CRM Short Courses

The volumes in the **CRM Short Courses** series have a primarily instructional aim, focusing on presenting topics of current interest to readers ranging from graduate students to experienced researchers in the mathematical sciences. Each text is aimed at bringing the reader to the forefront of research in a particular area or field, and can consist of one or several courses with a unified theme. The inclusion of exercises, while welcome, is not strictly required. Publications are largely but not exclusively, based on schools, instructional workshops and lecture series hosted by, or affiliated with, the *Centre de Researches Mathématiques* (CRM). Special emphasis is given to the quality of exposition and pedagogical value of each text.

More information about this series at http://www.springer.com/series/15360

Markus Heydenreich · Remco van der Hofstad

Progress in High-Dimensional Percolation and Random Graphs

CENTRE
DE RECHERCHES
MATHÉMATIQUES

Markus Heydenreich
Mathematisches Institut
Ludwig-Maximilians-Universität München
Munich, Bayern
Germany

Remco van der Hofstad
Department of Mathematics and
   Computer Science
Eindhoven University of Technology
Eindhoven
The Netherlands

CRM Short Courses
ISBN 978-3-319-87321-3   ISBN 978-3-319-62473-0 (eBook)
DOI 10.1007/978-3-319-62473-0

Mathematics Subject Classification (2010): 60K35, 60K37, 82B43

Printed on acid-free paper

This Springer imprint is published by Springer Nature
The registered company is Springer International Publishing AG
The registered company address is: Gewerbestrasse 11, 6330 Cham, Switzerland

Preface

This book focuses on percolation on high-dimensional lattices. We give a general introduction to percolation, stating the main results and defining the central objects. We assume no prior knowledge about percolation. This text is aimed at graduate students and researchers who wish to enter the wondrous world of high-dimensional percolation, with the aim to demystify the lace-expansion methodology that has been the key technique in high dimensions. This text can be used for reading seminars or advanced courses as well as for reference and individual study. The exposition is complemented with many exercises, and we invite readers to try them out and gain deeper understanding of the techniques presented here. Let us now summarize the content in more detail.

We describe mean-field results in high-dimensional percolation that make the intuition that "faraway critical percolation clusters are close to being independent" precise. We have two main purposes. The first main purpose is to give a self-contained proof of mean-field behavior for high-dimensional percolation, by proving that percolation in high dimensions has mean-field critical exponents $\beta = \gamma = 1, \delta = 2$ and $\eta = 0$, as for percolation on the tree. This proof is obtained by combining the Aizenman–Newman and Barsky–Aizenman differential inequalities, that rely on the *triangle condition*, with the lace-expansion proof of Hara and Slade of the infrared bound that, in turn, verifies the triangle condition. While there are expository texts discussing lace-expansion methodology, such as Slade's excellent Saint-Flour lecture notes, an introduction to high-dimensional percolation did not yet exist.

Aside from these classical results, that are now over 25 years old, our second main purpose is to discuss recent extensions and additions. We focus on (1) the recent proof that mean-field critical behavior holds for percolation in $d \geq 11$; (2) the proof of existence of arm exponents; (3) results on finite-size scaling and percolation on high-dimensional tori and their relationship to the Erdős–Rényi random graph; (4) extensions of these finite-size scaling results to hypercube percolation; (5) the existence of the *incipient infinite cluster* and its scaling properties, as well as the proof of the Alexander–Orbach conjecture for random walks on the high-dimensional incipient infinite cluster; (6) the novel lace expansion for the two-point function with a fixed number of pivotals; and (7) super-process limits of critical percolation clusters. The text is enriched with numerous open problems, which, we hope, will stimulate further research in the field.

This text is organized as follows. In Part I, consisting of Chaps. 1–3, we introduce percolation and prove its main properties such as the sharpness of the phase transition. In Part II, consisting of Chaps. 4–9, we discuss mean-field critical behavior by describing the two main techniques used, namely, differential inequalities and the lace expansion. In Parts I and Part II, all results are proved, making this the first self-contained text discussing high-dimensional percolation. In Part III, consisting of Chaps. 10–13, we describe recent progress in high-dimensional percolation. We provide partial proofs and give substantial overview and heuristics about how the proofs are obtained. In many of these results, the lace expansion and differential inequalities or their discrete analogues are central. In Part IV, consisting of Chaps. 14–16, we discuss related models and further open problems. Here we only provide heuristics and few details of the proofs, thus focussing on the overview and big picture.

This text could not have been written without help from many. We are grateful to Kilian Matzke, Andrea Schmidbauer, Gordon Slade, Si Tang, Sebastian Ziesche, as well as the reading groups in Geneva and Sapporo, for valuable comments and pointing out typos and omissions in an earlier version of the manuscript. Special thanks go to Robert Fitzner, who kindly prepared the graphics in this text.

This work would not have been possible without the generous support of various institutions. The work of MH is supported by the Netherlands Organisation for Scientific Research (NWO) through VENI grant 639.031.035. The work of RvdH is supported by the Netherlands Organisation for Scientific Research (NWO) through VICI grant 639.033.806 and the Gravitation NETWORKS grant 024.002.003.

The content of this book has been presented by RvdH at the CRM–PIMS Summer School in Probability 2015 in Montréal. He warmly thanks the organizers Louigi Addario-Berry, Omer Angel, Louis-Pierre Arguin, Martin Barlow, Ed Perkins and Lea Popovic for this opportunity, as well as the CRM for generous support.

Both of us have been working for many years on percolation and the lace expansion, and we thank our colleagues for joyful and inspiring collaborations that led to many joint articles. Most notably, we thank Christian Borgs, Jennifer Chayes, Robert Fitzner, Takashi Hara, Frank den Hollander, Mark Holmes, Tim Hulshof, Asaf Nachmias, Akira Sakai, Gordon Slade and Joel Spencer. Thank you for participating in the wonderful journeys in this beautiful branch of mathematical research!

Munich, Germany   Markus Heydenreich
Eindhoven, The Netherlands   Remco van der Hofstad
March 2017

Contents

Symbols

$E \circ F$	disjoint occurrence of events E and F; cf. (1.3.3)
$x \leftrightarrow y$	vertices x and y are connected; cf. p. 3
$x \Leftrightarrow y$	$= \{x \leftrightarrow y\} \circ \{x \leftrightarrow y\}$, vertices x and y are doubly connected; cf. Def. 6.3(a)
$\mathcal{C}(x)$	$= \{y : x \leftrightarrow y\}$, cluster of vertex x; cf. p. 3
$\widetilde{\mathcal{C}}^b(x)$	restricted cluster without using bond b; cf. Def. 6.3(b)
$\theta(p)$	$= \mathbb{P}_p(|\mathcal{C}(x)| = \infty)$, cluster density / percolation function; cf. (1.1.1)
p_c	$= \inf\{p : \theta(p) > 0\}$, percolation critical threshold; cf. (1.1.2)
$\chi(p)$	$= \mathbb{E}_p[|\mathcal{C}(x)|]$, expected cluster size / susceptibility; cf. (1.1.3)
$\chi^f(p)$	$= \mathbb{E}_p\big[|\mathcal{C}(x)|\mathbb{1}_{\{|\mathcal{C}(x)|<\infty\}}\big]$, truncated expected cluster size; cf. (1.1.5)
$\tau_p(x, y)$	$= \mathbb{P}_p(x \leftrightarrow y)$, two-point function; cf. (1.1.6)
$\tau_p^f(x, y)$	$= \mathbb{P}_p(x \leftrightarrow y, |\mathcal{C}(x)| < \infty)$, truncated two-point function; cf. (1.1.7)
$\xi(p)$	$= -\lim_{n\to\infty} n/\log \tau_p^f(ne_1)$, correlation length; cf. (1.1.11)
$\xi_2(p)$	$= \sqrt{(1/\chi^f(p))\sum_{x\in\mathbb{Z}^d} |x|^2 \tau_p^f(x)}$, average radius of gyration; cf. (1.1.12)
β	percolation function crit. exponent; cf. (1.2.1)
γ	susceptibility crit. exponent; cf. (1.2.5)
ν	correlation length crit. exponent; cf. (1.2.7)
ν_2	av. radius of gyration crit. exp; cf. (1.2.8)
Δ	gap crit. exponent; cf. (1.2.9)
δ	cluster tail crit. exponent; cf. (1.2.10)
ρ_{ex}	extrinsic one-arm crit. exponent; cf. (1.2.11)
ρ_{in}	intrinsic one-arm crit. exponent; cf. (1.2.12)
η	two-point function crit. exponent; cf. (1.2.13)
$D(x)$	$= \mathbb{1}_{\{|x|=1\}}/(2d)$, simple random walk step distribution; cf. (1.2.18)
$C_\mu(x)$	$= \sum_{n\geq 0} \mu^n D^{\star n}(x)$, random walk Green's function; cf. (2.2.8)
$J(x)$	$= p\mathbb{1}_{\{|x|=1\}} = 2dpD(x)$; cf. (6.2.1)
$x \overset{A}{\leftrightarrow} y$	connection *through* A; cf. Def. 6.2

$x \leftrightarrow y$ in A	connection *in* A; cf. Def. 6.2
$\tau^A(x)$	$= \mathbb{P}(x \leftrightarrow y$ in $\mathbb{Z}^d \setminus A)$, connection *off* A; cf. Def. 6.2
$\tilde{\tau}(x)$	$= 2dp(D \star \tau_p)(x)$, connection with extra bond; cf. (7.2.3)
$\Delta_p(x)$	$= (\tau_p \star \tau_p \star \tau_p)(x)$, open triangle diagram; cf. (4.2.1)
Δ_p	$= \max_x \Delta_p(x) = \Delta_p(0)$, triangle diagram; cf. (4.1.1), (7.2.1)
$\tilde{\Delta}_p(x)$	$= (\tau_p \star \tau_p \star \tilde{\tau}_p)(x)$, modified open triangle diagram; cf. (7.2.2)
$\tilde{\Delta}_p$	$= \max_x \tilde{\Delta}_p(x)$, modified triangle diagram; cf. (7.2.4)
$\tilde{\tilde{\Delta}}_p$	$= (D * D * \tau_p * \tau_p * \tau_p)(0)$, doubly modified triangle; cf. (8.3.8)
$\Delta_p^{\max}$	$= \sup_{x \in \mathbb{Z}^d}[\Delta_p(x) - \delta_{x,0}]$, another triangle variant; cf. (9.1.2)
$\Pi_p(x)$	irreducible two-point function; cf. (6.1.2), (8.5.2)
$E'(v, y; A)$	lace-expansion event; cf. (6.2.11)
$E(x, u, v, y; A)$	cutting-bond event; cf. (6.2.12)
$\Pi_M(x)$	$= \sum_{N=0}^M (-1)^N \Pi^{(N)}(x)$, finite approx. to $\Pi_p(x)$; cf. (6.2.3)
$\Pi^{(0)}(x)$	cf. (6.2.7)
$\Pi^{(1)}(x)$	cf. (6.2.25)
$\Pi^{(N)}(x)$	cf. (6.2.27)
$R_0(x)$	first inclusion-exclusion error; cf. (6.2.20)
$R_1(x)$	second inclusion-exclusion error; cf. (6.2.26)
$R_M(x)$	general inclusion-exclusion error; cf. (6.2.28)
$W_p(y; k)$	simple bounding diagram; cf. (7.2.5)
$H_p(k)$	simple bounding diagram; cf. (7.5.2)
$B_1(s, t, u, v)$	auxiliary bounding diagram; see Fig. 7.3; cf. (7.4.3)
$B_2(u, v, s, t)$	auxiliary bounding diagram; see Fig. 7.3; cf. (7.4.4)
$A_3(s, u, v)$	auxiliary bounding diagram; see Fig. 7.3; cf. (7.4.2)
Λ_n	$= \{-n, \ldots, n\}^d$, cube of width n in d dimensions; cf. p. 9
$\partial \Lambda_n$	$= \Lambda_n \setminus \Lambda_{n-1}$, inner vertex boundary of cube Λ_n; cf. p. 9
$\hat{f}(k)$	$= \sum_{x \in \mathbb{Z}^d} f(x) \, e^{ik \cdot x}$, Fourier transform; cf. (1.2.15)
$M(p, \gamma)$	$= \mathbb{E}_p[1 - (1 - \gamma)^{|\mathcal{C}(0)|}]$, percolation magnetization; cf. (3.4.1)
$\tau_{p,\gamma}(x)$	$= \mathbb{P}_{p,\gamma}(x \leftrightarrow y, x \not\leftrightarrow \mathcal{G})$, green-free two-point function; cf. (9.4.2)
$\chi(p, \gamma)$	$= \mathbb{E}_{p,\gamma}[|\mathcal{C}(0)| \mathbb{1}_{\{\mathcal{C}(0) \cap \mathcal{G} = \varnothing\}}]$, expected green-free cluster size; cf. (3.4.3)
$G_p(x)$	$= \sum_{v \in \mathcal{C}(o)} \mathbb{P}_p(\phi(v) = x)$, BRW two-point function; cf. (2.2.2)
$G_p^{\mathrm{f}}(x)$	truncated BRW two-point fct; cf. (2.2.3)
$b_n^{\mathrm{SRW}}(x)$	$= (2d)^n D^{\star n}(x)$, # fixed length SRW paths; cf. (10.2.1)
$b_n(x), b_n^t(x)$	# fixed length NBW paths; cf. (10.2.3)
$\tau_p^t(x)$	two-point function avoiding first step; cf. (10.3.1)
$\tau_{\mathrm{T},p}(x)$	two-point function on torus $\mathbb{T}_{n,d}$; cf. p. 179
$\Delta_{\mathrm{T},p}$	triangle diagram on torus $\mathbb{T}_{n,d}$; cf. (13.3.2)
$\chi_{\mathrm{T}}(p)$	expected cluster size on torus $\mathbb{T}_{n,d}$; cf. (13.3.1)
$\mathsf{Bin}(n, p)$	Binomial distribution
$\mathsf{Poi}(\lambda)$	Poisson distribution
$x \wedge y$	minimum, $\min\{x, y\}$
$x \vee y$	maximum, $\max\{x, y\}$
$o, O, o_{\mathbb{P}}, O_{\mathbb{P}}$	Landau symbols; cf. p. 17

Part I
Introduction to Percolation

Chapter 1
Introduction and Motivation

Percolation is a paradigmatic model in statistical physics. It is one of the simplest models that displays a *phase transition*. Percolation originated in the physics community as a model for a porous medium and has since seen many other applications. It also provides a highly active area of research, one in which tremendous progress was made in the past decades, notably in two dimensions and in high dimensions. Our focus is on recent progress in *high-dimensional* percolation. In this first chapter, we start by presenting an introduction to percolation.

There are a number of textbooks available with percolation as their main topic, most notably Grimmett [122] as a general reference on the topic, Hughes [176] discussing also the connections with physics including a nice historic account, Bollobás and Riordan [58] with emphasis on two-dimensional models, and Kesten [194]. For an expository account with a long list of open problems, we refer to Kesten's survey [198]. The foundation of percolation as a mathematical discipline is generally ascribed to Broadbent and Hammersley [65].

1.1 Introduction of the Model

We consider the hypercubic lattice $\mathbb{Z}^d$, with bonds between vertices $x, y \in \mathbb{Z}^d$ precisely when x and y are nearest neighbors, i.e., when $|x - y| = 1$, where $|x| = (\sum_{i=1}^{d} x_i^2)^{1/2}$ denotes the Euclidean norm. The set of bonds of the lattice $\mathbb{Z}^d$ is denoted by $\mathcal{E}(\mathbb{Z}^d)$. In the language of graph theory, bonds are called *edges*, and we use both terms in this text.

We first define bond percolation informally. The model has a parameter $p \in [0, 1]$. Each bond is *occupied* with probability p, and *vacant* otherwise, and the edge statuses are independent random variables. Percolation studies the random sublattice consisting of the *occupied* bonds. Alternatively, we remove bonds independently with a fixed probability $1 - p$.

We sometimes generalize from the lattice setting to percolation on more general graphs $\mathcal{G} = (\mathcal{V}, \mathcal{E})$, finite or infinite. Here, $\mathcal{V}$ is the vertex set and $\mathcal{E} \subseteq \mathcal{V} \times \mathcal{V}$ denotes the edge set.

© Springer International Publishing Switzerland 2017
M. Heydenreich and R. van der Hofstad, *Progress in High-Dimensional Percolation and Random Graphs*, CRM Short Courses, DOI 10.1007/978-3-319-62473-0_1

In such cases, we often assume that $\mathcal{G}$ is *transitive*, i.e., the neighborhoods of all points are the same. More precisely, transitivity means that for every $x, y \in \mathcal{V}$, there exists a bijection $\phi \colon \mathcal{V} \to \mathcal{V}$ for which $\phi(x) = y$ and $\{\phi(u), \phi(v)\} \in \mathcal{E}$ precisely when $\{u, v\} \in \mathcal{E}$. Such a bijection $\phi \colon \mathcal{V} \to \mathcal{V}$ is called an *automorphism*. In particular, transitivity of a graph $\mathcal{G}$ implies that each vertex has the same degree.

We consider the probability space $\{0, 1\}^{\mathcal{E}}$ equipped with the product topology (i.e., the minimal topology that makes finite-dimensional projections continuous). For a percolation configuration $\omega \in \{0, 1\}^{\mathcal{E}}$, a bond $b \in \mathcal{E}$ is occupied whenever $\omega(b) = 1$, and it is vacant whenever $\omega(b) = 0$. We equip this space with a family of product measures $(\mathbb{P}_p)_{p \in [0,1]}$ chosen such that $\mathbb{P}_p(b \text{ occupied}) = p$ for any $b \in \mathcal{E}$ and $p \in [0, 1]$. Let $\mathbb{E}_p$ denote expectation with respect to (w.r.t.) $\mathbb{P}_p$.

We say that x is *connected* to y and write $x \leftrightarrow y$ when there exists a (finite) path of occupied bonds connecting x and y. Formally, $x \leftrightarrow y$ on a configuration $\omega \in \{0, 1\}^{\mathcal{E}}$ if there exist $m \in \mathbb{N}$, $x = v_0, v_1, \ldots, v_{m-1}, v_m = y \in \mathbb{Z}^d$ with the property that $\{v_{i-1}, v_i\} \in \mathcal{E}$ and $\omega(\{v_{i-1}, v_i\}) = 1$ for all $i = 1, \ldots, m$. We further write $\{x \leftrightarrow y\} = \{\omega : x \leftrightarrow y \text{ on the configuration } \omega\}$. We let the *cluster* of x be all the vertices that are connected to x, i.e., $\mathcal{C}(x) = \{y : x \leftrightarrow y\}$. By convention, $x \in \mathcal{C}(x)$.

We mostly restrict to the setting where the probability that an edge is occupied is fixed. In the literature, also the setting is studied where the vertex set $\mathcal{V}$ is given by $\mathcal{V} = \mathbb{Z}^d$, the edge set $\mathcal{E}$ is given by $\mathcal{E} = \mathbb{Z}^d \times \mathbb{Z}^d$ and, for $b \in \mathcal{E}$, the probability that b is occupied depends on b in a translation invariant way; we discuss such an example in Sect. 5.2 and Sect. 15.4.

Percolation Function. We define the *percolation function* $p \mapsto \theta(p)$ by

$$\theta(p) = \mathbb{P}_p(|\mathcal{C}(x)| = \infty), \tag{1.1.1}$$

where $x \in \mathbb{Z}^d$ is an arbitrary vertex and $|\mathcal{C}(x)|$ denotes the number of vertices in $\mathcal{C}(x)$. By translation invariance, the above probability does not depend on the choice of x. We, therefore, often investigate $\mathcal{C}(0)$ where $0 \in \mathbb{Z}^d$ denotes the origin.

When $\theta(p) = 0$, then the probability that the origin is inside an infinite connected component is 0, so that there is almost surely no infinite connected component. When $\theta(p) > 0$, on the other hand, by ergodicity, the proportion of vertices in infinite connected components equals $\theta(p) > 0$, and we say that the system *percolates*.

We define the *percolation critical value* by

$$p_c = p_c(\mathbb{Z}^d) = \inf\{p : \theta(p) > 0\}. \tag{1.1.2}$$

The above critical value is sometimes written as $p_c = p_H$ in honor of Hammersley, who defined it in [128]. The first interesting question about percolation is whether the critical value is nontrivial, i.e., whether $0 < p_c < 1$, and this is indeed the case whenever $d > 1$. When $\theta(p) > 0$, then on $\mathbb{Z}^d$, there is a *unique* infinite cluster as first proved by Aizenman, Kesten and Newman [12]. Burton and Keane [67] gave a beautiful geometric proof of this nontrivial fact. A nice account on the implications of the Burton–Keane proof is due to Meester [214]. For other graphs, $p_c(\mathcal{G})$ can be defined as in (1.1.2), but there can be infinitely many infinite components (see Sect. 15.6 for details).

The main interest in percolation lies in what happens at or close to the critical value $p = p_c$. The following open problem states that percolation has no infinite critical cluster, which arguably is the holy grail in percolation theory:

Open Problem 1.1 (Continuity of $p \mapsto \theta(p)$). Show that for percolation on $\mathbb{Z}^d$ with $d \geq 2$, there are no infinite clusters at criticality, i.e., $\theta(p_c) = 0$.

In $d = 1$, $p_c = 1$, so that $\theta(p_c) = 1$ and the statement in Open Problem 1.1 is false. Most embarrassingly, Open Problem 1.1 is still open in full generality. For dimension $d = 2$, $p_c = \frac{1}{2}$ and $\theta(p_c) = 0$ has been proven in a seminal paper by Kesten [193]. The verification of Open Problem 1.1 for high dimensions is one of the main aims of this text. See in particular Thm. 5.1 and Cor. 5.2 below to get a taste of the kind of results that we are after here.

Instead of considering bond percolation, one can also study *site percolation*, for which we independently and with fixed probability $1 - p$ remove the *vertices* in $\mathbb{Z}^d$ along with all bonds that are adjacent to the removed vertices. It may be seen that any bond percolation model can be reformulated as a site percolation model on a suitably adjusted lattice, but the reverse statement fails; see Grimmett [122]. In line with the majority of the literature, our main focus is on bond percolation (with an exception in Sect. 16.2).

Susceptibility. We are interested in several key functions that describe the connections in bond percolation. The *susceptibility* $p \mapsto \chi(p)$ is the expected cluster size given by

$$\chi(p) = \mathbb{E}_p |\mathcal{C}(0)| . \tag{1.1.3}$$

Clearly, $\chi(p) = \infty$ for $p > p_c$, since then $|\mathcal{C}(0)| = \infty$ with probability $\theta(p) > 0$. Further, $p \mapsto \chi(p)$ is clearly increasing. Define the critical value $p_T = p_T(\mathbb{Z}^d)$ by

$$p_T = \sup\{p : \chi(p) < \infty\} . \tag{1.1.4}$$

The subscript T in $p_T(\mathbb{Z}^d)$ is in honor of H. Temperley. A natural question is whether $p_T(\mathcal{G}) = p_c(\mathcal{G})$ for general graphs $\mathcal{G}$, i.e., is $\chi(p) < \infty$ for every $p < p_c$? The latter indeed turns out to be true on $\mathbb{Z}^d$, and this is the main result in Chap. 3.

The main aim of this text is to describe the behavior of percolation at, or close to, the critical value p_c. For $p \in [0, 1]$, let

$$\chi^f(p) = \mathbb{E}_p\big[|\mathcal{C}(0)| \mathbb{1}_{\{|\mathcal{C}(0)| < \infty\}}\big] \tag{1.1.5}$$

denote the mean finite cluster size. Clearly, $\chi^f(p) = \chi(p)$ for $p < p_T$, but for $p > p_T$, this may not be true (indeed, for $p > p_c$, it is false). We define the *two-point function* $\tau_p : \mathbb{Z}^d \times \mathbb{Z}^d \to [0, 1]$ by

$$\tau_p(x, y) = \mathbb{P}_p(x \leftrightarrow y) . \tag{1.1.6}$$

When $p > p_c$, it is natural to expect that $\tau_p(x, y) \to \theta(p)^2$ when $|x - y| \to \infty$. To investigate the speed of this convergence, we define the *truncated two-point function* $\tau_p^f \colon \mathbb{Z}^d \times \mathbb{Z}^d \to [0, 1]$ by

$$\tau_p^f(x, y) = \mathbb{P}_p(x \leftrightarrow y, |\mathcal{C}(x)| < \infty) , \tag{1.1.7}$$

and note that $\tau_p^f(x, y) = \tau_p(x, y)$ whenever $|\mathcal{C}(x)| < \infty$ occurs a.s. On $\mathbb{Z}^d$, when the model is translation invariant, we have that $\tau_p^f(x, y) = \tau_p^f(y - x, 0) \equiv \tau_p^f(y - x)$.

In terms of $\tau_p^f(x)$, we can identify $\chi^f(p)$ as

$$\chi^f(p) = \mathbb{E}_p \left[\sum_{x \in \mathbb{Z}^d} \mathbb{1}_{\{0 \leftrightarrow x, |\mathcal{C}(0)| < \infty\}} \right] = \sum_{x \in \mathbb{Z}^d} \tau_p^f(x) . \tag{1.1.8}$$

Similarly,

$$\chi(p) = \mathbb{E}_p \left[\sum_{x \in \mathbb{Z}^d} \mathbb{1}_{\{0 \leftrightarrow x\}} \right] = \sum_{x \in \mathbb{Z}^d} \tau_p(x) , \tag{1.1.9}$$

where

$$\tau_p(x) = \tau_p(0, x) = \mathbb{P}_p(0 \leftrightarrow x) . \tag{1.1.10}$$

Correlation Length. An important measure of the *spatial extent* of clusters is the *correlation length* $\xi(p)$ defined by

$$\xi(p) = - \left(\lim_{n \to \infty} \frac{\log \tau_p^f(n e_1)}{n} \right)^{-1} , \tag{1.1.11}$$

where $e_1 = (1, 0, \ldots, 0)$. Existence of the limit is due to the FKG inequality, see (1.3.1) below, and a subadditivity argument; cf. [122, Thm. 6.44]. The correlation length thus characterizes exponential decay of the two-point function along coordinate axes. Exponential decay in *arbitrary* direction can be derived from this, see [122, Sect. 6.2] and references therein.

The correlation length $\xi(p)$ is closely related to (and should not be mistaken with) the *correlation length of order 2*, also known as the *average radius of gyration*, given by

$$\xi_2(p) = \sqrt{\frac{1}{\chi^f(p)} \sum_{x \in \mathbb{Z}^d} |x|^2 \tau_p^f(x)} , \tag{1.1.12}$$

where we recall that $|x|$ denotes the Euclidean norm of x. Indeed, there are several ways of defining the correlation length (or rather *some* correlation length), all of them presumably equivalent in the sense that they are bounded above and below by finite and positive constants times $\xi(p)$ defined in (1.1.11). However, these equivalences are often not known rigorously.

The correlation length measures the *dependence* between finite clusters at a given distance. If $|x - y| \gg \xi(p)$ and, for $p > p_c$, x and y are in finite clusters, then we can think of $\mathcal{C}(x)$ and $\mathcal{C}(y)$ as being close to independent, while if $|x - y| \ll \xi(p)$, then $\mathcal{C}(x)$ and $\mathcal{C}(y)$ are quite dependent. Another loosely formulated yet pedagogical way of interpreting the correlation length is the following. When looking at a percolation configuration in a window of length ℓ, the exponential decay of the correlation between clusters is clearly visible as soon as $\ell \gg \xi(p)$. On the other hand, when $\ell \ll \xi(p)$, the percolation configuration appears "indistinguishable" from a critical percolation configuration (where we bear in mind that $\lim_{p \to p_c} \xi(p) = \infty$).

1.2 Critical Behavior

The behavior of percolation models is most interesting and richest for p values that are close to the critical value. Clearly, the precise value of $p_c(\mathcal{G})$ depends sensitively on the nature of the underlying graph $\mathcal{G}$. By drawing an analogy to physical systems, physicists predict that the behavior of percolative systems close to criticality is rather insensitive to the precise details of the model, and it is only characterized by the macroscopic behavior. Thus, percolation is expected to behave in a *universal* manner. For example, it is predicted that the critical nature of finite-range percolation systems on $\mathbb{Z}^d$, under suitable symmetry conditions, is similar. While this prediction is far from rigorous, it does offer us a way of summarizing percolation models by only looking at their simplest examples. One of the key challenges of percolation theory is to make rigorous sense of this universality paradigm.

We now make the notion of universality more tangible, by discussing *critical exponents*. The critical nature of many physical systems is believed to be characterized by the validity of *power laws*, the exponent of which is a robust or universal measure of the underlying critical behavior. We start by giving an example of a critical exponent. It is predicted that

$$\theta(p) \sim (p - p_c)^\beta \quad \text{as } p \searrow p_c , \tag{1.2.1}$$

for some $\beta > 0$. The value of β is expected to be different for $\mathbb{Z}^d$ with different d, but (1.2.1) remains valid. The symbol $\sim$ in (1.2.1) can have several meanings, which we now elaborate on. We say that the critical exponent β exists in the *logarithmic form* if

$$\lim_{p \downarrow p_c} \frac{\log \theta(p)}{\log (p - p_c)} = \beta . \tag{1.2.2}$$

We say that β exists in the *bounded-ratio sense*, which we write as $\theta(p) \asymp (p - p_c)^\beta$, if there exist $0 < c_1 < c_2 < \infty$ such that, uniformly for $p \geq p_c$,

$$c_1(p - p_c)^\beta \leq \theta(p) \leq c_2(p - p_c)^\beta , \tag{1.2.3}$$

Finally, we say that β exists in the *asymptotic sense* if, as $p \searrow p_c$, there exists a $c > 0$ such that

$$\theta(p) = c(p - p_c)^\beta (1 + o(1)) . \tag{1.2.4}$$

The existence of a critical exponent is a priori unclear and needs a mathematical proof. Unfortunately, in general such a proof is missing, and we can only give proofs of the existence in special cases. Indeed, the existence of the critical exponent $\beta > 0$ is stronger than continuity of $p \mapsto \theta(p)$, which is unknown in general (recall Open Problem 1.1). Indeed, $p \mapsto \theta(p)$ is clearly continuous on $[0, p_c)$, and it is also continuous (and even infinitely differentiable) on $(p_c, 1]$ by the results of van den Berg and Keane [43] (for infinite differentiability of $p \mapsto \theta(p)$ for $p \in (p_c, 1]$, see Russo's paper [228]). Thus, continuity of $p \mapsto \theta(p)$ is equivalent to the statement that $\theta(p_c(\mathbb{Z}^d)) = 0$.

We now introduce several more critical exponents. The critical exponent γ for the expected cluster size is given by[1]

$$\chi^f(p) \sim |p - p_c|^{-\gamma} , \quad p \to p_c . \tag{1.2.5}$$

More precisely, we can think of (1.2.5) as defining the *two* critical exponents $\gamma, \gamma' > 0$ defined by

$$\begin{aligned}
\chi(p) &\sim (p_c - p)^{-\gamma} , & p \nearrow p_c , \\
\chi^f(p) &\sim (p - p_c)^{-\gamma'} , & p \searrow p_c ,
\end{aligned} \tag{1.2.6}$$

with the predicted equality $\gamma = \gamma'$.

Further, ν, ν' are defined by

$$\begin{aligned}
\xi(p) &\sim (p_c - p)^{-\nu} , & p \nearrow p_c , \\
\xi(p) &\sim (p - p_c)^{-\nu'} , & p \searrow p_c ,
\end{aligned} \tag{1.2.7}$$

again with the prediction that $\nu = \nu'$. In the same way, ν_2, ν_2' are defined by

$$\begin{aligned}
\xi_2(p) &\sim (p_c - p)^{-\nu_2} , & p \nearrow p_c , \\
\xi_2(p) &\sim (p - p_c)^{-\nu_2'} , & p \searrow p_c ,
\end{aligned} \tag{1.2.8}$$

and supposedly $\nu_2 = \nu_2' = \nu = \nu'$.

The *gap exponent* $\Delta > 0$ is defined by

$$\frac{\mathbb{E}_p\left[|\mathcal{C}(0)|^{k+1} \mathbb{1}_{\{|\mathcal{C}(0)| < \infty\}}\right]}{\mathbb{E}_p\left[|\mathcal{C}(0)|^{k} \mathbb{1}_{\{|\mathcal{C}(0)| < \infty\}}\right]} \sim |p - p_c|^{-\Delta} , \quad \text{where } k = 1, 2, 3, \dots , \tag{1.2.9}$$

with the unwritten assumption that Δ is independent of k. Also Δ can be defined, similarly to (1.2.6), as an exponent Δ for $p \nearrow p_c$ and another Δ' for $p \searrow p_c$, the values being equal.

---

[1]The careful reader may notice that we are anticipating here that $p_c = p_T$.

As mentioned before, it is highly unclear that these critical exponents are well defined, and that the value of $\Delta > 0$ does not depend on k. However, there are good physical reasons why these exponents are defined as they are.

The exponents $\beta, \gamma, \nu, \Delta$ can be thought of as *approach exponents* that measure the blow-up of various aspects of the cluster size as p approaches the critical value $p = p_c$. We finally define three critical exponents *at* criticality. The exponent $\delta \geq 1$ measures the power-law exponent of the critical cluster tail, i.e.,

$$\mathbb{P}_{p_c}(|\mathcal{C}(0)| \geq n) \sim n^{-1/\delta} , \quad n \to \infty , \tag{1.2.10}$$

the assumption that $\delta \geq 1$ following from the prediction that $\chi(p_c) = \infty$. See Chap. 3 for a proof of this fact. Further, we define the *extrinsic* arm exponent $\rho_{\mathrm{ex}} > 0$ by

$$\mathbb{P}_{p_c}(0 \leftrightarrow \partial \Lambda_n) \sim n^{-1/\rho_{\mathrm{ex}}} , \quad n \to \infty , \tag{1.2.11}$$

where $\Lambda_n = \{-n, \ldots, n\}^d$ denotes the $(\ell^\infty\text{-})$ball of radius n and $\partial \Lambda_n = \Lambda_n \setminus \Lambda_{n-1}$ its (vertex) boundary. In contrast to this, there is also the *intrinsic* arm exponent $\rho_{\mathrm{in}} > 0$, where, instead of the extrinsic ℓ^∞-metric, we use the *intrinsic* metric (also known as graph distance or hop-count distance) $d_{\mathcal{C}(0)}$ on the cluster $\mathcal{C}(0)$ interpreted as a graph,

$$\mathbb{P}_{p_c}(\exists x \in \mathbb{Z}^d : d_{\mathcal{C}(0)}(0, x) = n) \sim n^{-1/\rho_{\mathrm{in}}} , \quad n \to \infty , \tag{1.2.12}$$

Finally, η is defined by

$$\mathbb{E}_{p_c}\big[|\mathcal{C}(0) \cap \Lambda_n|\big] = \sum_{x \in \Lambda_n} \tau_{p_c}(x) \sim n^{2-\eta} , \quad n \to \infty . \tag{1.2.13}$$

There are several closely related versions of η that have been considered in the literature. First, we can define η in x-space by assuming that

$$\tau_{p_c}(x) = \mathbb{P}_{p_c}(0 \leftrightarrow x) \sim |x|^{-(d-2+\eta)} , \quad |x| \to \infty , \tag{1.2.14}$$

which is clearly stronger than (1.2.13) by summing over $x \in \Lambda_n$:

Exercise 1.1 (Versions for η). Prove that (1.2.14) implies (1.2.13).

For the second alternative definition of η, we rely on *Fourier theory*. Fourier transforms play a central role in high-dimensional percolation, so we take some time to introduce them now.

Unless specified otherwise, k always denotes an arbitrary element from the Fourier dual of $\mathbb{Z}^d$, which is the torus $(-\pi, \pi]^d$. The Fourier transform of a function $f : \mathbb{Z}^d \to \mathbb{C}$ is defined by

$$\hat{f}(k) = \sum_{x \in \mathbb{Z}^d} f(x) e^{ik \cdot x} , \quad k \in (-\pi, \pi]^d . \tag{1.2.15}$$

Even though the Fourier transforms have values in $\mathbb{C}$, most of our functions are symmetric w.r.t. reflection, that is, $f(x) = f(-x)$ for all $x \in \mathbb{Z}^d$. For such functions, the Fourier transform becomes

$$\hat{f}(k) = \sum_{x \in \mathbb{Z}^d} \cos(k \cdot x) f(x), \quad k \in (-\pi, \pi]^d, \tag{1.2.16}$$

and is in fact $\mathbb{R}$-valued. For two summable functions $f, g \colon \mathbb{Z}^d \to \mathbb{R}$, we let $f \star g$ denote their *convolution*, i.e.,

$$(f \star g)(x) = \sum_{x \in \mathbb{Z}^d} f(y) g(x - y). \tag{1.2.17}$$

We note that the Fourier transform of $f \star g$ is given by the product of $\hat{f}$ and $\hat{g}$.
We next discuss an example that is crucial throughout this text. Let

$$D(x) = \mathbb{1}_{\{|x|=1\}}/(2d) \tag{1.2.18}$$

be the transition probability of simple random walk, for which we compute

$$\hat{D}(k) = \frac{1}{2d} \sum_{x : |x|=1} e^{ik \cdot x} = \frac{1}{d} \sum_{i=1}^{d} \cos(k_i), \quad k = (k_1, \ldots, k_d) \in [-\pi, \pi)^d. \tag{1.2.19}$$

The simple random walk step distribution D and its Fourier transform $\hat{D}$ will play a central role in this text.

The Fourier transform of the two-point function is given by

$$\hat{\tau}_p(k) = \sum_{x \in \mathbb{Z}^d} e^{ik \cdot x} \tau_p(x), \tag{1.2.20}$$

so that, recalling (1.1.9),

$$\chi(p) = \hat{\tau}_p(0). \tag{1.2.21}$$

Then, we can define η in k-space by assuming that

$$\hat{\tau}_{p_c}(k) \sim |k|^{-2+\eta} \quad \text{as } |k| \to 0. \tag{1.2.22}$$

It can be seen that (1.2.22) again implies (1.2.13), but this is less obvious. See our paper with Hulshof [145] or the following exercise for more details:

Exercise 1.2 (Versions for η (cont.)). Prove that also (1.2.22) implies (1.2.13) (or look it up in [145]).

We emphasize here that (1.2.22) is a highly nontrivial result, since $\hat{\tau}_{p_c}(0) = \chi(p_c) = \infty$, so that $\hat{\tau}_{p_c}(k)$ is the Fourier transform of a nonsummable function. In particular, taking $k \neq 0$

Table 1.1 Conjectured or proved critical exponents for percolation. The values for dimension 2 are rigorously proven for site percolation on a triangular lattice [249], and the values for $d > 6$ are discussed in detail in this text.

dimension	β	γ	ν	δ	η	ρ_{ex}	ρ_{in}
2	$\dfrac{5}{36}$	$\dfrac{43}{18}$	$\dfrac{4}{3}$	$\dfrac{91}{5}$	$\dfrac{5}{24}$	$\dfrac{48}{5}$	?
3–6	?	?	?	?	?	?	?
$d > 6$	1	1	$\dfrac{1}{2}$	2	0	$\dfrac{1}{2}$	1

makes the sum in (1.2.20) finite. We argue in Chap. 4 that the same is true for the random walk Green's function, a statement that is quite relevant in high dimensions.

The existence of the percolation critical exponents is a deep result and has provided the source of enormous research in the past decades. Only for site percolation on the two-dimensional triangular lattice and for percolation in sufficiently high dimensions, these critical exponents are known to exist and their values have been identified rigorously. In Table 1.1, we list the conjectured and proved results for the percolation critical exponents. There are extensive numerical studies in the physics literature yielding approximations for the critical exponents in intermediate dimensions, see, for example, Stauffer and Aharony [250, Table 2.2].

The values for $d = 2$ are believed to be valid for any finite-range two-dimensional percolation model. At the moment, they are only proved to exist for site percolation on the two-dimensional triangular lattice and some related lattices. The history for the two-dimensional case is that Schramm [240] first identified a family of continuous models driven by one-dimensional Brownian motion, so-called Schramm–Loewner evolution (SLE). These are conformally invariant models in the plane of which the properties depend on the value of the parameter $\kappa > 0$ that describes the variance of the Brownian motion. Schramm continued by noting that *if* the scaling limit of two-dimensional percolation would be conformally invariant, *then* it must be equal to SLE with parameter $\kappa = 6$. A celebrated result by Smirnov [247] shows that indeed the scaling limit of critical percolation on the triangular lattice is conformally invariant. Schramm already noted that SLE with parameter $\kappa = 6$ has similar critical exponents as in Table 1.1, when defined in an appropriate way. Smirnov and Werner [249] identified the critical exponents relying on the conformal invariance proved by Smirnov. Recently, Grimmett and Manolescu [125] show that if the critical exponents exist for one member of a certain family of models, then they exist for all members of that family of isoradial graph. This family includes bond percolation on the triangular lattice as well as bond percolation on the square lattice. Since site percolation on the triangular lattice is not a member of this family, we are still waiting for the proof that any one of them is conformally invariant, as is widely believed. The critical exponent ρ_{in} has not yet been identified mathematically in dimension $d = 2$; for bounds on intrinsic distances in the planar case, we refer to Damron [81] and references therein.

The aim of this text is to describe the results in high dimensions. For percolation in sufficiently high dimensions, the critical exponents are now known to be equal to those in the last row of Table 1.1. The main works in this direction are the papers by Aizenman and

Newman [13] and Barsky and Aizenman [28], who proved that γ, β and δ take on these values when the so-called *triangle condition* is valid, and that by Hara and Slade [132], who proved the so-called infrared bound that, in turn, verifies that $\eta = 0$ and that the triangle condition holds. In this text, we explain and prove these results, as well as their more recent extensions. For example, we discuss the recent proof by Fitzner and the second author that shows that the mean-field critical exponents are valid for $d \geq 11$.

Percolation is a paradigmatic model in statistical physics, and a central notion in this field is *universality*. An example of universality in the setting of percolation is the prediction that any finite-range percolation model on $\mathbb{Z}^d$ has the *same* critical exponents. Indeed, the mean-field critical exponents in Table 1.1 *are* proved for *all* $d > 6$ when we consider percolation on a spread-out lattice consisting of all bonds $\{x, y\}$ with $\|x - y\|_\infty \leq L$ when $L \geq L_0(d)$ is chosen to be sufficiently large. Here, $\|x\|_\infty = \max_{i=1}^{d} |x_i|$ denotes the supremum norm of x. In particular, in this setting, the critical exponents do not depend on the value of L (even though the value of p_c is sensitive to L), an example of universality. While universality is quite plausible when describing real physical systems from the viewpoint of statistical physics, and while universality is a very useful notion since it allows us to study only the simplest finite-range model available, there are very few examples where universality can be rigorously proved. High-dimensional percolation provides one crucial example.

In the next section, we start by setting the stage. We describe three key technical tools that are used throughout this text.

1.3 Russo, FKG, and BKR

We mention two inequalities that play a profound role in percolation theory, namely the *FKG and BK(R) inequalities*. We also discuss *Russo's formula*, which is useful to describe derivatives of probabilities of events. Particularly, the BK inequality and Russo's formula are essential in analyzing high-dimensional percolation.

The *Fortuin–Kasteleyn–Ginibre* or *FKG inequality* in the context of percolation is called the *Harris inequality* and was first proved by Harris in [141]. The more general FKG inequality, which, for example, also applies to the Ising model, was derived by Fortuin, Kasteleyn and Ginibre in [111]. We say that an event E is *increasing* when, if E occurs for a percolation configuration ω and ω' denotes a configuration for which $\omega'(b) = 1$ for every bond b for which $\omega(b) = 1$, then E continues to hold for ω'. In other words, the event E remains to hold when we turn more bonds from vacant to occupied. The Harris inequality states that for two increasing events E and F that depend only on a finite number of bonds,

$$\mathbb{P}_p(E \cap F) \geq \mathbb{P}_p(E)\,\mathbb{P}_p(F)\,, \tag{1.3.1}$$

the FKG inequality gives the same conclusion under weaker assumptions on the measure involved. In words, for increasing events E and F, the occurrence of E makes the simultaneous occurrence of F more likely. The intuition for the FKG inequality is that if the increasing event E holds, then this makes it more likely for edges to be occupied, and, therefore, it becomes more likely that the increasing event F also holds. Thus, $\mathbb{P}_p(F \mid E) \geq \mathbb{P}_p(F)$, which is equivalent to (1.3.1). See Häggström [180] for a Markov chain proof of the FKG inequality or Grimmett [122, Sect. 2.2] for a proof using induction on the number of edges involved.

As an example, the FKG inequality yields that, for every $x, y, u, v \in \mathbb{Z}^d$,

$$\mathbb{P}_p(x \leftrightarrow y, u \leftrightarrow v) \geq \mathbb{P}_p(x \leftrightarrow y) \mathbb{P}_p(u \leftrightarrow v) . \tag{1.3.2}$$

While the events in (1.3.2) do not depend on finitely many bonds, the inequality can be obtained by an appropriate truncation argument.

The *van den Berg–Kesten or BK inequality* gives, in a certain sense, an opposite inequality. We first state it in the case of increasing events, for which it was proved by van den Berg and Kesten in [44]. The most general version is proved by Reimer in [226] and goes under the name of BKR inequality. For $K \subseteq \mathcal{E}$ and $\omega \in \{0, 1\}^{\mathcal{E}}$, we write $\omega_K(e) = \omega(e)$ for $e \in K$, and $\omega_K(e) = 0$ otherwise. Let E and F again be increasing events depending on finitely many bonds, and write

$$E \circ F = \{\omega : \exists K \subseteq \mathcal{E} \text{ such that } \omega_K \in E, \omega_{K^c} \in F\} . \tag{1.3.3}$$

Then, the van den Berg–Kesten (BK) inequality states that

$$\mathbb{P}_p(E \circ F) \leq \mathbb{P}_p(E) \mathbb{P}_p(F) . \tag{1.3.4}$$

For example, the event $\{x \leftrightarrow y\} \circ \{u \leftrightarrow v\}$ is the event that there are edge disjoint occupied paths from x to y and from u to v, and (1.3.3) implies that[2]

$$\mathbb{P}_p(\{x \leftrightarrow y\} \circ \{u \leftrightarrow v\}) \leq \mathbb{P}_p(x \leftrightarrow y) \mathbb{P}_p(u \leftrightarrow v) . \tag{1.3.5}$$

Intuitively, this can be understood by noting that, if $x \leftrightarrow y$ and $u \leftrightarrow v$ must occur disjointly, then we can first fix an occupied path connecting x and y in a certain arbitrary manner, and remove the occupied edges used in this path. Then, $\{x \leftrightarrow y\} \circ \{u \leftrightarrow v\}$ occurs when in the configuration with the edges removed, we still have that $u \leftrightarrow v$. Since we have removed the edges in the occupied path from x to y, this event now has smaller probability than $\mathbb{P}_p(u \leftrightarrow v)$. See Grimmett [122, Sect. 2.3] for a proof of the BK inequality.

Exercise 1.3 (BK for connection events). Use a restriction to finite domains and monotone convergence to prove that (1.3.5) follows from (1.3.4).

We continue to define a generalization of the BK inequality, allowing for events that are possibly nonmonotone, called the van den Berg–Kesten–Reimer (BKR) inequality, see [44, 62, 226]. For a set of edges B, we say that an event E *occurs in* B if and only if it occurs independently of the status of the edges not in B, i.e., it is the event

$$E_{|B} = \{\omega \in E : \forall \omega' \text{ such that } \omega' = \omega \text{ on } B \text{ also } \omega' \in E\} . \tag{1.3.6}$$

---

[2]These events again do not depend on finitely many bonds, but they can be approximated by events depending on finitely many bonds, so that our conclusion remains to hold. See Exerc. 1.3.

For two events E, F, we let $E \circ F$ denote the event

$$E \circ F = \{\omega : \exists B_1, B_2 \subseteq \mathcal{E}, B_1 \cap B_2 = \varnothing, \omega \in E_{|B_1} \cap F_{|B_2}\} . \tag{1.3.7}$$

We refer to the random sets of edges B_1, B_2 as *witnesses* for the events E and F, respectively. The BKR inequality states that

$$\mathbb{P}_p(E \circ F) \le \mathbb{P}_p(E) \, \mathbb{P}_p(F) . \tag{1.3.8}$$

While the BKR inequality in (1.3.8) is in its formulation identical to the BK inequality in (1.3.4), we do state the two inequalities separately for ease of reference.[3]

Exercise 1.4 (BKR for increasing events). Prove that (1.3.7) reduces to (1.3.3) when E and F are increasing.

Exercise 1.5 (BKR and FKG). Prove that the BKR inequality implies the Harris inequality when applied to an increasing event E and a decreasing event F, for which it states that $\mathbb{P}_p(E \cap F) \le \mathbb{P}_p(E) \, \mathbb{P}_p(F)$.

We finally discuss an important tool to study probabilities which goes under the name of *Russo's formula* [229] even though it appeared earlier in a similar form by Margulis [212]. Let E be an increasing event that is determined by the occupation status of *finitely* many bonds. Then, we say that the bond $\{u, v\}$ is *pivotal* for the event E when E occurs when the status of $\{u, v\}$ in the (possibly modified) configuration where $\{u, v\}$ is turned occupied, while E does not occur in the (possibly modified) configuration where $\{u, v\}$ is turned vacant. Thus, the bond $\{u, v\}$ is essential for the occurrence of the event E. The set of pivotal bonds for an event is *random*, as it depends on which other bonds are occupied and vacant in the configuration. Russo's formula states that for every increasing event E that depends on a *finite number of bonds*,

$$\frac{\mathrm{d}}{\mathrm{d}p} \mathbb{P}_p(E) = \sum_{e \in \mathcal{E}} \mathbb{P}_p(e \text{ pivotal for } E) . \tag{1.3.9}$$

Russo's formula allows us to study how the probability of an event changes as p varies. The fact that (1.3.9) is only valid for events that depend on a finite number of bonds is a nuisance, and there are many settings in which Russo's formula can be extended to events depending on infinitely many bonds by an appropriate limiting procedure. We tread lightly on this issue and refer to the literature for precise proofs.

To explain the intuition behind Russo's formula, we consider

$$\mathbb{P}_{p+\varepsilon}(E) - \mathbb{P}_p(E) , \tag{1.3.10}$$

and use the *Harris coupling*. In the Harris coupling, we assign a uniform random variable U_b to every bond b independently across the bonds and say that b is *p-occupied* when $U_b \le p$.

---

[3]Grimmett [122] uses $E \square F$ for the event in (1.3.7).

For an event E, we write E_p when E occurs for the p-occupied configuration. Then, we can write

$$\mathbb{P}_{p+\varepsilon}(E) - \mathbb{P}_p(E) = \mathbb{P}(E_{p+\varepsilon}) - \mathbb{P}(E_p) = \mathbb{P}(E_{p+\varepsilon} \cap E_p^c), \qquad (1.3.11)$$

since the event E is increasing. In the Harris coupling, every p-vacant bond is independently occupied with probability $\varepsilon/(1-p)$ to obtain the $(p+\varepsilon)$-configuration. Since E depends on the status of only finitely many bonds, if $\varepsilon > 0$ is very small, then most probably there is at most one bond that is different in the $(p+\varepsilon)$- and the p-configuration. Thus, when $E_{p+\varepsilon}$ occurs, but E_p does not, this bond must be pivotal for the occurrence of E_p. Furthermore, since E is increasing, this bond needs to be p-vacant and $(p+\varepsilon)$-occupied. Mind that the event $\{b$ is pivotal$\}$ is independent of the occupation of b. Thus, since the factors of $1-p$ cancel,

$$\mathbb{P}_{p+\varepsilon}(E) - \mathbb{P}_p(E) = \varepsilon \sum_b \mathbb{E}[\mathbb{1}_{\{b \text{ pivotal for } E_p\}}] + O(\varepsilon^2). \qquad (1.3.12)$$

Dividing by ε and taking the limit $\varepsilon \searrow 0$ gives (1.3.9).

1.4 Aim of This Book and What Is New?

We have now discussed the basic concepts in percolation. In this section, we present the main aim of this text, which is twofold:

First Main Aim of This Text

Give a self-contained proof for mean-field behavior of high-dimensional percolation.

We state and prove the fact that percolation in high dimensions has mean-field critical exponents $\beta = \gamma = 1$, $\delta = 2$, and $\eta = 0$. This proof is obtained by combining the results in three seminal papers:
(1) First, the two papers by Aizenman and Newman [13] and Barsky and Aizenman [28] that prove that these critical values exist and take their mean-field values assuming a certain geometrical condition called the triangle condition. The proofs of these facts rely on clever *differential inequalities*, some of which apply in general dimensions and imply mean-field bounds;
(2) Second, the paper by Hara and Slade [132] that proves the infrared bound, which, in turn, proves that the triangle condition holds. The proof relies on the *lace-expansion method*, a key tool in high-dimensional statistical mechanics. The lace expansion is a combinatorial expansion identity for the two-point function $\tau_p(x)$ that relates it to the random walk Green's function.

We perform the proofs needed for many of these results completely and thus provide a first self-contained expository version of these facts. The proofs include recent simplifications, for example, the use of trigonometric functions due to Borgs, Chayes, the second author, Slade and Spencer in [60], and the simplified analysis in joint work with Sakai [148]. See also the excellent Saint-Flour notes by Slade [246] for a related analysis; however, the main focus of that text is on self-avoiding walk instead.

Aside from these classical results, that are now approximate 25 years old, we also discuss recent extensions and additions. These results show that high-dimensional percolation is a highly active research field, in which still substantial progress is being made. Further, many important problems have not yet been resolved. Most prominently, we discuss the following:

Second Main Aim of This Text

Describe further mean-field results for percolation in high dimensions.

(1) The recent proof of Fitzner and the second author [107, 108] that mean-field behavior holds for percolation in dimensions $d \geq 11$;

(2) The proof of existence of arm exponents by Kozma and Nachmias [201, 202], and the relevance of $\eta = 0$ in x-space proved by Hara, the second author and Slade [131] and Hara [130];

(3) Finite-size scaling and percolation on high-dimensional tori and their relationship to the Erdős–Rényi random graph in work of Borgs, Chayes, the second author, Slade and Spencer [59, 60] and the authors [143, 144];

(4) Extensions of these finite-size scaling results to hypercube percolation by Borgs, Chayes, the second author, Slade and Spencer [61] and by Nachmias and the second author [164, 165];

(5) The existence of the *incipient infinite cluster* (IIC) as proved by Jarái and the second author [161], and with Hulshof in [145], and some of the scaling properties of the IIC in recent work with Hulshof and Miermont [147]. We further highlight the proof of the Alexander–Orbach conjecture for random walks on the high-dimensional IIC by Kozma and Nachmias [201], and the Euclidean scaling of random walks on the high-dimensional IIC by the authors with Hulshof [146];

(6) Super-process limits of critical percolation clusters, as proved by Hara and Slade [139, 140], and for oriented percolation by Slade and the second author [170].

While describing the results and proofs, we state many open problems along the way.

1.5 Organization and Notation

In Chap. 2, we start by analyzing percolation on the tree as well as its geometric embedding, which goes under the name branching random walk (BRW). We focus on its critical behavior. In Chap. 3, we show that the percolation phase transition is unique by proving that $p_c = p_T$ (recall (1.1.2) and (1.1.4)). This completes Part I of this text.

In Chap. 4, we describe a geometric condition, the so-called *triangle condition*, that implies that some of the critical exponents in $\mathbb{Z}^d$ are equal to the mean-field exponents for percolation on a tree identified in Sect. 2.1 and for branching random walk as identified in Sect. 2.2. This result lies at the heart of the methods to study high-dimensional percolation and formed the inspiration for Gordon Slade and Takashi Hara to study the two-point function in detail in [132]. We explain the relevance of the triangle condition by showing that it implies that $\gamma = 1$. In Chap. 5, we describe the main result in high-dimensional percolation, the so-called *infrared bound* that implies the triangle condition. In Chap. 6, we explain the philosophy behind the proof of the infrared bound. This proof is based on a combinatorial expansion that is often called the *lace expansion*, even though laces do not appear in it. We derive this expansion in Chap. 6, bound its coefficients in Chap. 7, and analyze its asymptotics in Chap. 8. This completes the proof of the infrared bound. In Chap. 9, we explain how the other critical exponents can be identified by using differential inequalities. Chapters 4–9 describe the classical mean-field results in high-dimensional percolation that constitute Part II of this text. Proofs in Parts I and II are given completely, so that the discussion of mean-field results is self-contained.

In Chap. 10, we discuss the nonbacktracking lace expansion that is used to prove that mean-field critical exponents exist for $d \geq 11$. In Chap. 11, we discuss some of the more recently investigated critical exponents, such as ν, η, ρ_{in} and ρ_{ex}. In Chap. 12, we define the so-called *incipient infinite cluster*, an infinite critical percolation cluster. In Chap. 13, we discuss finite-size effects in percolation by describing percolation on high-dimensional tori, and their relations to random graphs. We argue that the Erdős–Rényi random graph is the mean-field model in this setting, and state results that confirm this prediction. These results form Part III of this text, in which partial proofs are given and overviews in cases where proofs are incomplete.

In the remaining Part IV of this text, we discuss related and open problems. In Chap. 14, we investigate random walks on large percolation clusters. We close this text in Chap. 15 with several related results, ranging from relations between percolation and super-processes, oriented percolation, to long-range percolation and percolation on nonamenable graphs, as well as a list of open problems on related models in Chap. 16.

Notation. Let us introduce some standard notation. We say that a sequence of events E_n occurs *with high probability* (whp) when $\lim_{n\to\infty} \mathbb{P}(E_n) = 1$. We further write $f(n) = O(g(n))$ if $|f(n)|/|g(n)|$ is uniformly bounded from above by a positive constant as $n \to \infty$, $f(n) = \Theta(g(n))$ if $f(n) = O(g(n))$ and $g(n) = O(f(n))$, $f(n) = \Omega(g(n))$ if $1/f(n) = O(1/g(n))$, and $f(n) = o(g(n))$ if $f(n)/g(n)$ tends to 0 as $n \to \infty$. We say that $f(n) \gg g(n)$ when $g(n) = o(f(n))$.

We write $\| \cdot \|_\infty$ for the supremum (or ℓ^∞-) norm on $\mathbb{Z}^d$ or $\mathbb{R}^d$, $\| \cdot \|_1$ for the ℓ^1-norm, and $| \cdot |$ for the Euclidean (or ℓ^2-) norm.

For sequences of random variables $(X_n)_{n \geq 1}$, we let $X_n \xrightarrow{d} X$ denote that X_n converges in distribution to X, while $X_n \xrightarrow{\mathbb{P}} X$ denotes that X_n converges in probability to X and $X_n \xrightarrow{\text{a.s.}} X$ denotes that X_n converges almost surely to X. We write that $X_n = O_\mathbb{P}(Y_n)$ when $|X_n|/Y_n$ is a tight sequence of random variables and $X_n = o_\mathbb{P}(Y_n)$ when $|X_n|/Y_n \xrightarrow{\mathbb{P}} 0$.

Chapter 2
Fixing Ideas: Percolation on a Tree and Branching Random Walk

This text discusses percolation in high dimensions. When the dimension is high, space is so vast that faraway pieces of percolation clusters are close to being independent. The main purpose of this text is to make this imprecise statement precise. One reflection of this is that critical percolation clusters in high dimensions have relatively few cycles. On a tree, they are *precisely* independent, so that the above heuristic suggests that percolation on the high-dimensional hypercubic lattice is close to percolation on a tree. However, a tree does not have a Euclidean structure, and after discussing percolation on a tree, we discuss branching random walk, which we consider to be the proper mean-field model for percolation in high dimensions.

2.1 Percolation on a Tree

We start by studying percolation on the regular tree. In particular, we identify the critical exponents for percolation on a tree. We follow Grimmett [122, Sect. 10.1] or the second author's lecture notes [153, Sect. 1.2.2]. Let $\mathbb{T}_r$ denote the r-regular tree of degree r. The advantage of trees is that they do not contain cycles, which makes explicit computations possible. We first prove that the critical exponents for percolation on a regular tree exist and identify their values in the following theorem:

Theorem 2.1 (Critical behavior on the r-regular tree). *On the r-regular tree $\mathbb{T}_r$, $p_c(\mathbb{T}_r)$* $= p_T(\mathbb{T}_r) = 1/(r-1)$*, and* $\beta = \gamma = \gamma' = \rho_{\text{in}} = 1$ *and* $\delta = \Delta = \Delta' = 2$ *in the asymptotic sense.*

Proof. We make substantial use of the fact that percolation on a tree can be described in terms of *branching processes*. Let o denote a distinguished vertex that we call the *root* of the tree. For vertices $x, y \in \mathbb{T}_r$, we write $x \sim y$ whenever x and y are linked by an edge in the tree $\mathbb{T}_r$ and denote by $(I_{x,y})_{x\sim y}$ an i.i.d. family of Bernoulli random variables with parameter p indicating

© Springer International Publishing Switzerland 2017
M. Heydenreich and R. van der Hofstad, *Progress in High-Dimensional Percolation and Random Graphs*, CRM Short Courses, DOI 10.1007/978-3-319-62473-0_2

whether the edge $\{x, y\}$ is occupied or not. For $x \neq o$, we write $\mathcal{C}C_{BP}(x)$ for the forward cluster of x in $\mathbb{T}_r$, i.e., those vertices $y \in \mathbb{T}_r$ that are connected to x and for which the unique path from x to y only moves away from the root o. Then, clearly,

$$|\mathcal{C}(o)| = 1 + \sum_{e \sim o} I_{o,e} |\mathcal{C}_{BP}(e)| , \qquad (2.1.1)$$

where the sum is over all neighbors e of o, and $(|\mathcal{C}_{BP}(e)|)_{e \sim o}$ is an i.i.d. sequence independent of $(I_{o,e})_{e \sim o}$. Equation (2.1.1) allows us to deduce all information concerning $|\mathcal{C}(o)|$ from the information of $|\mathcal{C}_{BP}(e)|$. Also, for each $x \neq o$, $|\mathcal{C}_{BP}(x)|$ satisfies the formula

$$|\mathcal{C}_{BP}(x)| = 1 + \sum_{v \sim x : h(v) > h(x)} I_{x,v} |\mathcal{C}_{BP}(v)| , \qquad (2.1.2)$$

where $h(x)$ is the distance to the root (or *height*) of x in $\mathbb{T}_r$, and $(|\mathcal{C}_{BP}(v)|)_{v \sim x : h(v) > h(x)}$ is a set of $r - 1$ independent copies of $|\mathcal{C}_{BP}(x)|$. Thus, $|\mathcal{C}_{BP}(x)|$ is the *total population size* of a branching process, also known as its *total progeny*. We now derive the critical exponents one by one, in the order γ, Δ, β, γ', Δ', δ, and ρ_{in}.

Proof that $(\gamma = 1)$ *on the Tree.* We use (2.1.2) to conclude that

$$\chi_{BP}(p) := \mathbb{E}_p |\mathcal{C}_{BP}(x)| = 1 + (r-1) p \, \mathbb{E}_p |\mathcal{C}_{BP}(x)|$$
$$= 1 + (r-1) p \chi_{BP}(p) , \qquad (2.1.3)$$

so that

$$\chi_{BP}(p) = \mathbb{E}_p |\mathcal{C}_{BP}(x)| = \frac{1}{1 - (r-1)p} . \qquad (2.1.4)$$

From (2.1.1), we then obtain that, for $p < 1/(r-1)$,

$$\chi(p) = 1 + rp\chi_{BP}(p) = 1 + \frac{rp}{1 - (r-1)p} = \frac{1+p}{1 - (r-1)p} , \qquad (2.1.5)$$

while, for $p \geq 1/(r-1)$, $\chi(p) = \infty$. In particular, $p_T = 1/(r-1)$ and $\gamma = 1$ in the asymptotic sense. The computation of $\chi(p)$ can also be performed without the use of (2.1.2), by noting that, for $p \in [0, 1]$,

$$\tau_p(x) = p^{h(x)} , \qquad (2.1.6)$$

and the fact that, for $n \geq 1$, there are $r(r-1)^{n-1}$ vertices in $\mathbb{T}_r$ at height n, so that, for $p < 1/(r-1)$,

$$\chi(p) = 1 + \sum_{n=1}^{\infty} r(r-1)^{n-1} p^n = 1 + \frac{rp}{1 - (r-1)p} = \frac{1+p}{1 - (r-1)p} . \qquad (2.1.7)$$

However, for related results for percolation on a tree, the connection to branching processes in (2.1.2) is vital.

Proof that ($\Delta = 2$) on the Tree. You do it:

Exercise 2.1 ($\Delta = 2$ on the tree). Prove that $\Delta = 2$ on the tree.

Proof that ($\beta = 1$) on the Tree. We continue to investigate the critical exponent β for the percolation function on the tree. Let $\theta_{BP}(p) = \mathbb{P}_p(|\mathcal{C}_{BP}(x)| = \infty)$. Then $\theta_{BP}(p)$ is the survival probability of a branching process with a binomial offspring distribution with parameters $r - 1$ and p. Thus, $\theta_{BP}(p)$ satisfies the equation

$$\theta_{BP}(p) = 1 - \left(1 - p + p(1 - \theta_{BP}(p))\right)^{r-1} = 1 - \left(1 - p\theta_{BP}(p)\right)^{r-1} . \tag{2.1.8}$$

To compute $\theta_{BP}(p)$, it is more convenient to work with the extinction probability $\zeta_{BP}(p) = 1 - \theta_{BP}(p)$, which is the probability that the branching process dies out. The extinction probability $\zeta_{BP}(p)$ satisfies

$$\zeta_{BP}(p) = \left(1 - p + p\zeta_{BP}(p)\right)^{r-1} . \tag{2.1.9}$$

This equation can be seen by noting that each of the $r - 1$ possible children of the root needs to die out for the process to go extinct. By the absence of cycles, these events are independent and have the same probability, which explains the power $r - 1$ in (2.1.9). Further, the probability that a child of the root dies out equals $1 - p + p\zeta_{BP}(p)$, since either the edge leading to it is vacant, or the edge leading to it is occupied and then the branching process generated from this child needs to die out as well.

Equation (2.1.9) can be solved explicitly when $r = 2$ (the 'line graph'), where the unique solution is $\zeta_{BP}(p) = 1$ for $p \in [0, 1)$ and $\zeta_{BP}(1) = 0$. As a result, $\theta_{BP}(p) = 0$ for $p \in [0, 1)$ and $\theta_{BP}(1) = 1$, so that $p_c(\mathbb{T}_2) = 1$. Having dealt with $r = 2$, we henceforth assume $r \geq 3$.
When $r = 3$, (2.1.9) reduces to

$$p^2\zeta_{BP}(p)^2 + (2p(1 - p) - 1)\zeta_{BP}(p) + (1 - p)^2 = 0 , \tag{2.1.10}$$

so that

$$\zeta_{BP}(p) = \frac{1 - 2p(1 - p) \pm |2p - 1|}{2p^2} . \tag{2.1.11}$$

Since $\zeta_{BP}(1) = 0$, we must have that

$$\zeta_{BP}(p) = \frac{1 - 2p(1 - p) - |2p - 1|}{2p^2} , \tag{2.1.12}$$

so that $\zeta_{BP}(p) = 1$ for $p \in [0, \frac{1}{2}]$, while, for $p \in [\frac{1}{2}, 1]$,

$$\zeta_{\mathrm{BP}}(p) = \frac{1 - 2p(1-p) + (1-2p)}{2p^2} = \frac{2 - 4p + 2p^2}{2p^2} = \left(\frac{1-p}{p}\right)^2 . \qquad (2.1.13)$$

As a result, we have the explicit form $\theta_{\mathrm{BP}}(p) = 0$ for $p \in [0, \frac{1}{2}]$ and

$$\theta_{\mathrm{BP}}(p) = 1 - \left(\frac{1-p}{p}\right)^2 = \frac{2p-1}{p^2} , \qquad (2.1.14)$$

for $p \in [\frac{1}{2}, 1]$, so that $p_c(\mathbb{T}_3) = \frac{1}{2}$. In particular, $p \mapsto \theta_{\mathrm{BP}}(p)$ is continuous, and

$$\theta_{\mathrm{BP}}(p) = 8(p - p_c)\big(1 + o(1)\big) \quad \text{as } p \searrow p_c . \qquad (2.1.15)$$

It is not hard to see that (2.1.15) together with (2.1.1) implies that

$$\theta(p) = 12(p - p_c)\big(1 + o(1)\big) \quad \text{as } p \searrow p_c . \qquad (2.1.16)$$

Thus, for $r = 3$, the percolation function is continuous and $\beta = 1$ in the asymptotic sense.

One can easily extend the asymptotic analysis in (2.1.15)–(2.1.16) to $r \geq 4$, for which $p_c(\mathbb{T}_r) = p_T(\mathbb{T}_r) = 1/(r-1)$. We leave this as an exercise:

Exercise 2.2 (Asymptotics $p \mapsto \theta(p)$ for $\mathbb{T}_r$ with $r \geq 4$). Prove that, on $\mathbb{T}_r$ with $r \geq 4$,

$$\theta_{\mathrm{BP}}(p) = \frac{2(r-1)^2}{r-2}(p - p_c)\big(1 + o(1)\big) , \qquad (2.1.17)$$

and

$$\theta(p) = \frac{2r(r-1)}{r-2}(p - p_c)\big(1 + o(1)\big) . \qquad (2.1.18)$$

Proof that $\gamma' = 1$ on the Tree. In order to study $\chi^{\mathrm{f}}_{\mathrm{BP}}(p) = \mathbb{E}_p[|\mathcal{C}_{\mathrm{BP}}(x)|\mathbb{1}_{\{|\mathcal{C}_{\mathrm{BP}}(x)|<\infty\}}]$ for $p > p_c = 1/(r-1)$, we make use of the fact that

$$\chi^{\mathrm{f}}_{\mathrm{BP}}(p) = \big(1 - \theta_{\mathrm{BP}}(p)\big)\,\mathbb{E}_p[|\mathcal{C}_{\mathrm{BP}}(x)| \mid |\mathcal{C}_{\mathrm{BP}}(x)| < \infty] , \qquad (2.1.19)$$

and the conditional law of percolation on the tree given that $|\mathcal{C}_{\mathrm{BP}}(x)| < \infty$ is percolation on a tree with p replaced by the *dual* percolation probability p_d given by

$$p_d = p\big(1 - \theta_{\mathrm{BP}}(p)\big) . \qquad (2.1.20)$$

Indeed, each of the edges incident to the root that is occupied needs to be leading to a vertex that dies out itself, and all these events are independent. This explains that the offspring distribution of the root is binomial with parameter p_d as in (2.1.20) and $r - 1$. But then each of the children of the root is again conditioned to go extinct, so that also their offspring distribution is binomial with parameters p_d and $r - 1$.

The crucial fact is that $p_d < p_c(\mathbb{T}_r)$, which follows from the equality $1 - \theta_{BP}(p) = \zeta_{BP}(p)$, (2.1.9) and the fact that

$$
\begin{aligned}
(r-1)p\zeta_{BP}(p) &= (r-1)p\big(1 - p + p\zeta_{BP}(p)\big)^{r-1} \\
&< (r-1)p\big(1 - p + p\zeta_{BP}(p)\big)^{r-2} \\
&= \frac{d}{ds}(1 - p + ps)^{r-1}\Big|_{s=\zeta_{BP}(p)}.
\end{aligned}
\tag{2.1.21}
$$

Since $\zeta_{BP}(p)$ is the smallest solution of $(1 - p + ps)^{r-1} = s$, this implies that the derivative of $(1 - p + ps)^{r-1}$ at $s = \zeta_{BP}(p)$ is strictly bounded above by 1 for $p > p_c(\mathbb{T}_r)$. Thus, by conditioning a supercritical cluster in percolation on a tree to die out, we obtain a subcritical cluster at an appropriate subcritical p_d which is related to the original percolation parameter. This fact is sometimes called the *discrete duality principle*.

We use (2.1.7) to conclude that

$$
\begin{aligned}
\chi_{BP}^f(p) &= \big(1 - \theta_{BP}(p)\big)\, \mathbb{E}_p[|\mathscr{C}_{BP}(x)| \mid |\mathscr{C}_{BP}(x)| < \infty] \\
&= \big(1 - \theta_{BP}(p)\big)\frac{1}{1 - (r-1)p\big(1 - \theta_{BP}(p)\big)}.
\end{aligned}
\tag{2.1.22}
$$

Using that $\theta_{BP}(p) = C(p - p_c)\big(1 + o(1)\big)$ for $C > 1$, cf. (2.1.17), in the asymptotic sense then gives that

$$
\chi_{BP}^f(p) = \frac{C_{\gamma'} + o(1)}{p - p_c}.
\tag{2.1.23}
$$

By (2.1.1), this can easily be transferred to $\chi^f(p)$, so that also $\gamma' = 1$ in the asymptotic sense. *Proof that* $(\Delta' = 2)$ *on the Tree.* The above analysis can be extended to $\Delta' = 2$, by looking at higher moments of the cluster size conditioned to be finite. By the duality described above, this follows from the fact that $\Delta = 2$ and $\beta = 1$.

Proof that $(\delta = 2)$ *on the Tree.* We can compute δ by using the *random walk hitting time theorem*, see Grimmett's percolation book [122, Prop. 10.22] or the more recent proof by the second author and Keane in [162], where a simple proof is given for general branching processes. This result yields that

$$
\mathbb{P}_p(|\mathscr{C}_{BP}(x)| = k) = \frac{1}{k}\,\mathbb{P}(X_1 + \cdots + X_k = k - 1),
\tag{2.1.24}
$$

where $(X_i)_{i \geq 1}$ is an i.i.d. sequence of binomial random variables with parameter $r - 1$ and success probability p.

Exercise 2.3 (Branching processes and random walks). Prove that

$$\mathbb{P}_p(|\mathcal{C}_{\mathrm{BP}}(x)| = k) = \mathbb{P}_1(S_k = 0 \text{ for the first time}), \tag{2.1.25}$$

where $S_k = 1 + \sum_{i=1}^{k}(X_i - 1)$.

Exercise 2.4 (Random walk hitting time theorem). Prove (2.1.24) using induction on the equality

$$\mathbb{P}_m(S_k = 0 \text{ for first time}) = \frac{m}{k}\mathbb{P}_1(X_1 + \cdots + X_k = k - 1), \tag{2.1.26}$$

where $S_k = m + \sum_{i=1}^{k}(X_i - 1)$, and the previous exercise.

From (2.1.24), we conclude that

$$\mathbb{P}_p(|\mathcal{C}_{\mathrm{BP}}(x)| = k) = \frac{1}{k}\binom{k(r-1)}{k-1}p^{k-1}(1-p)^{k(r-1)-(k-1)}. \tag{2.1.27}$$

To prove that $\delta = 2$, we note that for $p = p_c(\mathbb{T}_r) = 1/(r-1)$, by a local limit theorem and for some C_δ,

$$\mathbb{P}_{p_c}(|\mathcal{C}_{\mathrm{BP}}(x)| = k) = \left(C_\delta/2 + o(1)\right)\frac{1}{\sqrt{k^3}}. \tag{2.1.28}$$

Summing over $k \geq n$, we obtain

$$\mathbb{P}_{p_c}(|\mathcal{C}_{\mathrm{BP}}(x)| \geq n) = \sum_{k \geq n}\left(C_\delta/2 + o(1)\right)\frac{1}{\sqrt{k^3}} = \frac{C_\delta + o(1)}{\sqrt{n}}. \tag{2.1.29}$$

This proves that $\delta = 2$ in an asymptotic sense on the tree.

Proof that $(\rho_{in} = 1)$ *on the Tree.* We can compute ρ_{in} by noting that

$$\theta_n = \mathbb{P}_{p_c}(\exists v \in \mathcal{C}_{\mathrm{BP}}(o) \text{ such that } h(v) = n) \tag{2.1.30}$$

satisfies the recursion relation

$$1 - \theta_n = (1 - p_c\theta_{n-1})^{r-1}. \tag{2.1.31}$$

It is not hard to see that (2.1.31) together with $p_c(\mathbb{T}_r) = 1/(r-1)$ implies that $\theta_n = (C_{in} + o(1))/n$, so that $\rho_{in} = 1$. This is left as Exerc. 2.5 below. $\square$

Exercise 2.5 (Proof of (2.1.31) and its asymptotics). (a) Prove (2.1.31).

(b) Prove that (2.1.31) implies that $\theta_n = 2/(\sigma^2 n)(1 + O(1/n))$, where $\sigma^2 = (r-2)/(r-1)$ is the variance of the offspring distribution. *Hint*: Perform induction on n for $v_n = 1/\theta_n$.

Exercise 2.6 (A lower bound on p_c for general graphs). (a) Use the percolation critical values $p_c(\mathbb{T}_r) = p_T(\mathbb{T}_r) = 1/(r-1)$ on the r-regular tree $\mathbb{T}_r$ in Thm. 2.1 to show that $p_c(\mathscr{G}) \geq 1/r$ when $\mathscr{G}$ is a transitive graph with degree r.

(b) Improve the bound in part (a) to $p_c(\mathscr{G}) \geq 1/(r-1)$ when $\mathscr{G}$ is a transitive graph with degree r.

The computation of the key objects for percolation on a tree is feasible due to the close relationship to branching processes, a topic which has attracted substantial interest in the probability community. See the books by Athreya and Ney [23], Harris [142] and Jagers [183] for detailed discussions about branching processes.

2.2 Branching Random Walk as the Percolation Mean-Field Model

We next argue that *branching random walk,* henceforth abbreviated by BRW, can be viewed as the mean-field model for percolation, and we shall see that the critical behavior of percolation in high dimensions is closely related to the critical behavior of BRW. Of course, percolation on the tree lacks a geometric embedding into (Euclidean) space. Therefore, the critical exponents $\rho_{\text{ex}}, \nu, \nu'$,and η are not so easily defined on the tree.

BRW is a *random embedding* of a branching process with $\text{Bin}(2d, p)$-offspring distribution into $\mathbb{Z}^d$. This can be intuitively understood as follows. Every vertex x in percolation on $\mathbb{Z}^d$ has a binomial number of neighbors with parameters p and $2d$ for which the edge leading to it is occupied. Thus, one could imagine *exploring* a cluster vertex by vertex. In high dimensions, space is quite vast, so that it is *relatively rare* to close a cycle. Cycles form the difference between percolation on a tree and percolation in $\mathbb{Z}^d$. BRW is precisely the process in which we *ignore cycles.* Thus, one might hope that BRW is closely related to percolation in sufficiently high dimensions. One of the main aims of this text is to make this intuition precise. Before starting with that, though, let us first investigate BRW in more detail, so as to obtain insight in the kind of results that we might be able to show for high-dimensional percolation.

We may think of branching random walk as percolation on the $2d$-ary tree that is embedded into the Euclidean lattice $\mathbb{Z}^d$, and we explain this *BRW embedding* now. For every $v \in \mathbb{T}_r$, we associate a spatial location $\phi(v) \in \mathbb{Z}^d$ in the following (random) way. We let $\phi(o) = 0$, so that the root in $\mathbb{T}_r$ is mapped to the origin in $\mathbb{Z}^d$. Further, for $v \in \mathbb{T}_r$, we let $p(v) \in \mathbb{T}_r$ denote the unique parent of v, i.e., the neighbor of v that is on the unique path to the root o in the tree $\mathbb{T}_r$. Then, for every $v \in \mathbb{T}_r$ having parent $p(v) \in \mathbb{T}_r$, we let $\phi(v) = \phi(p(v)) + Y_v$, where $Y_v \in \mathbb{Z}^d$ is a random neighbor of the origin, i.e., for every e with $|e| = 1$,

$$\mathbb{P}(Y_v = e) = 1/(2d) . \tag{2.2.1}$$

The random variables $(Y_v)_{v \in \mathbb{T}_r \setminus \{o\}}$ form a collection of i.i.d. random variables. We denote the *BRW two-point function* $G_p(x)$ by

$$G_p(x) = \mathbb{E}_p \left[\sum_{v \in \mathcal{C}(o)} \mathbb{1}_{\{\phi(v)=x\}} \right], \quad x \in \mathbb{Z}^d, \tag{2.2.2}$$

and its truncated version by

$$G_p^{\mathrm{f}}(x) = \mathbb{E}_p \left[\sum_{v \in \mathcal{C}(o)} \mathbb{1}_{\{\phi(v)=x\}} \mathbb{1}_{\{|\mathcal{C}(o)|<\infty\}} \right], \quad x \in \mathbb{Z}^d. \tag{2.2.3}$$

The BRW two-point functions $G_p(x)$ and $G_p^{\mathrm{f}}(x)$ have similar interpretations as the percolation two-point functions $\tau_p(x)$ and $\tau_p^{\mathrm{f}}(x)$ in (1.2.6)–(1.2.7).

Alternatively, denoting by $N(x)$ the total number of particles in $\mathcal{C}(o)$ that are mapped to $x \in \mathbb{Z}^d$ by ϕ,

$$\begin{aligned}
G_p^{\mathrm{f}}(x) &= \mathbb{E}_p[N(x)\mathbb{1}_{\{|\mathcal{C}(o)|<\infty\}}] \\
&= \mathbb{E}_p \left[\sum_{v \in \mathbb{T}_r} \mathbb{1}_{\{\phi(v)=x\}} \mathbb{1}_{\{v \in \mathcal{C}(o)\}} \mathbb{1}_{\{|\mathcal{C}(o)|<\infty\}} \right] \\
&= \sum_{v \in \mathbb{T}_r} \mathbb{P}_p \left(\phi(v) = x, v \in \mathcal{C}(o), |\mathcal{C}(o)| < \infty \right) \\
&= \sum_{v \in \mathbb{T}_r} \mathbb{P}(\phi(v) = x) \, \mathbb{P}_p(v \in \mathcal{C}(o), |\mathcal{C}(o)| < \infty),
\end{aligned} \tag{2.2.4}$$

the latter by the independence of the embedding of the tree and the occupation statuses of the bonds.

We now turn to $p \leq p_c(\mathbb{T}_r) = 1/(r-1)$, in which case we can remove the condition that $|\mathcal{C}(o)| < \infty$. When $h(v) = n$, in order for $\{v \in \mathcal{C}(o)\}$ to occur, all the n edges on the path between o and v have to be occupied, so that $\mathbb{P}_p(v \in \mathcal{C}(o)) = p^n$. Further,

$$\mathbb{P}(\phi(v) = x) = \mathbb{P} \left(\sum_{u \in \pi_v} Y_u = x \right), \tag{2.2.5}$$

where π_v contains all the vertices on the unique path between o and v. Again, when $h(v) = n$, and by the independence of the random variables $(Y_v)_{v \in \mathbb{T}_r \setminus \{o\}}$,

$$\mathbb{P}(\phi(v) = x) = \mathbb{P} \left(\sum_{i=1}^{n} Y_i = x \right) = D^{*n}(x), \tag{2.2.6}$$

where $D^{\star n}$ denotes the n-fold convolution of D with itself, and we recall that D denotes the simple random walk transition probability defined in (1.2.18). As a result, we obtain that

$$G_p(x) = \sum_{n \geq 0} r(r-1)^{n-1} p^n D^{\star n}(x) = \frac{r}{r-1} C_{(r-1)p}(x) , \qquad (2.2.7)$$

where $C_\mu(x)$ denotes the random walk Green's function given by

$$C_\mu(x) = \sum_{n \geq 0} \mu^n D^{\star n}(x) . \qquad (2.2.8)$$

It is well known that, for any $d \geq 3$, $\mu = 1$ serves as a critical value for the simple random walk Green's function $C_\mu(x)$ and that there exists a constant $A > 0$ such that

$$C_1(x) = \frac{A}{|x|^{d-2}} (1 + o(1)), \qquad (2.2.9)$$

cf. Uchiyama [256]. Probabilistically, $C_1(x)$ describes the expected number of visits to the site x of a random walk starting at the origin. We take $p = p_c(\mathbb{T}_r) = 1/(r-1)$, so that

$$G_{p_c}(x) = \frac{r}{r-1} C_1(x) . \qquad (2.2.10)$$

Thus, (2.2.9) implies that $\eta = 0$ in x-space for BRW.

The connection to BRW yields a powerful intuitive way to predict properties of percolation in high dimensions. Further, it yields a powerful relation between BRW and the random walk Green's function that will prove to be extremely useful later on.

In Fourier language, the fact that $\eta = 0$ is much simpler. Indeed, taking the Fourier transform of (2.2.8) leads to

$$\widehat{C}_\mu(k) = \sum_{n \geq 0} \mu^n \widehat{D}^n(k) = \frac{1}{1 - \mu \widehat{D}(k)} . \qquad (2.2.11)$$

Since $|\widehat{D}(k)| \leq 1$ with $\widehat{D}(0) = 1$, we again see that $\mu = 1$ serves as a critical value. For $\mu = 1$, we obtain

$$\widehat{C}_1(k) = \sum_{n \geq 0} \widehat{D}^n(k) = \frac{1}{1 - \widehat{D}(k)} . \qquad (2.2.12)$$

Using a series expansion of cosine, we obtain for $k \to 0$ that

$$1 - \widehat{D}(k) = \frac{1}{d} \sum_{i=1}^{d} [1 - \cos(k_i)]$$

$$= \frac{1}{2d} \sum_{i=1}^{d} k_i^2 (1 + o(1)) = \frac{1}{2d} |k|^2 (1 + o(1)), \qquad (2.2.13)$$

and arrive at

$$\widehat{C}_1(k) = \frac{2d}{|k|^2} (1 + o(1)) \quad \text{as } k \to 0. \qquad (2.2.14)$$

This proves that $\eta = 0$ in k-space.

BRW also allows us to define ρ_{ex} and ν_2, ν_2', ν, ν'. For ρ_{ex}, we write

$$\mathbb{P}_{p_c}(\exists v \in \mathcal{C}(o): \phi(v) \in \partial \Lambda_n) \sim n^{-1/\rho_{\mathrm{ex}}}, \qquad (2.2.15)$$

recalling that $\partial \Lambda_n$ consists of lattice sites at ℓ^∞-distance n from the origin. The exponents ν and ν' are the critical exponents of the BRW correlation length

$$\xi^{\mathrm{BRW}}(p) = -\lim_{n \to \infty} \left(\frac{\log(G_p^{\mathrm{f}}(ne))}{n} \right)^{-1}, \qquad (2.2.16)$$

where we recall (1.2.7) and (2.2.3). Similarly, the exponents ν_2 and ν_2' are the critical exponents of

$$\xi_2^{\mathrm{BRW}}(p) = \sqrt{\frac{1}{\chi^{\mathrm{f}}(p)} \sum_{x \in \mathbb{Z}^d} |x|^2 G_p^{\mathrm{f}}(x)}, \qquad (2.2.17)$$

where we recall (1.2.8) and (2.2.3). The following theorem collects results for the critical behavior of BRW:

Theorem 2.2 (Critical behavior branching random walk). *For BRW with a binomial offspring distribution with parameters $r - 1$ and p, the critical value equals $p_c = p_T = 1/(r - 1)$, and $\beta = \gamma = \gamma' = 1, \delta = \Delta = \Delta' = 2, \eta = 0$, and $\nu_2 = \nu_2' = \nu = \nu' = \rho_{\mathrm{ex}} = \frac{1}{2}$ in the asymptotic sense.*

Proof. All these critical exponents follow from Thm. 2.1, except $\eta, \nu_2, \nu_2', \nu, \nu'$, and ρ_{ex}. The fact that $\eta = 0$ follows from (2.2.9) and (2.2.10).

For ν_2 and ν_2', we note that

$$G_p^{\mathrm{f}}(x) = \sum_{v \in \mathbb{T}_r} D^{\star h(v)}(x) \tau_p^{\mathrm{f}}(v). \qquad (2.2.18)$$

Fix $p < p_c = 1/(r - 1)$, so that $\tau_p^{\mathrm{f}}(v) = \tau_p(v) = p^{h(v)}$. Then, using the simple random walk variance given by $\sum_{x \in \mathbb{Z}^d} |x|^2 D^{\star n}(x) = n$, we compute

$$\sum_{x \in \mathbb{Z}^d} |x|^2 G_p^{\mathrm{f}}(x) = \sum_v \sum_{x \in \mathbb{Z}^d} |x|^2 D^{\star h(v)}(x) \tau_p^{\mathrm{f}}(v) = \sum_v h(v) \tau_p^{\mathrm{f}}(v)$$

$$= \sum_{n \geq 1} n r (r-1)^{n-1} p^n = \frac{rp}{[1-(r-1)p]^2} \, , \tag{2.2.19}$$

so that $v_2 = \frac{1}{2}$ for BRW. This can be extended to $v_2' = \frac{1}{2}$ by using the duality between supercritical BRW conditioned to go extinct and subcritical BRW as discussed around (2.1.20). The fact that $v = \frac{1}{2}$ for BRW follows from a careful analysis of $C_\mu(ne_1)$ when $n \to \infty$, using large deviations for random walks. Again this can be extended to $v' = \frac{1}{2}$ by using the duality between supercritical BRW conditioned to go extinct and subcritical BRW.

The proof that $\rho_{\mathrm{ex}} = \frac{1}{2}$ is more involved and is therefore omitted here. $\qquad\square$

Chapter 3
Uniqueness of the Phase Transition

In this chapter, we state and prove the celebrated result that $p_c = p_T$, which was independently proved by Menshikov [215] and by Aizenman and Barsky [9]. This theorem is the starting point of the investigation of high-dimensional percolation. We first state the result in Sect. 3.1. We then give two separate proofs of the uniqueness of the phase transition. In Sect. 3.2, we first give the recent and beautiful proof by Duminil-Copin and Tassion [94]. After this, we give the proof by Aizenman and Barsky [9] by first stating its key result in Sect. 3.3, defining its central quantity, the *percolation magnetization*, in Sect. 3.4, as well as two differential inequalities that the magnetization satisfies. In Sect. 3.5, we infer a lower bound on the magnetization for $p = p_c$. In Sect. 3.6, we complete the Aizenman–Barsky uniqueness proof. Finally, in Sect. 3.7, we prove the crucial Aizenman–Barsky differential inequalities. These differential inequalities also play a pivotal role in the identification of mean-field critical exponents for percolation in high dimensions.

3.1 Main Result

The main result of this chapter is the following theorem:

Theorem 3.1 (Uniqueness of the phase transition [9, 215]). *For every $p < p_c$,*

$$\mathbb{E}_p|\mathcal{C}(0)| < \infty. \tag{3.1.1}$$

As a result, $p_c = p_T$.

Three fairly different proofs of this theorem exist. The one by Menshikov [215] (also presented by Grimmett [122, Sect. 5.2]) shows that the probability of the event A_n that there exists a vertex x with $\|x\|_1 \geq n$ such that 0 is connected to it decays exponentially for any $p < p_c(\mathbb{Z}^d)$. This immediately implies that $\mathbb{E}_p|\mathcal{C}(0)| < \infty$ for all $p < p_c(\mathbb{Z}^d)$. The proof relies on Russo's formula, as well as a lower bound on the expected number of pivotals for the event A_n in terms of the probability of A_n itself, and then bootstraps this to exponential

© Springer International Publishing Switzerland 2017

M. Heydenreich and R. van der Hofstad, *Progress in High-Dimensional Percolation and Random Graphs*, CRM Short Courses, DOI 10.1007/978-3-319-62473-0_3

decay of $\mathbb{P}_p(A_n)$ for n large. The second proof is by Aizenman and Barsky [9], and a recent proof is given by Duminil-Copin and Tassion [94]. We start with the latter, as it is quite simple and insightful. After this, we give the proof by Aizenman and Barsky [9], which relies on the magnetization and differential inequalities for it. The proof by Aizenman and Barsky turns out to be instrumental also to prove the existence of the critical exponents β and δ in high dimensions in Chap. 9.

3.2   The Duminil-Copin and Tassion Uniqueness Proof

We consider percolation on $\mathbb{Z}^d$ with $d \geq 2$. We start with some notation. Recall that the Λ_n denotes the cube with radius n, and $\partial \Lambda_n = \Lambda_n \setminus \Lambda_{n-1}$.

Theorem 3.2 (Equality of critical values and exponential decay [94]). *For any $d \geq 2$,*

(1) *For $p < p_c$, there exists $c = c(p, d) > 0$ such that for every $n \geq 1$,*

$$\mathbb{P}_p(0 \leftrightarrow \partial \Lambda_n) \leq e^{-cn} . \tag{3.2.1}$$

(2) *For $p > p_c$,*

$$\theta(p) \geq \frac{p - p_c}{p(1 - p_c)} . \tag{3.2.2}$$

(3) $\chi(p_c) = \infty$.

Indeed, (3.2.1) implies (3.1.1), so that Thm. 3.1 follows. The ingenious idea of Duminil-Copin and Tassion is yet another characterization for the critical value p_c, and we introduce this value now. For any (finite) vertex set $S \subset \mathbb{Z}^d$, we let ΔS denote its *edge boundary*, that is, all the (directed) edges (x, y) with $x \in S$ and $y \notin S$.

We say that x *is connected to* y *in* S when there exists $k \in \mathbb{N}$ and a path $\{v_i\}_{i=0}^k \subseteq S$ such that $v_0 = x$ and $v_k = y$, for which $\{v_{i-1}, v_i\}$ is occupied for every $i = 1, \ldots, k$. We write this as $\{x \leftrightarrow y \text{ in } S\}$.

From now on, we let S always stand for a finite set of vertices containing the origin, and for such S we define

$$\varphi_p(S) = p \sum_{(x,y) \in \Delta S} \mathbb{P}_p(0 \leftrightarrow x \text{ in } S) . \tag{3.2.3}$$

In words, $\varphi_p(S)$ is the expected number of occupied edges (x, y) in the boundary of S for which there is a path of bonds inside S that connects the origin to x. The novel critical value is then

$$\tilde{p}_c(\mathbb{Z}^d) = \sup\{p \in [0, 1] : \exists \text{ a finite set } S \text{ with } 0 \in S \subset \mathbb{Z}^d$$
$$\text{such that } \varphi_p(S) < 1\} . \tag{3.2.4}$$

The proof of Thm. 3.2 proceeds by showing assertion (1)–(3) of the theorem for $\tilde{p}_c$ rather than for p_c. This shows that there is exponential decay of $\mathbb{P}_p(0 \leftrightarrow \partial\Lambda_n)$ for $p < \tilde{p}_c$ and proves a linear lower bound on $\theta(p)$ for $p > \tilde{p}_c$. Consequently, $\tilde{p}_c = p_c$, and the theorem follows.

The argument uses an exploration like the one by Hammersley [127], but it is simplified significantly by using the BK inequality, which was not yet known to Hammersley.

Proof of subcritical exponential decay in Thm. 3.2(1). We start by proving Thm. 3.2(1). Let $p < \tilde{p}_c$, then there exists a finite set $S \subset \mathbb{Z}^d$ containing the origin with $\varphi_p(S) < 1$. Let L be so big that $S \subseteq \Lambda_{L-1}$. Assume that the event $0 \leftrightarrow \partial\Lambda_{kL}$ holds. We now prove that $\mathbb{P}_p(0 \leftrightarrow \partial\Lambda_{kL}) \leq \varphi_p(S)^k$ for every $k \geq 1$. To this end, we fix $k \geq 1$ and define

$$\mathcal{C}_S = \{z \in S : 0 \leftrightarrow z \text{ in } S\} \tag{3.2.5}$$

to be the cluster of 0 when restricting attention to the bonds in S. Since $S \cap \partial\Lambda_{kL} = \varnothing$, there must exist at least one edge $(x, y) \in \Delta S$ such that

(a) 0 is connected to x in S;
(b) (x, y) is occupied;
(c) y is connected to $\partial\Lambda_{kL}$ in $\mathcal{C}_S^c$.

Therefore, by a union bound followed by a decomposition with respect to the possible values of $\mathcal{C}_S$, we see that

$$\mathbb{P}_p(0 \leftrightarrow \partial\Lambda_{kL})$$
$$\leq \sum_{(x,y)\in\Delta S} \sum_{A\subseteq S} \mathbb{P}_p(\{0 \leftrightarrow x \text{ in } S, \mathcal{C}_S = A\} \cap \{(x, y) \text{ occ.}\}$$
$$\cap \{y \leftrightarrow \partial\Lambda_{kL} \text{ in } \mathbb{Z}^d \setminus A\})$$
$$= p \sum_{(x,y)\in\Delta S} \sum_{A\subseteq S} \mathbb{P}_p(0 \leftrightarrow x \text{ in } S, \mathcal{C}_S = A)$$
$$\times \mathbb{P}_p(y \leftrightarrow \partial\Lambda_{kL} \text{ in } \mathbb{Z}^d \setminus A) . \tag{3.2.6}$$

Here we abbreviate $\{(x, y) \text{ occ.}\}$ for the event that the directed bond (x, y) is occupied, and we use the fact that the events $\{0 \leftrightarrow x \text{ in } S, \mathcal{C}_S = A\}$, $\{(x, y) \text{ occ.}\}$ and $\{y \leftrightarrow \partial\Lambda_{kL} \text{ in } \mathbb{Z}^d \setminus A\}$ are *independent*. This follows since the first event depends only on bonds with both end points in $A \cap S$ or bonds in $\Delta A \setminus \Delta S$ (recall that ΔS denotes the edge boundary of S), the bond (x, y) is in ΔS, while the third event only depends on bonds with both end points outside A:

Exercise 3.1 (Check independence). Prove the independence in (3.2.6).

Since $\{y \leftrightarrow \partial\Lambda_{kL}\}$ is an increasing event and $y \in \Lambda_L$,

$$\mathbb{P}_p(y \leftrightarrow \partial\Lambda_{kL} \text{ in } \mathbb{Z}^d \setminus \mathcal{C}) \leq \mathbb{P}_p(y \leftrightarrow \partial\Lambda_{kL})$$
$$\leq \mathbb{P}_p(0 \leftrightarrow \partial\Lambda_{(k-1)L}) , \tag{3.2.7}$$

which leads us to

$$\mathbb{P}_p(0 \leftrightarrow \partial \Lambda_{kL})$$

$$\leq p \sum_{(x,y) \in \Delta S} \sum_{A \subseteq S} \mathbb{P}_p(0 \leftrightarrow x \text{ in } S, \mathcal{C}_S = A) \, \mathbb{P}_p(0 \leftrightarrow \partial \Lambda_{(k-1)L})$$

$$= \varphi_p(S) \, \mathbb{P}_p(0 \leftrightarrow \partial \Lambda_{(k-1)L}) \,. \tag{3.2.8}$$

Iterating this inequality leads to

$$\mathbb{P}_p(0 \leftrightarrow \partial \Lambda_{kL}) \leq \varphi_p(S)^k \,. \tag{3.2.9}$$

This proves Thm. 3.2(1) with $c = (1/L)\log(1/\varphi_p(S)) > 0$, since $\varphi_p(S) < 1$. □

Proof of supercritical mean-field lower bound on $\theta(p)$ in Thm. 3.2(2). For the proof of Thm. 3.2(2), we prove a *differential inequality* for $p \mapsto \mathbb{P}_p(0 \leftrightarrow \partial \Lambda_{kL})$:

Lemma 3.3 (Differential inequality for the one-arm probability [94]). *For every* $p \in [0, 1]$ *and* $n \geq 1$,

$$\frac{\mathrm{d}}{\mathrm{d}p} \mathbb{P}_p(0 \leftrightarrow \partial \Lambda_n) \geq \frac{1}{p(1-p)} \cdot \left(\inf_{0 \in S \subseteq \Lambda_n} \varphi_p(S) \right) [1 - \mathbb{P}_p(0 \leftrightarrow \partial \Lambda_n)] \,. \tag{3.2.10}$$

Before proving Lem. 3.3, we show how it implies Thm. 3.2(2). By assumption, for $p > \tilde{p}_c$, $\inf_{0 \in S \subseteq \Lambda_n} \varphi_p(S) \geq 1$. Then, let $f_n(p) = \mathbb{P}_p(0 \leftrightarrow \partial \Lambda_n)$, so that

$$\frac{\mathrm{d}}{\mathrm{d}p} \log[1/(1 - f_n(p))] = \frac{\frac{\mathrm{d}}{\mathrm{d}p} f_n(p)}{1 - f_n(p)} \geq \frac{1}{p(1-p)} = \frac{1}{p} + \frac{1}{1-p} \,. \tag{3.2.11}$$

We integrate (3.2.11) between $\tilde{p}_c$ and p to obtain

$$\log[1/(1 - f_n(p))] - \log[1/(1 - f_n(\tilde{p}_c))]$$
$$\geq \log(p/\tilde{p}_c) - \log[(1-p)/(1-\tilde{p}_c)] \,, \tag{3.2.12}$$

so that also

$$\log[1/(1 - f_n(p))] \geq \log(p/\tilde{p}_c) - \log[(1-p)/(1-\tilde{p}_c)] \,, \tag{3.2.13}$$

which implies

$$\frac{1}{1 - f_n(p)} \geq \frac{p(1-\tilde{p}_c)}{\tilde{p}_c(1-p)} \,. \tag{3.2.14}$$

We conclude that

$$\mathbb{P}_p(0 \leftrightarrow \partial \Lambda_n) = f_n(p) \geq 1 - \frac{\tilde{p}_c(1-p)}{p(1-\tilde{p}_c)}$$
$$= \frac{p(1-\tilde{p}_c) - \tilde{p}_c(1-p)}{p(1-\tilde{p}_c)} = \frac{p - \tilde{p}_c}{p(1-\tilde{p}_c)} \,. \tag{3.2.15}$$

Since this lower bound is independent of n, we can take $n \to \infty$ to obtain

$$\theta(p) \geq \frac{p - \tilde{p}_c}{p(1 - \tilde{p}_c)} ,$$

(3.2.16)

as required.

□

Proof of Lem. 3.3. We note that the event $\{0 \leftrightarrow \partial \Lambda_n\}$ depends only on the occupation status of the bonds inside Λ_n, thus we may apply Russo's formula (1.3.9) to obtain

$$\frac{d}{dp} \mathbb{P}_p(0 \leftrightarrow \partial \Lambda_n)$$

$$= \sum_{(x,y)} \mathbb{P}_p((x, y) \text{ pivotal for } 0 \leftrightarrow \partial \Lambda_n)$$

$$= \frac{1}{1 - p} \sum_{(x,y)} \mathbb{P}_p((x, y) \text{ pivotal for } 0 \leftrightarrow \partial \Lambda_n, 0 \nleftrightarrow \partial \Lambda_n) .$$

(3.2.17)

In the second line, we have used that the occupation of the bond (x, y) is independent of it being pivotal. Define

$$\mathcal{S} = \{x \in \Lambda_n \text{ such that } x \nleftrightarrow \partial \Lambda_n\} .$$

(3.2.18)

We note that $\mathcal{S}$ can be obtained by exploring clusters from the vertices in the boundary $\partial \Lambda_n$. When $0 \nleftrightarrow \partial \Lambda_n$, the set $\mathcal{S}$ is a subset of Λ_n containing the origin. We sum over all possible choices of $\mathcal{S}$ to obtain

$$\frac{d}{dp} \mathbb{P}_p(0 \leftrightarrow \partial \Lambda_n)$$

$$= \frac{1}{1 - p} \sum_{0 \in S \subseteq \Lambda_n} \sum_{(x,y)} \mathbb{P}_p((x, y) \text{ pivotal for } 0 \leftrightarrow \partial \Lambda_n, \mathcal{S} = S) .$$

(3.2.19)

Observe that on the event $\{\mathcal{S} = S\}$ with $0 \in S$, the pivotal edges (x, y) for $0 \leftrightarrow \partial \Lambda_n$ are those in ΔS for which $0 \leftrightarrow x$ in S:

Exercise 3.2 (Characterization of pivotal edges). Prove that on the event $\{\mathcal{S} = S\}$, the pivotal edges b for $\{0 \leftrightarrow \partial \Lambda_n\}$ are given by $b = (x, y) \in \Delta S$ such that $\{0 \leftrightarrow x$ in $S\}$.

Therefore,

$$\frac{d}{dp} \mathbb{P}_p(0 \leftrightarrow \partial \Lambda_n) = \frac{1}{1 - p} \sum_{0 \in S \subseteq \Lambda_n} \sum_{(x,y) \in \Delta S} \mathbb{P}_p(0 \leftrightarrow x \text{ in } S, \mathcal{S} = S).$$

(3.2.20)

The event $\{\mathcal{S} = S\}$ is measurable with respect to the edges with at least one end point outside of S, while $\{0 \leftrightarrow x$ in $S\}$ is measurable with respect to the edges having both end points in S, and are therefore independent. We conclude that

$$\frac{d}{dp} \mathbb{P}_p(0 \leftrightarrow \partial \Lambda_n) = \frac{1}{1-p} \sum_{0 \in S \subseteq \Lambda_n} \sum_{(x,y) \in \Delta S} \mathbb{P}_p(0 \leftrightarrow x \text{ in } S) \mathbb{P}_p(\mathcal{S} = S)$$

$$= \frac{1}{p(1-p)} \sum_{0 \in S \subseteq \Lambda_n} \varphi_p(S) \mathbb{P}_p(\mathcal{S} = S)$$

$$\geq \frac{1}{p(1-p)} \left(\inf_{0 \in S \subseteq \Lambda_n} \varphi_p(S) \right) \sum_{0 \in S \subseteq \Lambda_n} \mathbb{P}_p(\mathcal{S} = S) \tag{3.2.21}$$

$$= \frac{1}{p(1-p)} \left(\inf_{0 \in S \subseteq \Lambda_n} \varphi_p(S) \right) \mathbb{P}_p(0 \nleftrightarrow \partial \Lambda_n),$$

as desired. Here we note that $\mathcal{S} = S$ for some S with $0 \in S \subseteq \Lambda_n$ precisely when $0 \nleftrightarrow \partial \Lambda_n$.

□

Proof that critical expected cluster size is infinite in Thm. 3.2(3). Assertions (1) and (2) of Thm. 3.2 readily imply that $p_c = \tilde{p}_c$, we are therefore left to prove that $\chi(\tilde{p}_c) = \infty$. We note that the set

$$\mathcal{P} := \{p : \exists S \subset \mathbb{Z}^d, S \ni 0, |S| < \infty \text{ with } \varphi_p(S) < 1\}$$

is an open subset of $[0, 1]$, and hence $p_c = \tilde{p}_c \notin \mathcal{P}$. Thus,

$$1 \leq \varphi_{\tilde{p}_c}(\Lambda_n) \leq 2d\, \tilde{p}_c \sum_{x \in \partial \Lambda_n} \mathbb{P}_{\tilde{p}_c}(0 \leftrightarrow x \text{ in } \Lambda_n)$$

for all $n \in \mathbb{N}$, and

$$\chi(\tilde{p}_c) = \sum_{x \in \mathbb{Z}^d} \mathbb{P}_{\tilde{p}_c}(0 \leftrightarrow x) = \sum_{n \geq 0} \sum_{x \in \partial \Lambda_n} \mathbb{P}_{\tilde{p}_c}(0 \leftrightarrow x)$$

$$\geq \sum_{n \geq 0} \sum_{x \in \partial \Lambda_n} \mathbb{P}_{\tilde{p}_c}(0 \leftrightarrow x \text{ in } \Lambda_n) \tag{3.2.22}$$

$$\geq \sum_{n \geq 1} \frac{1}{2d\, \tilde{p}_c} = \infty,$$

as desired.

□

3.3 The Aizenman–Barsky Uniqueness Proof

In the remainder of this chapter, we discuss the Aizenman–Barsky [9] proof of the uniqueness of the phase transition in Thm. 3.1. The main ingredient of the Aizenman–Barsky proof is the following theorem:

Theorem 3.4 (Bound on $\theta(p)$ when $\chi^f(p) = \infty$ [9]). *If $p \in [0, 1]$ is such that $\chi^f(p) = \infty$, then either $\theta(p) > 0$, or $\theta(p) = 0$ and $\theta(q) \geq (q - p)/(2q)$ for all $q > p$.*

Theorem 3.4 immediately implies the uniqueness of the phase transition in Thm. 3.1:

Proof of Thm. 3.1. Trivially, $p_T \leq p_c$. We argue by contradiction and assume that $p_T < p_c$. Then there exists a $p < p_c$ such that $\chi(p) = \infty$. Since $p < p_c$, we know that $\chi(p) = \chi^f(p)$ (since $|\mathcal{C}(0)| < \infty$ a.s.), so that also $\chi^f(p) = \infty$. By Thm. 3.4 and the fact that $\theta(p) = 0$, it then follows that $\theta(q) \geq (q - p)/(2q)$ for all $q > p$, so that $q \geq p_c$ for every $q > p$. Thus, $p \geq p_c$, which is a contradiction. $\qquad\square$

The remainder of this chapter is devoted to the proof of Thm. 3.4. We start by discussing the *magnetization*, which is an essential tool in studying percolation, particularly in high dimensions.

3.4 The Magnetization

For $\gamma \in [0, 1]$, the percolation *magnetization* is defined by

$$M(p, \gamma) = \sum_{k \geq 1} [1 - (1 - \gamma)^k] \, \mathbb{P}_p(|\mathcal{C}(0)| = k) = \mathbb{E}_p \left[1 - (1 - \gamma)^{|\mathcal{C}(0)|} \right]. \qquad (3.4.1)$$

Alternatively, $M(p, \gamma) = 1 - G_{|\mathcal{C}(0)|}(1 - \gamma)$, where $G_{|\mathcal{C}(0)|}(s) = \mathbb{E}_p[s^{|\mathcal{C}(0)|}]$ denotes the probability generating function of $|\mathcal{C}(0)|$ under the measure $\mathbb{P}_p$.

The quantity $M(p, \gamma)$ also has a useful probabilistic interpretation in terms of a combined bond and site percolation model. For this, we color each vertex $x \in \mathbb{Z}^d$ *green* with probability γ. Denote the resulting probability measure of bond and site variables by $\mathbb{P}_{p,\gamma}$, and denote the (random) set of green sites by $\mathcal{G}$. Then,

$$M(p, \gamma) = \mathbb{P}_{p,\gamma}(0 \leftrightarrow \mathcal{G}) = \mathbb{P}_{p,\gamma}(\mathcal{C}(0) \cap \mathcal{G} \neq \emptyset) \qquad (3.4.2)$$

is the probability that the cluster of the origin contains at least one green vertex.

We investigate the behavior of $M(p, \gamma)$ for p close (or equal) to the critical value p_c, and $\gamma > 0$ small. We note that if $\mathbb{P}_{p_c}(|\mathcal{C}(0)| \geq n) \sim n^{-1/\delta}$, then the generating function $M(p_c, \gamma)$ behaves as $M(p_c, \gamma) \sim \gamma^{1/\delta}$. Such relation go under the name of *Abelian Theorems*, and are fairly straightforward. The reverse statements, known as *Tauberian theorems*, are typically more delicate. Thus, we can use the magnetization to study δ.

Exercise 3.3 (Magnetization and cluster size distribution). Show that $\mathbb{P}_{p_c}(|\mathcal{C}(0)| \geq n) \sim n^{-1/\delta}$ as $n \to \infty$ for some $\delta > 0$ implies $M(p_c, \gamma) \sim \gamma^{1/\delta}$ as $\gamma \searrow 0$.

For fixed p, the function $\gamma \mapsto M(p, \gamma)$ is strictly increasing, with $M(p, 0) = 0$ and $M(p, 1) = 1$. In addition, for $\gamma \in (0, 1)$, $M(p, \gamma)$ is strictly increasing in p. Define the *green-free susceptibility* or (in the probabilistic interpretation) the expected size of a *green-free cluster* by

$$\chi(p, \gamma) = \mathbb{E}_p \left[|\mathcal{C}(0)| (1 - \gamma)^{|\mathcal{C}(0)|} \right] = \mathbb{E}_{p,\gamma} \left[|\mathcal{C}(0)| \mathbb{1}_{\{\mathcal{C}(0) \cap \mathcal{G} = \emptyset\}} \right]. \qquad (3.4.3)$$

Then, we note that $\gamma \mapsto \partial M(p,\gamma)/\partial \gamma = (1-\gamma)^{-1}\chi(p,\gamma)$ is monotone decreasing in $\gamma \in (0,1)$. That the magnetization is a useful quantity can also be seen in the following exercise:

Exercise 3.4 (Magnetization and percolation). Prove that $\lim_{\gamma \searrow 0} M(p,\gamma) = \theta(p)$, while $\lim_{\gamma \searrow 0} \chi(p,\gamma) = \chi^{\mathrm{f}}(p)$.

The probabilistic interpretations of $M(p,\gamma)$ and $\chi(p,\gamma)$ allow one to prove the following *differential inequalities* that lie at the heart of the proof of Thm. 3.4:

Lemma 3.5 (Aizenman–Barsky differential inequalities [9]). *If $0 < p < 1$ and $0 < \gamma < 1$, then*

$$(1-p)\frac{\partial M}{\partial p} \le 2d(1-\gamma)M\frac{\partial M}{\partial \gamma} , \tag{3.4.4}$$

$$M \le \gamma \frac{\partial M}{\partial \gamma} + M^2 + pM\frac{\partial M}{\partial p} . \tag{3.4.5}$$

Lemma 3.5 is proved in Sect. 3.7. We start by using the differential inequalities to derive bounds on the magnetization and related quantities.

3.5 The Lower Bound on the Magnetization

In this section, we prove the following lower bound on the magnetization $M(p,\gamma)$:

Proposition 3.6 (Lower bound on M). Let $p \in (0,1)$ be such that $\chi^{\mathrm{f}}(p) = \infty$, and $\gamma \in (0,1)$, and let $K = 1 + 2dp/(1-p)$. Then,

$$M(p,\gamma) \ge \sqrt{\gamma/K} . \tag{3.5.1}$$

Proof. We fix $p \in (0,1)$ such that $\chi^{\mathrm{f}}(p) = \infty$, and drop the p dependence from the notation. Inserting (3.4.4) into (3.4.5), defining $\widetilde{K} = 2dp/(1-p)$, and using $1 - \gamma \le 1$, we get

$$M \le \gamma \frac{dM}{d\gamma} + M^2 + \widetilde{K}M^2\frac{dM}{d\gamma} . \tag{3.5.2}$$

Since $M > 0$ as long as $\gamma > 0$, we can divide by $M^2 dM/d\gamma$ to get

$$\frac{1}{M}\frac{d\gamma}{dM} - \frac{1}{M^2}\gamma \le \widetilde{K} + \frac{d\gamma}{dM} , \tag{3.5.3}$$

where we are using the fact that $\gamma \mapsto M(p,\gamma)$ has a well-defined inverse function on $[0,1]$. Therefore,

$$\frac{d}{dM}\left(\frac{\gamma}{M}\right) \le \widetilde{K} + \frac{d\gamma}{dM} . \tag{3.5.4}$$

Next we integrate (3.5.4) between 0 and M and use that $\gamma(0) = 0$ and $\lim_{M \to 0} \gamma(M)/M = \gamma'(0) = 1/M'(0) = 1/\chi^{\mathrm{f}}(p) = 0$ to get

$$\frac{\gamma}{M} \leq \widetilde{K} M + \gamma . \tag{3.5.5}$$

Observing that $1 - (1 - \gamma)^k \geq 1 - (1 - \gamma) = \gamma$, we see from (3.4.1) that $\gamma \leq M$, which simplifies (3.5.5) to

$$\frac{\gamma}{M} \leq K M , \tag{3.5.6}$$

where $K = \widetilde{K} + 1$. This completes the proof of (3.5.1). □

Exercise 3.5 If we drop the assumption $\chi^{\mathrm{f}}(p) = \infty$ in Prop. 3.6, show that

$$M(p, \gamma) \geq \sqrt{\frac{\gamma}{K} + \left[\frac{1}{2K\chi^{\mathrm{f}}(p)}\right]^2} - \frac{1}{2K\chi^{\mathrm{f}}(p)} . \tag{3.5.7}$$

3.6 Aizenman–Barsky Proof of Mean-Field Lower Bound on $\theta(p)$

In this section, we prove Thm. 3.4. We follow [122, Proof of Thm. (5.48)]. Fix $\gamma, a \in (0, 1)$ and suppose that $\chi^{\mathrm{f}}(a) = \infty$. If $\theta(a) > 0$, then there is nothing to prove. Thus, we shall henceforth assume that $\theta(a) = 0$ and $\chi^{\mathrm{f}}(a) = \infty$. However, mind that $M(a, \gamma) > 0$, since $\gamma \in (0, 1)$.

We rewrite (3.4.5) as

$$\frac{1}{M(p, \gamma)} \frac{\partial M(p, \gamma)}{\partial \gamma} + \frac{1}{\gamma} \frac{\partial}{\partial p}[pM(p, \gamma) - p] \geq 0 . \tag{3.6.1}$$

We integrate over $a \leq p \leq b$ and $\delta \leq \gamma \leq \varepsilon$, where $0 < a < b < 1, 0 < \delta < \varepsilon$. On this rectangle, $M(p, \gamma)$ is maximal for $(p, \gamma) = (b, \varepsilon)$. Thus,

$$(b - a) \log\big(M(b, \varepsilon)/M(a, \delta)\big) + \log(\varepsilon/\delta)[bM(b, \varepsilon) - aM(a, \delta) - (b - a)] \geq 0 . \tag{3.6.2}$$

Bounding $aM(a, \delta) \geq 0$ leads further to

$$(b - a) \log(M(b, \varepsilon)/M(a, \delta)) + \log(\varepsilon/\delta)[bM(b, \varepsilon) - (b - a)] \geq 0 . \tag{3.6.3}$$

We divide by $\log(\varepsilon/\delta)$ and take the limit as $\delta \searrow 0$. By Prop. 3.6,

$$\limsup_{\delta \searrow 0} \frac{\log(M(a, \delta))}{\log(\delta)} \leq \frac{1}{2} , \tag{3.6.4}$$

so that we conclude

$$0 \leq \tfrac{1}{2}(b - a) + \left[bM(b, \varepsilon) - (b - a) \right] = bM(b, \varepsilon) - \tfrac{1}{2}(b - a) . \qquad (3.6.5)$$

Finally, let $\varepsilon \searrow 0$ and use that $M(b, \varepsilon) \searrow \theta(b)$ to arrive at

$$b\theta(b) - \tfrac{1}{2}(b - a) \geq 0. \qquad (3.6.6)$$

Dividing by b proves the claim in Thm. 3.4.   □

3.7 Proof of the Aizenman–Barsky Differential Inequalities

3.7.1 Proof of (3.4.4)

We follow the proof of [122, Lem. (5.51)]. We use (3.4.2) and Russo's formula (1.3.9) to obtain[1]

$$(1 - p)\frac{\partial}{\partial p}M(p, \gamma) = (1 - p) \sum_{\{x,y\}} \mathbb{P}_{p,\gamma}(\{x, y\} \text{ pivotal for } 0 \leftrightarrow \mathcal{G})$$

$$= \sum_{\{x,y\}} \mathbb{P}_{p,\gamma}(\{x, y\} \text{ vacant and pivotal for } 0 \leftrightarrow \mathcal{G}) , \qquad (3.7.1)$$

where the sum is over all bonds $\{x, y\} \in \mathcal{E}(\mathbb{Z}^d)$. Instead of summing over *undirected* bonds, we can instead sum over *directed* bonds, which we denote by (u, v), once we introduce a direction into the event. Indeed, we say that the (directed) bond (x, y) is vacant and pivotal for $0 \leftrightarrow \mathcal{G}$ precisely when the following three events occur:

(a) the origin is not connected to $\mathcal{G}$;
(b) x is connected to the origin by an occupied path;
(c) y is joined to $\mathcal{G}$ by an occupied path.

Therefore,

$$(1 - p)\frac{\partial}{\partial p}M(p, \gamma) = \sum_{(x,y)} \mathbb{P}_{p,\gamma}(0 \leftrightarrow x, \mathcal{C}(0) \cap \mathcal{G} = \varnothing, y \leftrightarrow \mathcal{G}) . \qquad (3.7.2)$$

We proceed by conditioning on $\mathcal{C}(0)$, to obtain

---

[1]Bear in mind that we are dealing with an event that relies on infinitely many bonds. The fact that Russo's formula can be applied follows by a truncation argument by first proving the result on a finite box, and then showing convergence when the width of the box grows large, cf. [22]. A similar argument is used for the susceptibility in the proof of Thm. 4.2 below. We refrain from giving the details here.

$$(1-p)\frac{\partial}{\partial p}M(p,\gamma)$$
$$= \sum_{(x,y)}\sum_S \mathbb{P}_p(\mathcal{C}(0)=S)\,\mathbb{P}_{p,\gamma}(S\cap\mathcal{G}=\varnothing, y\leftrightarrow\mathcal{G}\mid\mathcal{C}(0)=S)\,,\quad (3.7.3)$$

where the sum is over all finite[2] connected sets S of vertices that contain 0 and x but do not contain y. Conditionally on $\mathcal{C}(0)=S$, the two events $\{\mathcal{C}(0)\cap\mathcal{G}=\varnothing\}$ and $\{y\leftrightarrow\mathcal{G}\}$ are *independent*, since the first depends only on the status of vertices in S, while the second depends on the status of vertices outside of S and edges having no end point in S. Thus,

$$\mathbb{P}_{p,\gamma}(S\cap\mathcal{G}=\varnothing, y\leftrightarrow\mathcal{G}\mid\mathcal{C}(0)=S)$$
$$= \mathbb{P}_{p,\gamma}(S\cap\mathcal{G}=\varnothing\mid\mathcal{C}(0)=S)\,\mathbb{P}_{p,\gamma}(y\leftrightarrow\mathcal{G}\mid\mathcal{C}(0)=S)\,.\quad (3.7.4)$$

Now, since S does not contain y, the connection from y to $\mathcal{G}$ should avoid all vertices in $\mathcal{C}(0)$, so that

$$\mathbb{P}_{p,\gamma}(y\leftrightarrow\mathcal{G}\mid\mathcal{C}(0)=S)\le\mathbb{P}_{p,\gamma}(y\leftrightarrow\mathcal{G})=M(p,\gamma)\,.\quad (3.7.5)$$

Therefore,

$$(1-p)\frac{\partial}{\partial p}M(p,\gamma)$$
$$\le \sum_{(x,y)}\sum_S \mathbb{P}_p(\mathcal{C}(0)=S)\,\mathbb{P}_{p,\gamma}(S\cap\mathcal{G}=\varnothing\mid\mathcal{C}(0)=S)M(p,\gamma)$$
$$\le M(p,\gamma)\sum_{(x,y)}\mathbb{P}_{p,\gamma}(x\in\mathcal{C}(0),\mathcal{C}(0)\cap\mathcal{G}=\varnothing)\,,\quad (3.7.6)$$

where the inequality follows since we no longer restrict to S that do not contain y. The above probability is independent of y, and since there are $2d$ possible choices for y such that (x,y) is a bond, we obtain

$$(1-p)\frac{\partial}{\partial p}M(p,\gamma)\le 2dM(p,\gamma)\sum_x\mathbb{P}_{p,\gamma}(x\in\mathcal{C}(0),\mathcal{C}(0)\cap\mathcal{G}=\varnothing)\,.\quad (3.7.7)$$

Finally, we note that

$$\sum_{x\in\mathbb{Z}^d}\mathbb{P}_{p,\gamma}(x\in\mathcal{C}(0),\mathcal{C}(0)\cap\mathcal{G}=\varnothing)=\mathbb{E}_{p,\gamma}\big[|\mathcal{C}(0)|\mathbb{1}_{\{\mathcal{C}(0)\cap\mathcal{G}=\varnothing\}}\big]$$
$$= (1-\gamma)\frac{\partial}{\partial\gamma}M(p,\gamma)\,.\quad (3.7.8)$$

---

[2]The condition $\mathcal{C}(0)\cap\mathcal{G}=\varnothing$ for $\gamma>0$ implies $|\mathcal{C}(0)|<\infty$ a.s., thus we can restrict ourselves to finite clusters here.

This completes the proof of (3.4.4). □

Exercise 3.6 (Differential inequality for $\theta(p)$ [73]). Let $p > p_c$. Assume that the event $\{0 \leftrightarrow \infty\}$ a.s. has a finite number of pivotal bonds. Show that $\theta(p)$ satisfies the differential inequality

$$\theta(p) \leq \theta(p)^2 + p\theta(p)\frac{\partial}{\partial p}\theta(p) . \tag{3.7.9}$$

Exercise 3.7 (Finite number of pivotals for connection to infinity). Argue why it is reasonable that the event $\{0 \leftrightarrow \infty\}$ a.s. has a finite number of pivotal bonds. Check how Chayes and Chayes [73] prove this fact. Further, prove that this fact is indeed true when we assume that the *half-space* percolation function

$$\theta^{\text{half}}(p) := \mathbb{P}_p\big(0 \leftrightarrow \infty \text{ in } \{0, 1, 2, \ldots\} \times \mathbb{Z}^{d-1}\big)$$

satisfies $\theta^{\text{half}}(p) > 0$ for every $p > p_c$.

Exercise 3.8 (Mean-field bound $\theta(p)$ revisited [73]). Use the differential inequality in Exerc. 3.6 to prove that there exists a constant $a > 0$ such that $\theta(p) \geq a(p - p_c)$ for $p > p_c$.

3.7.2 Proof of (3.4.5)

We again follow the proof of [122, Lem. (5.53)] closely. The proof of (3.4.5) is very similar to that of (3.4.4). We use $M(p, \gamma) = \mathbb{P}_{p,\gamma}(\mathcal{C}(0) \cap \mathcal{G} \neq \varnothing)$ to split

$$M(p, \gamma) = \mathbb{P}_{p,\gamma}(|\mathcal{C}(0) \cap \mathcal{G}| = 1) + \mathbb{P}_{p,\gamma}(|\mathcal{C}(0) \cap \mathcal{G}| \geq 2) . \tag{3.7.10}$$

The first term on the r.h.s. of (3.7.10) equals $\gamma \partial M/\partial \gamma$, which is the first term in (3.4.5), since

$$\mathbb{P}_{p,\gamma}(|\mathcal{C}(0) \cap \mathcal{G}| = 1) = \sum_{n=1}^{\infty} n\gamma(1-\gamma)^{n-1}\mathbb{P}_p(|\mathcal{C}(0)| = n)$$

$$= \gamma\frac{\partial}{\partial\gamma}\sum_{n=1}^{\infty}[1 - (1-\gamma)^n]\mathbb{P}_p(|\mathcal{C}(0)| = n)$$

$$= \gamma\frac{\partial}{\partial\gamma}M(p, \gamma) . \tag{3.7.11}$$

For the second term, we define A_x to be the event that either $x \in \mathcal{G}$ or that x is connected by an occupied path to a vertex $g \in \mathcal{G}$. Then,

$$\mathbb{P}_{p,\gamma}(|\mathcal{C}(0) \cap \mathcal{G}| \geq 2) = \mathbb{P}_{p,\gamma}(A_0 \circ A_0)$$
$$+ \mathbb{P}_{p,\gamma}(|\mathcal{C}(0) \cap \mathcal{G}| \geq 2, A_0 \circ A_0 \text{ does not occur}) . \tag{3.7.12}$$

We apply the BK inequality to obtain

$$\mathbb{P}_{p,\gamma}(A_0 \circ A_0) \le \mathbb{P}_{p,\gamma}(A_0)^2 = M(p,\gamma)^2 , \qquad (3.7.13)$$

leading to the second term in (3.4.5). Thus, we are left to prove that

$$\mathbb{P}_{p,\gamma}(|\mathcal{C}(0) \cap \mathcal{G}| \ge 2, A_0 \circ A_0 \text{ does not occur}) \le pM \frac{\partial M}{\partial p} . \qquad (3.7.14)$$

To prove (3.7.14), we move on to investigate the event that $|\mathcal{C}(0) \cap \mathcal{G}| \ge 2$, but that $A_0 \circ A_0$ does not occur. This event is equivalent to the existence of a (directed) bond (x, y) for which the following occurs:

(i) the edge (x, y) is occupied; and
(ii) in the subgraph of $\mathbb{Z}^d$ obtained by deleting (x, y), the following events occur:

    (a) no vertex of $\mathcal{G}$ is joined to the origin by an open path;
    (b) x is joined to 0 by an occupied path;
    (c) the event $A_y \circ A_y$ occurs.

The events in (ii) are independent of the occupation status of the bond (x, y), so that

$$\mathbb{P}_{p,\gamma}(|\mathcal{C}(0) \cap \mathcal{G}| \ge 2, A_0 \circ A_0 \text{ does not occur})$$
$$= \frac{p}{1-p} \sum_{(x,y)} \mathbb{P}_{p,\gamma}((x, y) \text{ closed}, x \in \mathcal{C}(0), \mathcal{C}(0) \cap \mathcal{G} = \varnothing, A_y \circ A_y)$$
$$\le \frac{p}{1-p} \sum_{(x,y)} \mathbb{P}_{p,\gamma}(x \in \mathcal{C}(0), \mathcal{C}(0) \cap \mathcal{G} = \varnothing, A_y \circ A_y) . \qquad (3.7.15)$$

We condition on $\mathcal{C}(0)$ to obtain

$$\mathbb{P}_{p,\gamma}(\mathcal{C}(0) \cap \mathcal{G} = \varnothing, A_y \circ A_y)$$
$$= \frac{p}{1-p} \sum_{(x,y)} \sum_{S} \mathbb{P}_p(\mathcal{C}(0) = S)$$
$$\times \mathbb{P}_{p,\gamma}(\mathcal{C}(0) \cap \mathcal{G} = \varnothing, A_y \circ A_y \mid \mathcal{C}(0) = S) , \quad (3.7.16)$$

where the sum over S is over all finite and connected sets of vertices which contain 0 and x but not y. Since S does not contain y, we can write

$$\mathbb{P}_{p,\gamma}(\mathcal{C}(0) \cap \mathcal{G} = \varnothing, A_y \circ A_y \mid \mathcal{C}(0) = S)$$
$$= \mathbb{P}_{p,\gamma}(\{\mathcal{C}(0) \cap \mathcal{G} = \varnothing\} \cap \{A_y \circ A_y \text{ in } \mathbb{Z}^d \setminus S\} \mid \mathcal{C}(0) = S) , \qquad (3.7.17)$$

where we write $\{A_y \circ A_y \text{ in } \mathbb{Z}^d \setminus S\}$ when there exist two bond disjoint paths connecting y and $\mathcal{G}$ that do not contain any vertices in S.

Conditionally on $\mathcal{C}(0) = S$, the events $\{\mathcal{C}(0) \cap \mathcal{G} = \varnothing\}$ and $\{A_y \circ A_y$ in $\mathbb{Z}^d \setminus S\}$ are *independent*, since $\mathcal{C}(0) \cap \mathcal{G} = \varnothing$ depends on the occupation statuses of the vertices in S, while $\{A_y \circ A_y$ in $\mathbb{Z}^d \setminus S\}$ depends on the vertices in $\mathbb{Z}^d \setminus S$ and the edges between them. Thus,

$$\mathbb{P}_{p,\gamma}(\mathcal{C}(0) \cap \mathcal{G} = \varnothing, A_y \circ A_y \mid \mathcal{C}(0) = S)$$
$$= \mathbb{P}_{p,\gamma}(\mathcal{C}(0) \cap \mathcal{G} = \varnothing \mid \mathcal{C}(0) = S)\, \mathbb{P}_{p,\gamma}(A_y \circ A_y \text{ in } \mathbb{Z}^d \setminus S).$$

$$(3.7.18)$$

By the BK inequality,

$$\mathbb{P}_{p,\gamma}(A_y \circ A_y \text{ in } \mathbb{Z}^d \setminus S) \leq \mathbb{P}_{p,\gamma}(A_y \text{ in } \mathbb{Z}^d \setminus S)^2 \leq M(p,\gamma)\, \mathbb{P}_{p,\gamma}(A_y \text{ in } \mathbb{Z}^d \setminus S). \quad (3.7.19)$$

As a result,

$$\frac{p}{1-p} \sum_{(x,y)} \mathbb{P}_{p,\gamma}(x \in \mathcal{C}(0), \mathcal{C}(0) \cap \mathcal{G} = \varnothing, A_y \circ A_y)$$

$$\leq \frac{pM(p,\gamma)}{1-p} \sum_{(x,y)} \sum_S \mathbb{P}_p(\mathcal{C}(0) = S)$$
$$\times \mathbb{P}_{p,\gamma}(\mathcal{C}(0) \cap \mathcal{G} = \varnothing \mid \mathcal{C}(0) = S)\, \mathbb{P}_{p,\gamma}(A_y \text{ in } \mathbb{Z}^d \setminus S)$$

$$= \frac{pM(p,\gamma)}{1-p} \sum_{(x,y)} \sum_S \mathbb{P}_p(\mathcal{C}(0) = S)$$
$$\times \mathbb{P}_{p,\gamma}(\mathcal{C}(0) \cap \mathcal{G} = \varnothing, A_y \mid \mathcal{C}(0) = S)$$

$$= \frac{pM(p,\gamma)}{1-p} \sum_{(x,y)} \mathbb{P}_{p,\gamma}(x \in \mathcal{C}(0), \mathcal{C}(0) \cap \mathcal{G} = \varnothing, \mathcal{C}(y) \cap \mathcal{G} \neq \varnothing)$$

$$= pM(p,\gamma)\frac{\partial M}{\partial p}, \qquad (3.7.20)$$

where the first equality follows again by conditional independence, and the last equality by the fact that

$$(1-p)\frac{\partial M}{\partial p} = \sum_{(x,y)} \mathbb{P}_{p,\gamma}(x \in \mathcal{C}(0), \mathcal{C}(0) \cap \mathcal{G} = \varnothing, \mathcal{C}(y) \cap \mathcal{G} \neq \varnothing), \qquad (3.7.21)$$

where we recall (3.7.2). This completes the proof of (3.4.5). □

Part II
Mean-Field Behavior: Differential Inequalities and the Lace Expansion

Chapter 4
Critical Exponents and the Triangle Condition

In this chapter, we define an important condition, the so-called *triangle condition*, which implies mean-field behavior in percolation in the sense that the percolation critical exponents β, γ and δ take on their mean-field values $\beta = \gamma = 1, \delta = 2$. This chapter is organized as follows. We state the main result in Sect. 4.1. In Sect. 4.2, we illustrate the use of the triangle condition in its simplest setting by proving that it implies that $\gamma = 1$. The proof that it implies that $\beta = 1, \delta = 2$ is deferred to Chap. 9. We discuss one-sided mean-field bounds on the critical exponents δ and β in Sect. 4.3.

4.1 Definition of the Triangle Condition

We start by defining the triangle condition. Aizenman and Newman [13] were the first to establish the existence of mean-field critical exponents for percolation models under a condition called the *triangle condition*:

The triangle condition is the condition that

$$\triangle_{p_c} = \sum_{x,y \in \mathbb{Z}^d} \tau_{p_c}(0, x) \tau_{p_c}(x, y) \tau_{p_c}(y, 0) < \infty , \qquad (4.1.1)$$

where we recall that $\tau_p(x, y) = \mathbb{P}_p(x \leftrightarrow y)$ is the percolation two-point function.

The following theorem states that the critical exponents β, γ, δ exist under the triangle condition, and take on their mean-field values for percolation on the tree (recall Thm. 2.1):

© Springer International Publishing Switzerland 2017
M. Heydenreich and R. van der Hofstad, *Progress in High-Dimensional Percolation and Random Graphs*, CRM Short Courses, DOI 10.1007/978-3-319-62473-0_4

Theorem 4.1 (Mean-field critical exponents under the triangle condition [13, 28]). *For percolation on a transitive graph, if the triangle condition holds, then $\beta = \gamma = 1, \delta = 2$ in the bounded-ratio sense.*

It is a priori not obvious that the triangle condition is the appropriate condition for mean-field behavior, but the present theorem advocates that it is. Note that the triangle condition is also a strong condition. In particular, it implies that $\tau_{p_c}(x) \to 0$ as $|x| \to \infty$, which, in turn, implies that $\theta(p_c) = 0$. Therefore, the triangle condition implies that the percolation function is continuous.

We give the proof that $\gamma = 1$ under the triangle condition below. In [28], Barsky and Aizenman also proved that, under the same condition, $\beta = 1$ and $\delta = 2$. We give these proofs in Chap. 9 below. Needless to say, without the actual *verification* of the triangle condition, Thm. 4.1 would not prove anything, but it does highlight why the verification of the triangle condition is such an important result.

Exercise 4.1 (Random walk triangle condition). Prove that

$$\triangle^{(\text{RW})} = \sum_{x,y \in \mathbb{Z}^d} C_1(x)C_1(y-x)C_1(y) < \infty , \tag{4.1.2}$$

if and only if $d > 6$, where $C_\mu(x)$ denotes the random walk Green's function defined in (2.2.8).

4.2 The Susceptibility Critical Exponent γ

Here, we give the Aizenman–Newman argument [13] to identify the critical exponent $\gamma = 1$. In its statement, we use the following *open triangle* diagram, which, for $x \in \mathbb{Z}^d$, is defined as

$$\triangle_p(x) = \sum_{u,v} \tau_p(u)\tau_p(v-u)\tau_p(x-v) . \tag{4.2.1}$$

Then, the main result concerning γ is as follows:

Theorem 4.2 ($\gamma = 1$ under the triangle condition [13]). *For percolation on a transitive graph, if the triangle condition holds in the form that $\triangle_{p_c}(e) < 1$ for $e \sim 0$, then $\gamma = 1$ in the bounded-ratio sense.*

The proof of Thm. 4.2 consists of two parts, a lower bound on $\chi(p)$ for $p < p_c$ that is valid for *every* transitive percolation model, and an upper bound that relies on the triangle condition $\triangle_{p_c}(e) < 1$ in (4.2.1). We next present these two proofs.

Proof that $\gamma \geq 1$. We use Russo's formula (1.3.9), and for the time being ignoring the restriction to events depending on finitely many bonds, to obtain

$$\frac{\mathrm{d}}{\mathrm{d}p}\tau_p(x) = \sum_{(u,v)} \mathbb{P}_p\big((u, v) \text{ pivotal for } 0 \leftrightarrow x\big) . \tag{4.2.2}$$

Mind that Russo's formula is formulated for unoriented bonds; however, we can replace it by a sum over *oriented* bonds if we declare (u, v) to be pivotal for $0 \leftrightarrow x$ whenever $0 \leftrightarrow u$ and $v \leftrightarrow x$ occur, and there is no connection from 0 to x when the bond $\{u, v\}$ is made vacant. This yields the above formula. Summing over $x \in \mathbb{Z}^d$ and assuming differentiability of χ yield

$$\frac{d}{dp}\chi(p) = \sum_{x \in \mathbb{Z}^d} \sum_{(u,v)} \mathbb{P}_p\big((u, v) \text{ pivotal for } 0 \leftrightarrow x\big) . \qquad (4.2.3)$$

We next note that if (u, v) is pivotal for $0 \leftrightarrow x$, then there exist two *disjoint* paths of occupied bonds connecting 0 and u, and v and x, respectively. Thus, $\{0 \leftrightarrow u\} \circ \{v \leftrightarrow x\}$ occurs. By the BK inequality (1.3.4),

$$\frac{d}{dp}\chi(p) \leq \sum_{x} \sum_{(u,v)} \tau_p(u)\tau_p(x - v) = 2d\chi(p)^2 . \qquad (4.2.4)$$

To make (4.2.4) precise, one has to overcome a few technicalities, since Russo's formula in (1.3.9) can only be applied to events E depending on a *finite* number of bonds, which is not the case for $\{0 \leftrightarrow x\}$. We return to these issues below.

To complete the argument, note that we can rewrite (4.2.4) as

$$\frac{d}{dp}\chi(p)^{-1} \geq -2d . \qquad (4.2.5)$$

We assume for the time being that $\chi(p_c)^{-1} = 0$. Ignoring this subtle point, we can integrate (4.2.5) over $[p, p_c]$ to obtain

$$\chi(p_c)^{-1} - \chi(p)^{-1} = -\chi(p)^{-1} \geq -2d(p_c - p) , \qquad (4.2.6)$$

so that

$$\chi(p) \geq \frac{1}{2d(p_c - p)} . \qquad (4.2.7)$$

This proves that $\gamma \geq 1$ if γ exists. Of course, (4.2.7) is stronger than the existence of γ, as it provides a pointwise lower bound valid for every $p \in [0, p_c)$. $\qquad\square$

Finite-Size Corrections and $\chi(p_c) = \infty$. In the above proof, we neglected three subtleties. The first is that we have used Russo's formula (1.3.9) in a setting where the event actually depends on *infinitely* many bonds, the second is that we have assumed differentiability of $\chi(p)$ and $\chi(p)^{-1}$, the third is that we have assumed that $\chi(p_c) = \infty$. We have already shown that $\chi(p_c) = \infty$ in Thm. 3.2(3). All three can be resolved by performing an appropriate finite-size approximation, which we are presenting next. Along the way, we give an alternative proof of Thm. 3.2(3). We start by introducing the necessary notation.

Let $\tau_p^n(x, y)$ denote the probability that x is connected to y by only using the bonds in the cube $\Lambda_n = \{-n, \ldots, n\}^d$, and let $\mathbb{P}_p^n$ denote the percolation measure on the cube Λ_n. Let $\chi^n(p) = \max_{w \in \Lambda_n} \sum_{x \in \Lambda_n} \tau_p^n(w, x)$ be the maximal expected cluster size on the cube Λ_n. A technical ingredient to the proof is the following continuity statement:

Lemma 4.3 (Continuity of equicontinuous functions). *Let* $(f_\alpha)_{\alpha \in A}$ *be an equicontinuous family of functions on an interval* $[t_1, t_2]$, *i.e., for every given* $\varepsilon > 0$, *there is a* $\delta > 0$ *such that* $|f_\alpha(s) - f_\alpha(t)| < \varepsilon$ *whenever* $|s - t| < \delta$, *uniformly in* $\alpha \in A$. *Furthermore, suppose that* $\sup_{\alpha \in A} f_\alpha(t) < \infty$ *for each* $t \in [t_1, t_2]$. *Then* $t \mapsto \sup_{\alpha \in A} f_\alpha(t)$ *is continuous on* $[t_1, t_2]$.

A proof of this standard result can be found, for example, in [246, Lem. 5.12]. Note that $(f_\alpha)_{\alpha \in A}$ is equicontinuous whenever $\sup_{\alpha \in A} \sup_{t \in [t_1, t_2]} |f_\alpha'(t)| \leq K$ for some $K > 0$. Then, the finite-size corrections are investigated in the following lemma:

Lemma 4.4 (Continuity of $p \mapsto 1/\chi(p)$). *The functions* $p \mapsto 1/\chi^n(p)$ *are equicontinuous, and* $\chi^n(p) \to \chi(p)$ *for every* $p \in [0, 1]$. *As a result,* $p \mapsto 1/\chi(p)$ *is continuous, and in particular,* $\chi(p_c) = \infty$.

Proof. We start by proving that $\chi^n(p) \to \chi(p)$ for every $p \in [0, 1]$. Note that

$$\chi(p) \geq \chi^n(p) \geq \sum_{x \in \Lambda_n} \tau_p^n(0, x) . \tag{4.2.8}$$

Further, by the bounded convergence theorem, $\tau_p^n(0, y) \nearrow \tau_p(0, y)$ for every $y \in \mathbb{Z}^d$, so that monotone convergence implies

$$\chi^n(p) \nearrow \chi(p) . \tag{4.2.9}$$

In particular, also

$$\lim_{n \to \infty} 1/\chi^n(p) = 1/\chi(p) . \tag{4.2.10}$$

This proves the pointwise convergence of $\chi^n(p)$.

We continue with the claimed equicontinuity. As in (4.2.2),

$$\frac{d}{dp} \tau_p^n(w, x) = \sum_{(u,v)} \mathbb{P}_p^n \big((u, v) \text{ pivotal for } w \leftrightarrow x \big) , \tag{4.2.11}$$

where now the sum is over all edges with both endpoints in the cube Λ_n. Thus, summing over $x \in \Lambda_n$ yields, for fixed $w \in \Lambda_n$,

$$\frac{d}{dp} \sum_{x \in \Lambda_n} \tau_p^n(w, x) = \sum_{x \in \Lambda_n} \frac{d}{dp} \tau_p^n(w, x)$$

$$= \sum_{x \in \Lambda_n} \sum_{(u,v)} \mathbb{P}_p^n \big((u, v) \text{ pivotal for } w \leftrightarrow x \big) . \tag{4.2.12}$$

Following the steps above, we thus arrive at the statement that, for every $p \in [0, 1]$ and uniformly in $w \in \Lambda_n$,

$$\frac{d}{dp} \sum_{x \in \Lambda_n} \tau_p^n(w, x) \leq 2d\chi^n(p)^2 . \tag{4.2.13}$$

We next relate this to $(d/dp)\chi^n(p)$. The function $p \mapsto \chi^n(p)$ is continuous, as a maximum of increasing and continuous functions. Secondly, $\chi^n(p)$ is a finite maximum of differentiable functions, and therefore, $\chi^n(p)$ is almost everywhere differentiable. Finally, since $\chi^n(p)$ is a finite maximum of continuous functions, there is (at least one) maximizer w_p (when there are multiple, pick any one of them). Therefore,

$$
\begin{aligned}
\chi^n(p + \varepsilon) - \chi^n(p) &= \sum_{x \in \Lambda_n} [\tau_{p+\varepsilon}^n(w_{p+\varepsilon}, x) - \tau_p^n(w_p, x)] \\
&\leq \sum_{x \in \Lambda_n} [\tau_{p+\varepsilon}^n(w_{p+\varepsilon}, x) - \tau_p^n(w_{p+\varepsilon}, x)] \\
&\leq \max_{w \in \Lambda_n} \sum_{x \in \Lambda_n} [\tau_{p+\varepsilon}^n(w, x) - \tau_p^n(w, x)] .
\end{aligned}
\tag{4.2.14}
$$

Dividing by ε and letting $\varepsilon \searrow 0$ lead to

$$
\frac{d}{dp}\chi^n(p) \leq \max_{w \in \Lambda_n} \sum_{x \in \Lambda_n} \frac{d}{dp}\tau_p^n(w, x) .
\tag{4.2.15}
$$

As a result,

$$
\frac{d}{dp}\chi^n(p) \leq 2d\chi^n(p)^2 ,
\tag{4.2.16}
$$

so that

$$
0 \geq \frac{d}{dp}\frac{1}{\chi^n(p)} \geq -2d .
\tag{4.2.17}
$$

Thus, the functions $p \mapsto 1/\chi^n(p)$ form an equicontinuous nonincreasing family of functions, with uniformly bounded derivative. Further, as $n \to \infty$, the function $1/\chi^n(p)$ converges to $1/\chi(p)$ for every $p \in [0, 1]$ by (4.2.10). By equicontinuity, the limit must be continuous. Since $1/\chi(p) = 0$ for every $p > p_c$, we conclude that $\chi(p_c) = 0$ (which we already proved using other means in Thm. 3.2(3)). $\qquad\square$

We continue by integrating $1/\chi^n(p)$ as in (4.2.6) to deduce

$$
\chi^n(p_c)^{-1} - \chi^n(p)^{-1} \geq 2d(p_c - p) \quad \text{for all } n
$$

and the limit $n \to \infty$ (using $\chi^n(p_c)^{-1} \to 0$) obtains (4.2.7). This completes the proof of $\gamma \geq 1$. $\qquad\square$

In the above, we have been quite precise in our finite-size arguments. In the sequel, we often ignore such issues and refer to the literature for the precise proofs. We next turn to the upper bound $\gamma \leq 1$, for which we recall that we assume that $\triangle_{p_c}(e) < 1$ for $e \sim 0$, where, for $x \in \mathbb{Z}^d$, we define $\triangle_{p_c}(x)$ to be the open triangle in (4.2.1). We note that $\triangle_p = \triangle_p(0)$. We shall see that we require several related triangles in the sequel.

Exercise 4.2 (A derivative of a finite maximum of functions). Fix a finite set of indices I. For $i \in I$, let $p \mapsto f_i(p)$ be a differentiable function. Adapt the argument in (4.2.14) to show that

$$\frac{d}{dp} \max_{i \in I} f_i(p) \leq \max_{i \in I} \frac{d}{dp} f_i(p) . \tag{4.2.18}$$

Proof that $\gamma \leq 1$ *when* $\Delta_{p_c}(e) < 1$ *for* $e \sim 0$. The proof that $\gamma \leq 1$ when the triangle condition holds is a nice example of the general methodology in high dimension. The BK inequality gives an upper bound on (4.2.3), and, in order to prove that $\gamma = 1$, a matching lower bound on (4.2.3) needs to be obtained. For this, we need similar objects as in the lace expansion performed later in Sect. 6.2, and start by introducing a few objects that we define again, more formally, in Def. 6.2 below. We start by exploiting the independence of the occupation statuses of the bonds to explicitly write

$$\mathbb{P}_p\big((u,v) \text{ pivotal for } 0 \leftrightarrow x\big) = \mathbb{E}_p\big[\mathbb{1}_{\{0 \leftrightarrow u\}} \tau^{\widetilde{\mathcal{C}}^{(u,v)}(0)}(v,x)\big] , \tag{4.2.19}$$

where, for $A \subset \mathbb{Z}^d$, the restricted two-point function $\tau^A(v,x)$ is the probability that v is connected to x using only bonds with both endpoints outside A, and $\widetilde{\mathcal{C}}^{(u,v)}(0)$ consists of those sites that are connected to 0 without using the bond (u,v). We prove equalities such as (4.2.19) in more generality in Lem. 6.4 below and omit the proof here. Such explicit equalities are at the heart of the lace-expansion methodology in high-dimensional percolation.

Clearly, $\tau^{\widetilde{\mathcal{C}}^{(u,v)}(0)}(v,x) \leq \tau(v,x)$, and this reproves the upper bound obtained in (4.2.4) using the BK inequality. We continue to prove the lower bound. For $A \subseteq \mathbb{Z}^d$, we write $\{x \overset{A}{\leftrightarrow} y\}$ if either $x = y \in A$ or if $x \leftrightarrow y$, and every occupied path connecting x to y has at least one bond with an endpoint in A, and call this event "x is connected to y *through A*." We note that

$$\tau(v,x) - \tau^A(v,x) = \mathbb{P}_p\big(v \overset{A}{\leftrightarrow} x\big) . \tag{4.2.20}$$

Thus,

$$\frac{d}{dp}\chi(p) = \sum_x \sum_{(u,v)} \mathbb{E}_p[\mathbb{1}_{\{0 \leftrightarrow u\}}]\tau_p(v,x)$$
$$- \sum_x \sum_{(u,v)} \mathbb{E}_p\Big[\mathbb{1}_{\{0 \leftrightarrow u\}} \, \mathbb{P}_p\Big(v \xleftarrow{\widetilde{\mathcal{C}}^{(u,v)}(0)} x\Big)\Big]$$
$$= 2d\chi(p)^2 - \sum_x \sum_{(u,v)} \mathbb{E}_p\Big[\mathbb{1}_{\{0 \leftrightarrow u\}} \, \mathbb{P}_p\Big(v \xleftarrow{\widetilde{\mathcal{C}}^{(u,v)}(0)} x\Big)\Big] . \tag{4.2.21}$$

Now, for any $A \subseteq \mathbb{Z}^d$,

$$\mathbb{P}_p\big(v \overset{A}{\leftrightarrow} x\big) \leq \sum_{a \in A} \mathbb{P}_p(\{v \leftrightarrow a\} \circ \{a \leftrightarrow x\}) , \tag{4.2.22}$$

which, after resumming over x to retrieve a factor $\chi(p)$, leads to

$$\frac{d}{dp}\chi(p) \ge 2d\chi(p)^2 - \chi(p) \sum_{(u,v)} \sum_{a \in \mathbb{Z}^d} \mathbb{E}_p[\mathbb{1}_{\{0 \leftrightarrow u\}} \mathbb{1}_{\{a \in \widetilde{\mathcal{C}}^{(u,v)}(0)\}}] \, \mathbb{P}_p(v \leftrightarrow a)$$

$$\ge 2d\chi(p)^2 - \chi(p) \sum_{(u,v)} \sum_{a \in \mathbb{Z}^d} \mathbb{E}_p[\mathbb{1}_{\{0 \leftrightarrow u,a\}}] \, \mathbb{P}_p(v \leftrightarrow a) \,. \tag{4.2.23}$$

Applying a union bound combined with the BK inequality yields

$$\mathbb{P}_p(0 \leftrightarrow u, 0 \leftrightarrow a) \le \sum_z \mathbb{P}_p\big(\{0 \leftrightarrow z\} \circ \{z \leftrightarrow u\} \circ \{z \leftrightarrow a\}\big)$$

$$\le \sum_z \tau_p(z)\tau_p(u-z)\tau_p(a-z) \,, \tag{4.2.24}$$

so that

$$\sum_{(u,v)} \sum_a \mathbb{P}_p(0 \leftrightarrow u, 0 \leftrightarrow a) \, \mathbb{P}_p(v \leftrightarrow a) \le 2d\chi(p)\triangle_p(e)$$

$$\le 2d\chi(p)\triangle_{p_c}(e) \,, \tag{4.2.25}$$

implying

$$\frac{d}{dp}\chi(p) \ge 2d\chi(p)^2[1 - \triangle_{p_c}(e)] \tag{4.2.26}$$

for $e = v - u$. If we know that $\triangle_{p_c}(e) < 1$, then we can integrate the above equation in a similar fashion as around (4.2.5) to obtain $\gamma = 1$. When we only have the *finiteness* of the triangle, then some more work under the name of *ultraviolet regularization* is necessary to make the same conclusion, see [13, Lem. 6.3] for details. An alternative argument involving operator theory is given by Kozma [200]. We omit the details of this ultraviolet regularization argument.   □

The main ingredient in the above proof is (4.2.19). If we would ignore the interaction, which causes only a small quantitative error in high dimensions, then we could replace (4.2.19) by

$$\sum_{(u,v)} \mathbb{P}_p\big((u,v) \text{ pivotal for } 0 \leftrightarrow x\big) \approx \sum_{(u,v)} \mathbb{E}_p[\mathbb{1}_{\{0 \leftrightarrow u\}}\tau(v,x)]$$

$$= 2d(\tau_p \star D \star \tau_p)(x) \,. \tag{4.2.27}$$

We use *inclusion–exclusion* to make such approximations precise in the further sections, as this is the main idea behind the lace expansion.

4.3   Mean-Field Bounds on the Critical Exponents δ and β and Overview

It turns out that the differential inequalities imply one-sided bounds on various critical exponents. We have already seen in the proof of Thm. 4.2 that $\gamma \geq 1$ is valid in *any* dimension (and
even in every transitive percolation model). Such a bound in terms of a mean-field critical
exponent is called a *mean-field bound*. In this section, we discuss so-called mean-field bounds
on the critical exponents β and δ following from the analysis in Chap. 3.

The bound on δ is at the moment a little tricky, and we need to resort to a generating function
estimate. For this, we say that δ exists *for the generating function*, e.g., in the bounded-ratio
sense, when there exist constants $0 < c_\delta < C_\delta < \infty$ such that

$$c_\delta \gamma^{1/\delta} \leq M(p_c, \gamma) \leq C_\delta \gamma^{1/\delta} \ . \tag{4.3.1}$$

Then, we obtain the following mean-field bounds:

Corollary 4.5 (Mean-field bounds on β and δ). *It is the case that $\beta \leq 1$ when the critical
exponent β exists. Further, $\delta \geq 2$ when the critical exponent δ exists for the generating
function as in* (4.3.1).

Proof. The bound on β directly follows from Thm. 3.2(2). The bound on δ is Prop. 3.6. $\square$

Taking Stock of Where We Are Now. We have identified the *triangle condition* in (4.1.1)
as being the decisive factor for mean-field behavior in percolation. We have proved that it
implies the mean-field susceptibility critical exponent $\gamma = 1$, and we have further seen mean-
field lower bounds on β and δ. In the next chapter, we investigate the *infrared bound* for the
percolation two-point function, and we see that this implies that the triangle condition is
satisfied in sufficiently high dimensions. In Chap. 9, after having completed the proof of the
infrared bound, we perform the proofs that $\beta = 1$ and $\delta = 2$ under the triangle condition.

Chapter 5
Proof of Triangle Condition: The Infrared Bound

In this chapter, we state one of the key results in high-dimensional percolation, the so-called *infrared bound*. We state the result in Sect. 5.1, and in Sect. 5.2, we extend the result to a slightly modified percolation model known as *spread-out percolation*. In Sect. 5.3, we give an overview of the proof of the infrared bound relying on the *lace expansion*. Finally, in Sect. 5.4, we discuss the *random walk triangle*. The main idea is that if the random walk triangles is sufficiently small, then the infrared bound follows.

5.1 The Infrared Bound

The infrared bound was first proved for percolation in the seminal 1990 paper by Hara and Slade [132]. It has been the main workhorse for high-dimensional percolation in the past decades. The infrared bound provides an upper bound for the percolation two-point function in Fourier space, in terms of the random walk Green's function:

Theorem 5.1 (Infrared bound [132]). *For percolation, there exists $d_0 > 6$ such that for $d \geq d_0 > 6$, there exists a constant $A(d) < 3$ (the 'amplitude') such that, uniformly in $p \in [0, p_c)$ and $k \in (-\pi, \pi]^d$,*

$$\hat{\tau}_p(k) \leq \frac{A(d)}{1 - \widehat{D}(k)} . \tag{5.1.1}$$

Mind that Lem. 5.3 asserts that $\hat{\tau}_p(k)$ is nonnegative, so that (5.1.1) is even valid for the absolute value of $\hat{\tau}_p(k)$.

For the critical simple random walk Green's function, the inequality (5.1.1) in Thm. 5.1 holds with an equality and $A(d) = 1$, recall (2.2.12). Thus, we can think of Thm. 5.1 as stating that the percolation two-point function τ_{p_c} is a small perturbation of the simple random walk Green's function. In light of the relation between the simple random walk Green's function and the branching random walk two-point function in (2.2.7), we can alternatively think of Thm. 5.1 as stating that critical percolation in high dimensions is a small perturbation of

© Springer International Publishing Switzerland 2017
M. Heydenreich and R. van der Hofstad, *Progress in High-Dimensional Percolation and Random Graphs*, CRM Short Courses, DOI 10.1007/978-3-319-62473-0_5

critical branching random walk. In the next three chapters, we establish this connection and prove Thm. 5.1 in full detail.

Next, we establish the triangle condition as a consequence of the infrared bound:

Corollary 5.2 (Triangle condition and critical exponents). *For percolation, there exists $d_0 > 6$ such that for $d \geq d_0 > 6$, the triangle condition holds. Therefore, by Thm. 4.1, the critical exponents γ, β and δ exist in the bounded-ratio sense, and take on the mean-field values $\gamma = \beta = 1, \delta = 2$.*

In the proof of Cor. 5.2, the following nonnegativity result of $\hat{\tau}_p(k)$, proved by Aizenman and Newman [13, Lem. 3.3] and interesting in its own right, proves to be useful:

Lemma 5.3 (Nonnegativity $\hat{\tau}_p(k)$). *Let $p \in [0, 1]$ be such that $\chi(p) < \infty$. Then, for every $k \in (-\pi, \pi]^d$,*

$$\hat{\tau}_p(k) \geq 0 . \tag{5.1.2}$$

Proof. We view $\tau_p(x, y)$ as an operator and prove that it is of positive type, meaning that for every summable function $f: \mathbb{Z}^d \to \mathbb{R}$,

$$\sum_{x,y \in \mathbb{Z}^d} \bar{f}(x)\tau_p(x, y)f(y) \geq 0, \tag{5.1.3}$$

where $\bar{f}(x)$ denotes the complex conjugate of $f(x)$. We can write

$$\sum_{x,y \in \mathbb{Z}^d} \bar{f}(x)\tau_p(x, y)f(y) = \sum_{x,y \in \mathbb{Z}^d} \mathbb{E}_p[\bar{f}(x)\mathbb{1}_{\{x \longleftrightarrow y\}}f(y)]$$

$$= \mathbb{E}_p\left[\sum_{\mathscr{C}}\left|\sum_{x \in \mathscr{C}} f(x)\right|^2\right] \geq 0, \tag{5.1.4}$$

where the first sum is over all clusters $\mathscr{C}$ in a random partitioning of $\mathbb{Z}^d$. The claim now follows from Bochner's theorem, cf. [225, Thm. IX.9]. $\square$

Proof of Cor. 5.2. The equalities $\gamma = \beta = 1, \delta = 2$ in the bounded-ratio sense follow from the triangle condition and Thm. 4.1. We now establish that the triangle condition follows from Thm. 5.1. Note that, for every $p \in [0, 1]$,

$$\triangle_p = \sum_{x,y \in \mathbb{Z}^d} \tau_p(0, x)\tau_p(x, y)\tau_p(y, 0) = \tau_p^{*3}(0) , \tag{5.1.5}$$

where $\tau_p^{*3}(0)$ denotes the three-fold convolution of τ_p with itself evaluated at $x = 0$. We use the Fourier inversion theorem, which states that

$$f(x) = \int_{(-\pi,\pi]^d} e^{-ik \cdot x} \hat{f}(k) \frac{dk}{(2\pi)^d} \ , \tag{5.1.6}$$

whenever $|\hat{f}(k)|$ is integrable. We apply (5.1.6) to (5.1.5) with $p < p_c$ and use that

$$\widehat{\tau_p^{*3}}(k) = \hat{\tau}_p(k)^3 \ , \tag{5.1.7}$$

to arrive at

$$\triangle_p = \int_{(-\pi,\pi]^d} \hat{\tau}_p(k)^3 \frac{dk}{(2\pi)^d} \ . \tag{5.1.8}$$

By the infrared bound in Thm. 5.1,

$$\triangle_p \le A(d)^3 \int_{(-\pi,\pi]^d} \frac{1}{[1 - \widehat{D}(k)]^3} \frac{dk}{(2\pi)^d} \ . \tag{5.1.9}$$

The arising integral in (5.1.9) is bounded *uniformly in the dimension d* for $d > 6$, as we show next.

We claim that for any integer $n < d/2$, there is a constant $C = C(n) > 0$ independent of the dimension d such that

$$\int_{(-\pi,\pi]^d} \frac{1}{[1 - \widehat{D}(k)]^n} \frac{dk}{(2\pi)^d} \le C \ . \tag{5.1.10}$$

In order to prove this, we note that $1 - \cos(k) \ge \frac{2}{\pi^2} k^2$ for every $k \in (-\pi, \pi]$, and we thus get

$$1 - \widehat{D}(k) = \frac{1}{d} \sum_{i=1}^{d} [1 - \cos(k_i)] \ge \frac{2}{\pi^2 d} \sum_{i=1}^{d} k_i^2 = \frac{2}{\pi^2} \frac{|k|^2}{d} \ , \quad k \in (-k, k]^d \ . \tag{5.1.11}$$

This gives readily

$$\int_{(-\pi,\pi]^d} \frac{1}{[1 - \widehat{D}(k)]^n} \frac{dk}{(2\pi)^d} \le \frac{\pi^{2n}}{2^n} \int_{(-\pi,\pi]^d} \frac{d^n}{|k|^{2n}} \frac{dk}{(2\pi)^d} . \tag{5.1.12}$$

The right-hand side of (5.1.12) is finite if (and only if) $d > 2n$. For $a > 0$ and $n > 0$,

$$\frac{1}{a^n} = \frac{1}{\Gamma(n)} \int_0^{\infty} t^{n-1} e^{-ta} \, dt \ . \tag{5.1.13}$$

Applying this with $a = |k|^2/d$ yields

$$\frac{1}{\Gamma(2n)} \frac{\pi^{2n}}{2^n} \int_0^\infty t^{n-1} \left(\int_{-\pi}^\pi \left(e^{-t\theta^2} \right)^{1/d} \frac{d\theta}{2\pi} \right)^d dt \tag{5.1.14}$$

as an upper bound for (5.1.12). This is nonincreasing in d, because $\| f \|_p \le \| f \|_q$ for $0 < p \le q \le \infty$ on a probability space by Lyapunov's inequality, and the term in brackets is finite as a Gaussian integral. It remains to show that the integral over t is finite (for small d). In order to establish this, we bound further, for $d > 2n$,

$$\int_{-\pi}^\pi \left(e^{-t\theta^2} \right)^{1/d} \frac{d\theta}{2\pi} \le \min \left\{ 1, \int_{-\infty}^\infty \left(e^{-t\theta^2} \right)^{1/d} \frac{d\theta}{2\pi} \right\} = \sqrt{2\pi d/t} \wedge 1 , \tag{5.1.15}$$

which, upon substitution into (5.1.14), yields

$$\frac{1}{\Gamma(2n)} \frac{\pi^{2n}}{2^n} \int_0^\infty t^{n-1} \left(\sqrt{2\pi d/t} \wedge 1 \right)^d dt , \tag{5.1.16}$$

which is finite when $n - 1 - d/2 < -1$, i.e., when $d > 2n$. This proves (5.1.10).

Since $A(d)$ is uniformly bounded by 3, we get from (5.1.9) and (5.1.10) that $\triangle_p$ is bounded uniformly in $p < p_c$. Pointwise convergence of $\tau_p(x)$ to $\tau_{p_c}(x)$ implies that the claim also holds for $p = p_c$, since, by monotone convergence,

$$\begin{aligned} \triangle_{p_c} &= \sum_{x,y \in \mathbb{Z}^d} \tau_{p_c}(0,x) \tau_{p_c}(x,y) \tau_{p_c}(y,0) \\ &= \lim_{p \nearrow p_c} \sum_{x,y \in \mathbb{Z}^d} \tau_p(0,x) \tau_p(x,y) \tau_p(y,0) \\ &= \lim_{p \nearrow p_c} \triangle_p < \infty . \end{aligned} \tag{5.1.17}$$

$\square$

The Triangle $\triangle_{p_c}(e)$ with $e \sim 0$. It is useful to also consider $\triangle_{p_c}(e)$, as this is the quantity that appears for example in the proof of the upper bound $\gamma \le 1$. For this, we note that by symmetry,

$$\begin{aligned} \triangle_p(e) &= \frac{1}{2d} \sum_{e \sim 0} \tau_p^{\star 3}(e) = (D \star \tau_p^{\star 3})(0) \\ &= \int_{(-\pi,\pi]^d} \widehat{D}(k) \hat{\tau}_p(k)^3 \frac{dk}{(2\pi)^d} . \end{aligned} \tag{5.1.18}$$

By Cauchy–Schwarz,

$$\triangle_p(e)^2 \le \int_{(-\pi,\pi]^d} \widehat{D}(k)^2 \frac{dk}{(2\pi)^d} \times \int_{(-\pi,\pi]^d} \hat{\tau}_p(k)^6 \frac{dk}{(2\pi)^d} . \tag{5.1.19}$$

The second term is bounded as in (5.1.9) and (5.1.10), it is thus finite when $d > 12$, and its bound is uniform in the dimension d. The first term is small when d is large, due to the fact that

$$\int_{(-\pi,\pi]^d} \widehat{D}(k)^2 \frac{dk}{(2\pi)^d} = \mathbb{P}(S_2 = 0) = \frac{1}{2d}, \tag{5.1.20}$$

where $(S_m)_{m \geq 0}$ denotes simple random walk in d dimensions. Thus, for $d > 12$, $\triangle_p(e)^2 = O(1/d)$ uniformly in $p < p_c$, so that, for $d > 12$ sufficiently large, we indeed obtain that $\triangle_{p_c}(e) < 1$. $\qquad\qquad\qquad\qquad\qquad\qquad\qquad\qquad\qquad\qquad\qquad\qquad\qquad\qquad\qquad\quad$ $\square$

Exercise 5.1 (Proof of 1 − cos(k) bounds). Prove that $1 - \cos(k) \leq k^2/2$ and $1 - \cos(k) \geq (2/\pi^2)k^2$ for every $k \in (-\pi, \pi]$.

Exercise 5.2 (Bound on open triangle). Let $p < p_c$. Prove that $\triangle_p(x)$ is maximal for $x = 0$. *Hint:* Use the Fourier inversion theorem on $\triangle_p(x) = \tau_p^{\star 3}(x)$ and Lem. 5.3.

Kozma [200] proves the statement that the open triangle is finite when the closed triangle is for general vertex-transitive graphs. He uses completely different methods using operator theory, since Fourier theory is not applicable in such generality.

The above not only suggests that the triangle condition should be satisfied whenever $d > 6$, but there is also strong evidence that the critical exponents should not take on their mean-field values when $d < 6$. We return to this issue in Sect. 11.3, where we show that $\rho_{ex} = \frac{1}{2}$ and $\eta = 0$ in x-space imply that $d \geq 6$.

5.2 Spread-Out Models

In this section, we state another theorem that shows mean-field behavior, but now for the so-called *spread-out model*. We refer to our previously considered percolation model as the *nearest-neighbor* model. In the spread-out model, we consider vertex set $\mathbb{Z}^d$ with a different set of edges, namely

$$\mathcal{E} = \{\{x, y\} : \|x - y\|_\infty \leq L\}, \tag{5.2.1}$$

where $L \geq 1$ is a sufficiently large constant. The reason to study such, arguably somewhat esoteric, spread-out models is that we can make the triangle diagram small by taking L large, so that the dimension can remain unchanged. This makes the role of the upper critical dimension clearer, as indicated by the following theorem. In its statement, we abbreviate $p_c = p_c(L, \mathbb{Z}^d)$:

Theorem 5.4 (Infrared bound [132]). *For spread-out percolation with $d > 6$, there exists $L_0 = L_0(d)$ and $A = A(d, L)$ such that for every $L \geq L_0$ and uniformly in $p \in [0, p_c)$ and $k \in (-\pi, \pi]^d$,*

$$\hat{\tau}_p(k) \leq \frac{A(d, L)}{|k|^2}. \tag{5.2.2}$$

Theorem 5.4 is proved in an identical fashion as Thm. 5.1, and, alternatively, the result can also be formulated as the statement that there exists a constant $A(d, L)$ such that, uniformly in $p \in [0, p_c)$ and $k \in (-\pi, \pi]^d$,

$$\hat{\tau}_p(k) \leq \frac{A(d, L)}{1 - \widehat{D}_L(k)} , \tag{5.2.3}$$

where

$$D_L(x) = \frac{\mathbb{1}_{\{\|x\|_\infty \leq L\}}}{(2L + 1)^d} \tag{5.2.4}$$

is the random walk step distribution on the cube of width L. Here, for L large, $A(d, L) = 1 + O(L^{-d})$, so that (5.2.3) can really be seen as the perturbative statement that $\hat{\tau}_p(k)$ is approximately the Fourier transform of the random walk Green's function or BRW two-point function with step distributions D_L.

In Thm. 5.4, the so-called *upper critical dimension* appears as $d_c = 6$, and it is believed that mean-field results cannot be valid for $d < d_c$, while for $d = d_c$, logarithmic corrections appear to mean-field behavior. We return to this question in Sect. 11.3. In the next section, we explain the philosophy behind the proofs of Thms. 5.1 and 5.4, focusing on Thm. 5.1.

Exercise 5.3 (Nontriviality of percolation on spread-out lattices). Let $d \geq 2$. Prove that $p_c = p_c(L, \mathbb{Z}^d)$ for the spread-out model with edge set in (5.2.1) satisfies $p_c \in (0, 1)$.

5.3   Overview of Proof: A Lace-Expansion Analysis

The proof of the infrared bound in Thm. 5.1 makes use of a clever combinatorial expansion technique, somewhat like the cluster expansion in statistical mechanics. This expansion goes under the name of the *percolation lace expansion*. Lace-expansion techniques have been used to prove mean-field results for various models, including the self-avoiding walk, percolation, lattice trees and lattice animals, the Ising model, φ^4-models, and the contact process. In this text, we focus on the percolation setting. For an excellent survey of the lace-expansion method with a focus on self-avoiding walk, we refer the reader to the Saint-Flour lecture notes of Slade [246].

Any lace-expansion analysis consists of three main steps that are strongly intertwined:

(I) Derivation Lace Expansion. First, we need to *derive* the lace expansion. For percolation, the lace expansion takes the form

$$\tau_p(x) = \delta_{0,x} + 2dp(D \star \tau_p)(x) + 2dp(\Pi_p \star D \star \tau_p)(x) + \Pi_p(x) , \tag{5.3.1}$$

valid for all $p \leq p_c$ and with explicit, but quite complicated, formulas for the function $\Pi_p \colon \mathbb{Z}^d \to \mathbb{R}$, whose finiteness is a priori not obvious. The function Π_p and its properties form the key to the understanding of the two-point function, as can be seen by taking the Fourier transform and solving for $\hat{\tau}_p(k)$ to yield

$$\hat{\tau}_p(k) = \frac{1 + \hat{\Pi}_p(k)}{1 - 2dp\widehat{D}(k)[1 + \hat{\Pi}_p(k)]}.$$ (5.3.2)

Thus, the function Π_p uniquely identifies the two-point function and is sometimes called the *irreducible two-point function*. The derivation of the lace expansion is performed in Chap. 6.

(II) Bounds on the Lace-Expansion Coefficients. The second key ingredient in an analysis using the lace expansion are the bounds on the lace-expansion coefficients Π_p. These bounds facilitate the proof of the fact that the equations in (5.3.1) and (5.3.2) are small perturbations of simple random walk. The bounds on the lace-expansion coefficients are proved in Chap. 7.

(III) Bootstrap Analysis of the Lace Expansion. Now that we have completed the bounds on the lace-expansion coefficients, we are ready to analyze the lace expansion, with the aim to prove the infrared bound in Thm. 5.1. This is difficult, since the bounds on the lace-expansion coefficients are in terms of the two-point function τ_p itself, thus creating an apparent circularity in the reasoning. Therefore, we need a bootstrap argument to circumvent this problem. The bootstrap analysis is performed in Chap. 8.

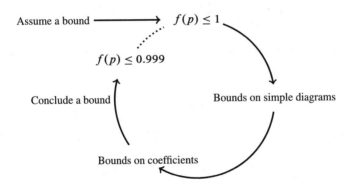

Fig. 5.1 Proof of claim (iii): $f(p) \le 1$ implies $f(p) \le 0.999$.

In the bootstrap, we assume that certain bounds hold for $p < p_c$ on a certain (finite) set of functions $f_i(p)$ and then conclude that a better bound must also hold. Here, we can think of $f_1(p) = 2dp$ or $f_2(p) = \sup_{k \in (-\pi,\pi)^d} \hat{\tau}_p(k)[1 - \widehat{D}(k)]$. These estimates can then be used to prove sharp bounds on the lace-expansion coefficients, which in turn imply better bounds on the bootstrap functions. When $f_i(0)$ obey the better bound for every i, and the functions $p \mapsto f_i(p)$ are continuous, this implies that the better bounds hold for *every* $p \in [0, p_c)$ and thus also *at* $p = p_c$. This is intuitively explained in Fig. 5.1.

Together, the above steps prove that when the dimension is such the *random walk triangle condition* holds, i.e., when

$$\triangle^{(\mathrm{RW})} = \sum_{x,y \in \mathbb{Z}^d} C_1(x)C_1(y - x)C_1(y) < \infty,$$ (5.3.3)

then also the percolation triangle condition in (4.1.1) is satisfied. For technical reasons, we need that $\triangle^{(RW)}$ is sufficiently close to 1, which is only true when d is sufficiently large. We see such restrictions on $\triangle_p$ also reflected in the requirement that $\triangle_{p_c}(e) < 1$ for $e \sim 0$ in the proof that $\gamma \leq 1$. In the next section, we investigate such random walk quantities in more detail.

5.4   The Random Walk Triangle

In this section, we prove random walk conditions very similar to the random walk triangle condition. These are formulated in the following proposition:

Proposition 5.5 (Random walk triangles). *For simple random walk, for any $l, n \geq 0$, there exist constants $c_{2l,n}^{(RW)}$ independent of d such that for $d > 4n$,*

$$\int_{(-\pi,\pi]^d} \frac{\widehat{D}(k)^{2l}}{[1 - \widehat{D}(k)]^n} \frac{dk}{(2\pi)^d} \leq c_{2l,n}^{(RW)} d^{-l} . \tag{5.4.1}$$

As a consequence, we get that (under the hypothesis of the theorem) for any $\lambda \in [0, 1]$,

$$\int_{(-\pi,\pi]^d} \frac{\widehat{D}(k)^{2l}}{[1 - \lambda\widehat{D}(k)]^n} \frac{dk}{(2\pi)^d} \leq c_{2l,n}^{(RW)} d^{-l} . \tag{5.4.2}$$

The reason for this is that we can represent the left-hand side in x-space and bound as

$$\sum_x (D^{*2l} * C_\lambda^{*n})(x) \leq \sum_x (D^{*2l} * C_1^{*n})(x) , \tag{5.4.3}$$

and then (after retransformation into Fourier space) applying (5.4.1).

Proof of Prop. 5.5. We follow [60, Sect. 2.2.2]. The Cauchy–Schwarz inequality[1] yields

$$\int_{(-\pi,\pi]^d} \frac{\widehat{D}(k)^{2l}}{[1 - \widehat{D}(k)]^n} \frac{dk}{(2\pi)^d}$$
$$\leq \left(\int_{(-\pi,\pi]^d} \widehat{D}(k)^{4l} \frac{dk}{(2\pi)^d} \right)^{1/2} \left(\int_{(-\pi,\pi]^d} \frac{1}{[1 - \widehat{D}(k)]^{2n}} \frac{dk}{(2\pi)^d} \right)^{1/2} . \tag{5.4.4}$$

The second term on the right-hand side is bounded uniformly by (5.1.10). Finally, we argue that the first term on the right-hand side of (5.4.4) is small if d is large. Note that

---

[1] The Hölder inequality gives better bounds here. In particular, it requires $d > 2n$ only, cf. (2.19) in [60]. Since we are working in high dimensions anyway, the extra dimensions needed here do not matter.

$$\int_{(-\pi,\pi]^d} \widehat{D}(k)^{4l} \frac{dk}{(2\pi)^d} = D^{\star 4l}(0) \tag{5.4.5}$$

is the probability that simple random walk returns to its starting point at the $4l$th step. This is bounded from above by $c_l(2d)^{-2l}$ with c_l being a well-chosen constant, because the first $2l$ steps must be compensated by the last $2l$. Finally, the square root yields the upper bound $O(d^{-l})$. $\qquad\qquad\qquad\qquad\qquad\qquad\qquad\qquad\qquad\qquad\qquad\qquad\qquad\qquad\qquad\qquad$ □

Fourier theory plays a very important role in this text, so we encourage you to make the following exercises:

Exercise 5.4 (Related triangles). Show that for $d > 6$ and simple random walk, uniformly for $\lambda \le 1$ and $k \in (-\pi, \pi]^d$,

$$\int_{(-\pi,\pi]^d} \widehat{D}(l)^2 \widehat{C}_\lambda(l)^2 \tfrac{1}{2}[\widehat{C}_\lambda(l+k) + \widehat{C}_\lambda(l-k)]\frac{dl}{(2\pi)^d} \le c_{2,3}^{(\mathrm{RW})}/d, \tag{5.4.6}$$

$$\int_{(-\pi,\pi]^d} \widehat{D}(l)^2 \widehat{C}_\lambda(l)\widehat{C}_\lambda(l-k)\widehat{C}_\lambda(l+k)\frac{dl}{(2\pi)^d} \le c_{2,3}^{(\mathrm{RW})}/d, \tag{5.4.7}$$

where $c_{2,3}^{(\mathrm{RW})}$ is the same as in Prop. 5.5. [*Hint:* For (5.4.7), show that

$$\widehat{C}_\lambda(l-k)\widehat{C}_\lambda(l+k) = \left(\sum_{x\in\mathbb{Z}^d} \cos(l\cdot x)\cos(k\cdot x)C_\lambda(x)\right)^2$$

$$-\left(\sum_{x\in\mathbb{Z}^d} \sin(l\cdot x)\sin(k\cdot x)C_\lambda(x)\right)^2, \tag{5.4.8}$$

and use that $l \mapsto \sum_{x\in\mathbb{Z}^d} \cos(l\cdot x)\cos(k\cdot x)C_\lambda(x)$ is the Fourier transform of $x \mapsto \cos(k\cdot x)C_\lambda(x)$.]

Exercise 5.5 (Percolation open polygons). Fix $p < p_c$. Use the Inverse Fourier Theorem and Lem. 5.3 to prove that $\tau_p^{\star s}(x) \le \tau_p^{\star s}(0)$ for every $x \in \mathbb{Z}^d$ and $s \ge 1$. Show further that $(D^{\star 2} \star \tau_p^{\star s})(x) \le (D^{\star 2} \star \tau_p^{\star s})(0)$ for every $x \in \mathbb{Z}^d$ and $s \ge 1$. Conclude that these bounds also hold for $p = p_c$.

Random Walk Triangle Condition for Spread-Out Models. We recall our discussion of spread-out models in Sect. 5.2. The major motivation for the spread-out model with D_L as in (5.2.4) stems from the fact that we can prove mean-field behaviour for $d \ge d_0 = 7$ provided that L is sufficiently large (in contrast to the nearest-neighbor model, where d_0 need to be sufficiently large). The reason for this difference is seen best when contrasting the assertion of Prop. 5.5 with the corresponding statement for spread-out models, which reads as follows:

Proposition 5.6 (Random walk triangles for spread-out model). *For the spread-out walk, there exist constants $c_{2l,n}^{(\mathrm{RW})} = c_{2l,n}^{(\mathrm{RW})}(d)$ independent of L such that for $d > 2n$ and $l \ge 1$,*

$$\int_{(-\pi,\pi]^d} \frac{\widehat{D}_L(k)^{2l}}{[1-\widehat{D}_L(k)]^n} \frac{dk}{(2\pi)^d} \leq c_{2l,n}^{(\mathrm{RW})} L^{-d}, \tag{5.4.9}$$

where D_L is defined in (5.2.4).

Recall from (5.3.3) that the *random walk triangle diagram* $\Delta^{(\mathrm{RW})}$ is obtained as the left-hand side of (5.4.9) with $l=0$ and $n=3$, and the proposition thus gives that not only $\Delta^{(\mathrm{RW})}$ is finite if $d>6$, but also that it is arbitrarily *small*.

Proof. We again follow [60, Sect. 2.2.2] and [148, Sect. 3]. We separately consider the regions $\|k\|_\infty \leq L^{-1}$ and $\|k\|_\infty > L^{-1}$. We note that for suitable constants $c_1, c_2 > 0$,

$$1-\widehat{D}_L(k) \geq c_1 L^2 |k|^2 \quad \text{if } \|k\|_\infty \leq L^{-1}, \tag{5.4.10}$$

$$1-\widehat{D}_L(k) > c_2 \qquad \text{if } \|k\|_\infty \geq L^{-1}, \tag{5.4.11}$$

both inequalities being fairly straightforward to prove.

Exercise 5.6 (Bounds on $1-\widehat{D}_L$). Prove (5.4.10) and (5.4.11).

By (5.4.10) and the bound $\widehat{D}_L(k)^2 \leq 1$, the corresponding contributions to the integral are

$$\int_{k:\|k\|_\infty \leq L^{-1}} \frac{\widehat{D}_L(k)^{2l}}{[1-\widehat{D}_L(k)]^n} \frac{dk}{(2\pi)^d}$$

$$\leq \frac{1}{c_1^s L^{2n}} \int_{k:\|k\|_\infty \leq L^{-1}} \frac{1}{|k|^{2n}} \frac{dk}{(2\pi)^d} \leq C_{d,c_1} L^{-d}, \tag{5.4.12}$$

if $d > 2n$, where C_{d,c_1} is a constant depending (only) on d and c_1, and by (5.4.11),

$$\int_{k:\|k\|_\infty > L^{-1}} \frac{\widehat{D}_L(k)^{2l}}{[1-\widehat{D}_L(k)]^n} \frac{dk}{(2\pi)^d}$$

$$\leq c_2^{-n} \int_{k:\|k\|_\infty > L^{-1}} \widehat{D}_L(k)^{2l} \frac{dk}{(2\pi)^d} \leq \text{const } L^{-d}, \tag{5.4.13}$$

for some positive constant and $l \geq 1$. In the last step, we have used that

$$\int_{k\in(-\pi,\pi]^d} \widehat{D}_L(k)^2 \frac{dk}{(2\pi)^d} = (D_L \star D_L)(0)$$

$$= \sum_{y\in\mathbb{Z}^d} D_L(y)^2 \leq \text{const } L^{-d}, \tag{5.4.14}$$

since $\max_x D_L(x) = O(L^{-d})$. $\square$

Chapter 6
The Derivation of the Lace Expansion via Inclusion–Exclusion

In this chapter, we perform the combinatorial expansion that relies on inclusion–exclusion for the percolation two-point function $\tau_p(x)$. We follow the classical Hara-Slade expansion [132]. An overview of this expansion can be found in the related paper by Hara and Slade [137]. We adapt it using ideas in [60] and [148]. We provide an overview to the expansion in Sect. 6.1. We give a self-contained and detailed derivation of the expansion in Sect. 6.2. We close this chapter in Sect. 6.3 by discussing the lace expansion, focussing in particular on the error term that arises in the inclusion–exclusion expansion. For an introduction to the lace expansion for various different models, see the Saint-Flour lecture notes by Slade [246]. We start by giving an overview of the expansion.

6.1 The Inclusion–Exclusion Lace Expansion

The term "lace" was used by Brydges and Spencer [66] for certain graphs that arose in the expansion that they invented to study the self-avoiding walk. Although the inclusion–exclusion expansion for percolation evolved from the lace expansion for the self-avoiding walk, this graphical construction does not occur for percolation, so that the term "lace" expansion is a misnomer in the percolation context. However, the name has stuck for historical reasons.

We start by giving a brief introduction to the inclusion–exclusion lace expansion, with an indication of how it is used to prove the infrared bound in Thm. 5.1 that implies the triangle condition. We restrict attention here to percolation on the hypercubic lattice $\mathbb{Z}^d$, with d large. In fact, the derivation of the expansion as well as obtaining the bounding diagrams in Chaps. 6 and 7 applies verbatim to any transitive graph. Only during the Fourier analysis in Chap. 8, we are limited to the lattice $\mathbb{Z}^d$, as Fourier techniques are not available in such general context.

Given a percolation cluster containing 0 and x, we call any bond whose removal would disconnect 0 from x a *pivotal* bond. The connected components that remain after removing all pivotal bonds are called *sausages*, see Fig. 6.1. Since, by definition, sausages are separated by at least one pivotal bond, no two sausages can share a vertex. Thus, the sausages are

© Springer International Publishing Switzerland 2017
M. Heydenreich and R. van der Hofstad, *Progress in High-Dimensional Percolation and Random Graphs*, CRM Short Courses, DOI 10.1007/978-3-319-62473-0_6

constrained to be *mutually avoiding*. In high dimensions, this is, at least intuitively, a relatively *weak* constraint, since sausage intersections require a cycle, and cycles are unlikely in high dimensions. In fact, for p asymptotically proportional to $1/(2d)$, and hence for $p = p_c$, the probability that the origin is in a cycle of length 4 is of order $d^2 d^{-4} = d^{-2}$, and larger cycles are even more rare. The fact that cycles are unlikely also means that clusters tend to be tree-like and percolation paths tend to be close to random walk paths. This makes it reasonable to attempt an inclusion–exclusion analysis, where the connection from 0 to x is treated as a random walk path, with correction terms taking into account cycles in sausages and intersections between sausages that appear through the use of inclusion–exclusion. With this in mind, it makes sense to relate $\tau_p(x)$ to the random walk Green's function as defined in (2.2.8).

For this, we start from (2.2.8) and notice that the random walk Green's function satisfies

$$C_\mu(x) = \delta_{0,x} + \mu(D \star C_\mu)(x) . \tag{6.1.1}$$

Indeed, the first term arises from the zero step walk, and the second term arises by conditioning on the first step of the simple random walk.

The lace expansion of Hara and Slade [132] is a combinatorial expansion for $\tau_p(x)$ that makes this inclusion–exclusion procedure precise. It produces a convolution equation of the form

$$\tau_p(x) = \delta_{0,x} + 2dp(D \star \tau_p)(x) + 2dp(\Pi_p \star D \star \tau_p)(x) + \Pi_p(x) \tag{6.1.2}$$

for the two-point function, valid for $p \le p_c$. The expansion gives explicit but complicated formulas for the function $\Pi_p \colon \mathbb{Z}^d \to \mathbb{R}$. It turns out that if $d \ge d_0$ and $d_0 > 6$ is sufficiently large, then $\widehat{\Pi}_p(k) = O(1/d)$ uniformly in $p \le p_c$. Putting $\Pi_p \equiv 0$ in (6.1.2) gives (6.1.1),

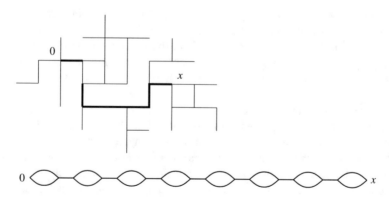

Fig. 6.1 Percolation cluster with a string of eight sausages joining **0** to x, and a schematic representation of the string of sausages. The 7 pivotal bonds are shown in bold.

and in this sense the percolation two-point function can be regarded as a small perturbation of the random walk Green's function.

Applying the Fourier transform, (6.1.2) can be solved to give

$$\hat{\tau}_p(k) = \frac{1 + \widehat{\Pi}_p(k)}{1 - 2dp\widehat{D}(k)[1 + \widehat{\Pi}_p(k)]} .$$ (6.1.3)

Therefore, $\widehat{\Pi}_p(k)$ is sometimes called the *irreducible two-point function*. It follows that $\widehat{\Pi}_p(k)$ uniquely identifies $\hat{\tau}_p(k)$ and vice versa:

Exercise 6.1 (Uniqueness lace-expansion coefficient). Take $p < p_c$. Show that (6.1.3) uniquely identifies $\widehat{\Pi}_p(k)$.

We show that when $d \geq d_0$ and $d_0 > 6$ is sufficiently large, then $\widehat{\Pi}_p(k)$ can be well approximated by $\widehat{\Pi}_p(0)$. Since $\widehat{\Pi}_p(k)$ is also small compared to 1, (6.1.3) suggests that the approximation

$$\hat{\tau}_p(k) \approx \frac{1}{1 - 2dp[1 + \widehat{\Pi}_p(0)]\widehat{D}(k)}$$ (6.1.4)

is reasonable (where $\approx$ denotes an uncontrolled approximation). Comparing with (2.2.11), this suggests that

$$\hat{\tau}_p(k) \approx \hat{C}_{\mu_p}(k) \quad \text{with } \mu_p = 2dp[1 + \widehat{\Pi}_p(0)] .$$ (6.1.5)

Since $\hat{\tau}_p(0) = \chi(p) \to \infty$ as $p \nearrow p_c$ (recall Thm. 3.2(3)), the critical threshold must satisfy the implicit equation

$$2dp_c[1 + \widehat{\Pi}_{p_c}(0)] = 1 .$$ (6.1.6)

In order to derive the infrared bound in Thm. 5.1, it then suffices to prove that $\hat{\tau}_p(k)$ in (6.1.3) can be bounded from above by $A(d)/[1 - \widehat{D}(k)]$. A problem for this is that in order to bound the expansion coefficients $\widehat{\Pi}_p(k)$, we need to rely on bounds on the two-point function $\hat{\tau}_p(k)$, which seems to lead to a circular argument. This circularity is avoided in Chap. 8 using an ingenious *bootstrap argument*. In the remainder of this chapter, we derive the expansion in (6.1.2) and identify the coefficients $\widehat{\Pi}_p(k)$.

6.2 Derivation of the Inclusion–Exclusion Lace Expansion

In this section, we derive a version of the inclusion–exclusion lace expansion (6.1.2) that contains a remainder term. We use the method of Hara and Slade [132], which applies directly in this general setting, and we follow the presentation of Borgs et al. in [60] closely. Parts of the exposition of the expansion are taken verbatim from [60, Sect. 3.2], where the expansion was derived for general graphs and further references are given.

We start by introducing some notation. Fix $p \in [0, 1]$. We define

$$J(x) = p \mathbb{1}_{\{x \sim 0\}} = 2dp D(x) , \qquad (6.2.1)$$

with D given by (1.2.18). We write $\tau(x) = \tau_p(x)$ for brevity, and generally drop subscripts indicating dependence on p. The result of this section is summarized in the following proposition:

Proposition 6.1 (Inclusion-exclusion lace expansion). *Let p be such that $\theta(p) = 0$. For each $M = 0, 1, 2, \ldots$, the expansion takes the form*

$$\tau(x) = \delta_{0,x} + (J \star \tau)(x) + (\Pi_M \star J \star \tau)(x) + \Pi_M(x) + R_M(x) , \qquad (6.2.2)$$

where the $\star$ denotes convolution. The function $\Pi_M : \mathbb{Z}^d \to \mathbb{R}$ is the key quantity in the expansion, and $R_M(x)$ is a remainder term defined in (6.2.20) for $M = 0$, in (6.2.26) for $M = 1$ and in (6.2.28) for $M \geq 2$. The dependence of Π_M on M is given by

$$\Pi_M(x) = \sum_{N=0}^{M} (-1)^N \Pi^{(N)}(x) , \qquad (6.2.3)$$

with $\Pi^{(N)}(x)$ independent of M. $\Pi^{(N)}(x)$ is defined in (6.2.7) for $N = 0$, in (6.2.25) for $N = 1$, and in (6.2.27) for $N \geq 2$.

The alternating sign in (6.2.3) arises via a repeated inclusion–exclusion argument. In Chap. 8, we prove that, for all $d > d_0$ with $d_0 \geq 6$ large enough and $p < p_c$,

$$\lim_{M \to \infty} \sum_{x \in \mathbb{Z}^d} |R_M(x)| = 0 . \qquad (6.2.4)$$

This leads to (6.1.2) with $\Pi = \Pi_\infty$ (see also Sect. 6.3 below). Convergence properties of (6.2.3) when $M = \infty$ are also established in Chap. 8. The remainder of this section gives the proof of Prop. 6.1.

Initiating the Expansion. To get started with the expansion, we need the following definitions.

Definition 6.2 (Connections in and through sets of vertices).
(a) Given a set of vertices $A \subseteq \mathbb{Z}^d$, we say that x and y are *connected in A* and write $\{x \leftrightarrow y \text{ in } A\}$, if $x = y \in A$ or if there is an occupied path from x to y having all its endpoints in A. We define the *restricted two-point function* by

$$\tau^A(x, y) = \mathbb{P}(x \leftrightarrow y \text{ in } \mathbb{Z}^d \backslash A) . \qquad (6.2.5)$$

(b) Given a bond configuration and a set of vertices $A \subseteq \mathbb{Z}^d$, we say that x and y are *connected through A*, if $x \leftrightarrow y$ and every occupied path connecting x to y has at least one bond with an endpoint in A, or if $x = y \in A$. This event is written as $\{x \overset{A}{\leftrightarrow} y\}$.

With this definition at hand, we can extend (4.2.20) to get for every $x, y \in \mathbb{Z}^d$, $A \subseteq \mathbb{Z}^d$,

$$\tau^A(x, y) = \tau(x, y) - \left(\tau(x, y) - \tau^A(x, y)\right) = \tau(x, y) - \mathbb{P}(x \overset{A}{\longleftrightarrow} y). \tag{6.2.6}$$

Definition 6.3 (Double connections and pivotal bonds).

(a) Given a bond configuration, we say that x is *doubly connected to* y, and we write $x \Longleftrightarrow y$, if $x = y$ or if there are at least two bond-disjoint paths from x to y consisting of occupied bonds.

(b) Given a bond configuration and a bond b, we define the *restricted cluster* $\widetilde{\mathcal{C}}^b(x)$ to be the set of vertices connected to x in the new configuration obtained by setting b to be vacant.

(c) Given a bond configuration, a bond $\{u, v\}$ (occupied or not) is called *pivotal* for the connection from x to y if (i) either $x \longleftrightarrow u$ and $y \longleftrightarrow v$, or $x \longleftrightarrow v$ and $y \longleftrightarrow u$, and (ii) $y \notin \widetilde{\mathcal{C}}^{\{u,v\}}(x)$. Bonds are not usually regarded as directed. However, it is convenient at times to regard a bond $\{u, v\}$ as directed from u to v, and we emphasize this point of view with the notation (u, v). A directed bond (u, v) is pivotal for the connection from x to y if $x \longleftrightarrow u$, $v \longleftrightarrow y$, and $y \notin \widetilde{\mathcal{C}}^{\{u,v\}}(x)$. We denote by $\text{Piv}(x, y)$ the set of directed pivotal bonds for the connection from x to y.

Start of the Expansion for $M = 0$. To begin the expansion, we define

$$\Pi^{(0)}(x) = \mathbb{P}(0 \Longleftrightarrow x) - \delta_{0,x} \tag{6.2.7}$$

and distinguish configurations with $0 \longleftrightarrow x$ according to whether or not there is a double connection, to obtain

$$\tau(x) = \delta_{0,x} + \Pi^{(0)}(x) + \mathbb{P}(0 \longleftrightarrow x, 0 \not\Longleftrightarrow x). \tag{6.2.8}$$

If 0 is connected to x, but not doubly, then $\text{Piv}(0, x)$ is nonempty. There is therefore a unique element $(u, v) \in \text{Piv}(0, x)$ (the *first* pivotal bond, sometimes also called the *cutting bond*) such that $0 \Longleftrightarrow u$, and we can write

$$\mathbb{P}(0 \longleftrightarrow x, 0 \not\Longleftrightarrow x)$$
$$= \sum_{(u,v)} \mathbb{P}(0 \Longleftrightarrow u \text{ and } (u, v) \text{ is occupied and pivotal for } 0 \longleftrightarrow x). \tag{6.2.9}$$

Now comes the essential part of the expansion. Ideally, we would like to *factor* the probability on the right side of (6.2.9) as

$$\mathbb{P}(0 \Longleftrightarrow u) \, \mathbb{P}((u, v) \text{ is occupied}) \, \mathbb{P}(v \longleftrightarrow x)$$
$$= \left(\delta_{0,u} + \Pi^{(0)}(u)\right) J(u, v) \tau(x - v), \tag{6.2.10}$$

where $J(u, v) = J(v - u)$. The expression (6.2.10) is the same as (6.2.2) with $\Pi_M = \Pi^{(0)}$ and $R_M = 0$. However, (6.2.9) does not factor in this way because the cluster $\widetilde{\mathcal{C}}^{(u,v)}(u)$ is constrained not to intersect the cluster $\widetilde{\mathcal{C}}^{(u,v)}(v)$, since (u, v) is pivotal. What we can do is approximate the probability on the right-hand side of (6.2.9) by (6.2.10) using inclusion–exclusion and then attempt to deal with the error term.

The Cutting-Bond Lemma. In order to use inclusion–exclusion, we use the next lemma, which gives an identity for the probability on the right-hand side of (6.2.9). In fact, we also need a more general identity, involving the following generalizations of the event appearing on the right-hand side of (6.2.9). Let $x, u, v, y \in \mathbb{Z}^d$ with $u \sim v$, and $A \subseteq \mathbb{Z}^d$ be nonempty. Then we define the events

$$E'(v, y; A) = \left\{ v \overset{A}{\leftrightarrow} y \right\} \cap \left\{ \nexists (u', v') \in \text{Piv}(v, y) \text{ such that } v \overset{A}{\leftrightarrow} u' \right\}, \tag{6.2.11}$$

and

$$E(x, u, v, y; A) = E'(x, u; A)$$
$$\cap \{(u, v) \text{ is occupied and pivotal for } x \leftrightarrow y\} . \tag{6.2.12}$$

Note that $\{x \Longleftrightarrow y\} = E'(x, y; \mathbb{Z}^d)$, while $E(0, u, v, x; \mathbb{Z}^d)$ is the event appearing on the right-hand side of (6.2.9). A version of Lem. 6.4, with $E'(x, u; A)$ replaced by $\{0 \leftrightarrow u\}$ on both sides of (6.2.13), appeared as (4.2.19) in the Aizenman–Newman proof of $\gamma = 1$. See also Exerc. 6.2 below.

Lemma 6.4 (The cutting-bond lemma). *Let $p \in [0, 1]$ be such that $\theta(p) = 0$, let $x, u, v, y \in \mathbb{Z}^d$, and let $A \subseteq \mathbb{Z}^d$ be nonempty. Then*

$$\mathbb{E}\left(\mathbb{1}_{E(x,u,v,y;A)}\right) = p \, \mathbb{E}\left(\mathbb{1}_{E'(x,u;A)} \, \tau^{\widetilde{\mathscr{C}}^{\{u,v\}}(x)}(v, y)\right) . \tag{6.2.13}$$

Proof. The event appearing in the left side of (6.2.13) is depicted in Fig. 6.2. We first observe that the event $E'(x, u; A) \cap \{(u, v) \in \text{Piv}(x, y)\}$ is independent of the occupation status of the bond (u, v). This is true by definition for $\{(u, v) \in \text{Piv}(x, y)\}$, and when (u, v) is pivotal, the occurrence or not of $E'(x, u; A)$ cannot be affected by $\{u, v\}$ since in this case $E'(x, u; A)$ is determined by the occupied paths from x to u and no such path uses the bond $\{u, v\}$. Therefore, the left side of (6.2.13) is equal to

$$p \, \mathbb{E}\left(\mathbb{1}_{E'(x,u;A) \cap \{(u,v) \in \text{Piv}(x,y)\}}\right) . \tag{6.2.14}$$

By conditioning on $\widetilde{\mathscr{C}}^{\{u,v\}}(x)$, (6.2.14) is equal to

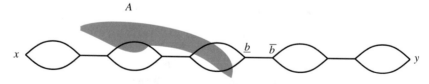

Fig. 6.2 Event $E(x, b, y; A)$ of Lem. 6.4. The shaded regions represent the vertices in A. There is no restriction on intersections between A and $\widetilde{\mathscr{C}}^b(y)$.

$$p \sum_{S:S \ni x} \mathbb{E}\left(\mathbb{1}_{E'(x,u;A)\cap\{(u,v)\in\mathrm{Piv}(x,y)\}\cap\{\widetilde{\mathcal{C}}^{\{u,v\}}(x)=S\}}\right), \tag{6.2.15}$$

where the sum is over all finite connected sets of vertices S containing x. This is possible, since by assumption $\theta(p) = 0$, so that, in particular, $|\widetilde{\mathcal{C}}^{\{u,v\}}(x)| < \infty$ a.s.

In (6.2.15), we make the replacement, valid for all connected S such that $x \leftrightarrow u$ in S,

$$\{(u, v) \in \mathrm{Piv}(x, y)\} \cap \{\widetilde{\mathcal{C}}^{\{u,v\}}(x) = S\}$$
$$= \{v \leftrightarrow y \text{ in } \mathbb{Z}^d \backslash S\} \cap \{\widetilde{\mathcal{C}}^{\{u,v\}}(x) = S\}. \tag{6.2.16}$$

Indeed, on the one hand, $\{v \leftrightarrow y \text{ in } \mathbb{Z}^d \setminus \widetilde{\mathcal{C}}^{\{u,v\}}(x)\} = \{v \leftrightarrow y \text{ in } \mathbb{Z}^d \setminus S\}$ must occur when $(u, v) \in \mathrm{Piv}(x, y)$ and when $\widetilde{\mathcal{C}}^{\{u,v\}}(x) = S$. On the other hand, when $\{v \leftrightarrow y \text{ in } \mathbb{Z}^d \setminus \widetilde{\mathcal{C}}^{\{u,v\}}(x)\} = \{v \leftrightarrow y \text{ in } \mathbb{Z}^d \setminus S\}$ occurs, then also $(u, v) \in \mathrm{Piv}(x, y)$ (recall Def. 6.3(c)).

The event $\{v \leftrightarrow y \text{ in } \mathbb{Z}^d \backslash S\}$ depends only on the occupation status of bonds that do not have an endpoint in S. Further, given that $\{v \leftrightarrow y \text{ in } \mathbb{Z}^d \backslash S\} \cap \{\widetilde{\mathcal{C}}^{\{u,v\}}(x) = S\}$ occurs, the event $E'(x, u; A)$ is determined by the occupation status of bonds that do have an endpoint in $S = \widetilde{\mathcal{C}}^{\{u,v\}}(x)$. Similarly, the event $\{\widetilde{\mathcal{C}}^{\{u,v\}}(x) = S\}$ depends on bonds that have one or both endpoints in S. Hence, given S, the event $E'(x, u; A) \cap \{\widetilde{\mathcal{C}}^{\{u,v\}}(x) = S\}$ is independent of the event that $\{v \leftrightarrow y \text{ in } \mathbb{Z}^d \backslash S\}$, and therefore (6.2.15) is equal to

$$p \sum_{S:S \ni x} \mathbb{E}\left(\mathbb{1}_{E'(x,u;A)\cap\{\widetilde{\mathcal{C}}^{\{u,v\}}(x)=S\}}\right) \tau^S(v, y). \tag{6.2.17}$$

Bringing the restricted two-point function inside the expectation, replacing the superscript S by $\widetilde{\mathcal{C}}^{\{u,v\}}(x)$, and performing the sum over S give the desired result. $\qquad \square$

Exercise 6.2 (The Aizenman–Newman identity (4.2.19)). Prove (4.2.19) by adapting the proof of Lem. 6.4.

Completion expansion for $M = 0$. It follows from (6.2.9) and Lem. 6.4 that

$$\mathbb{P}(0 \leftrightarrow x, 0 \not\Leftrightarrow x) = \sum_{(u,v)} J(u, v)\,\mathbb{E}\left(\mathbb{1}_{\{0 \Leftrightarrow u\}} \tau^{\widetilde{\mathcal{C}}^{(u,v)}(0)}(v, x)\right), \tag{6.2.18}$$

where we write $\widetilde{\mathcal{C}}^{(u,v)}(0)$ in place of $\widetilde{\mathcal{C}}^{\{u,v\}}(0)$ to emphasize the directed nature of the bond (u, v). On the right side, $\tau^{\widetilde{\mathcal{C}}^{(u,v)}(0)}(v, x)$ is the restricted two-point function *given* the cluster $\widetilde{\mathcal{C}}^{(u,v)}(0)$ of the outer expectation, so that in the expectation defining $\tau^{\widetilde{\mathcal{C}}^{(u,v)}(0)}(v, x)$, $\widetilde{\mathcal{C}}^{(u,v)}(0)$ should be regarded as a *fixed* set. We stress this delicate point here, as it is crucial also in the rest of the expansion. The inner expectation on the right-hand side of (6.2.18) effectively introduces a second percolation model on a second graph, which is coupled to the original percolation model via the set $\widetilde{\mathcal{C}}^{(u,v)}(0)$.

We write, using (6.2.6) with $A = \widetilde{\mathcal{C}}^{(u,v)}(0)$,

$$\tau^{\widetilde{\mathcal{C}}^{(u,v)}(0)}(v,x) = \tau(v,x) - \left(\tau(v,x) - \tau^{\widetilde{\mathcal{C}}^{(u,v)}(0)}(v,x)\right)$$

$$= \tau(x-v) - \mathbb{P}\left(v \xleftarrow{\widetilde{\mathcal{C}}^{(u,v)}(0)} x\right), \qquad (6.2.19)$$

insert this into (6.2.18), and use (6.2.8) and (6.2.7) to obtain

$$\tau(x) = \delta_{0,x} + \Pi^{(0)}(x) + \sum_{(u,v)} \left(\delta_{0,u} + \Pi^{(0)}(u)\right) J(u,v)\tau(x-v)$$

$$- \sum_{(u,v)} J(u,v)\, \mathbb{E}_0\left(\mathbb{1}_{\{0 \Longleftrightarrow u\}}\, \mathbb{P}_1(v \xleftarrow{\widetilde{\mathcal{C}}_0^{(u,v)}(0)} x)\right). \qquad (6.2.20)$$

Here, we have introduced the subscripts 0 and 1 to denote the two percolation configurations. Thus, the law of $\widetilde{\mathcal{C}}_0^{(u,v)}(0)$ is described by $\mathbb{P}_0$ and $\mathbb{E}_0$, and $\widetilde{\mathcal{C}}_0^{(u,v)}(0)$ should be considered a *fixed* set inside the expectation $\mathbb{E}_1$. With $R_0(x)$ equal to the last term on the right side of (6.2.20) (including the minus sign), this proves (6.2.2) for $M = 0$.

Expansion for $M = 1$ and Cutting-Bond Partition. To continue the expansion, we would like to rewrite $R_0(x)$, i.e., the final term of (6.2.20), in terms of a product with the two-point function. A configuration contributing to the expectation in the final term of (6.2.20) is illustrated schematically in Fig. 6.3, in which the bonds drawn with heavy lines should be regarded as living on a different graph than the bonds drawn with lighter lines, as explained previously. Our goal is to extract a factor $\tau(y-v')$, where v' is the top of the bond b' shown in Fig. 6.3.

Given a configuration in which $v \xleftrightarrow{A} x$, the *cutting bond* (u',v') is defined to be the first pivotal bond for $v \leftrightarrow x$ such that $v \xleftrightarrow{A} u'$. It is possible that no such bond exists, as, for example, would be the case in Fig. 6.3 if only the leftmost four sausages were included in the figure (using the terminology of Sect. 6.1), with x in the location currently occupied by u', which is the bottom of the bond b'. Recall the definitions of $E'(v,x;A)$ and $E(x,u,v,y;A)$ in (6.2.11) and (6.2.12). By partitioning $\{v \xleftrightarrow{A} x\}$ according to the location of the cutting bond (or the lack of a cutting bond), we obtain the following partition:

Lemma 6.5 (Cutting-bond partition). *The event $\{v \xleftrightarrow{A} x\}$ can be partitioned as*

$$\{v \xleftrightarrow{A} x\} = E'(v,x;A) \bigcup \bigcup_{(u',v')} E(v,u',v',x;A). \qquad (6.2.21)$$

Exercise 6.3 Prove Lem. 6.5.

Lemma. 6.5 implies that

$$\mathbb{P}\left(v \xleftrightarrow{A} x\right) = \mathbb{P}\left(E'(v,x;A)\right) + \sum_{(u',v')} \mathbb{P}\left(E(v,u',v',x;A)\right). \qquad (6.2.22)$$

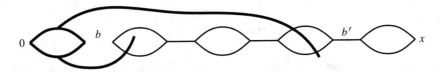

Fig. 6.3 Possible configuration appearing in the second stage of the expansion.

Using the Cutting-Bond Lemma 6.4, this gives

$$\mathbb{P}\left(v \overset{A}{\leftrightarrow} x\right) = \mathbb{P}\left(E'(v, x; A)\right)$$
$$+ \sum_{(u',v')} J(u', v')\,\mathbb{E}\left(\mathbb{1}_{E'(v,u';A)}\tau^{\widetilde{\mathcal{C}}^{(u',v')}(v)}(v', x)\right). \tag{6.2.23}$$

Inserting the identity (6.2.19) into (6.2.23), we obtain

$$\mathbb{P}\left(v \overset{A}{\leftrightarrow} x\right) = \mathbb{P}\left(E'(v, x; A)\right) + \sum_{(u',v')} J(u', v')\,\mathbb{P}\left(E'(v, u'; A)\right)\tau(x - v')$$
$$- \sum_{(u',v')} J(u', v')\,\mathbb{E}_1\left(\mathbb{1}_{E'(v,u';A)}\,\mathbb{P}_2\left(v' \xleftarrow{\widetilde{\mathcal{C}}_1^{(u',v')}(v)} x\right)\right). \tag{6.2.24}$$

In the last term on the right-hand side, we have again introduced subscripts for $\widetilde{\mathcal{C}}$ and the expectations, to indicate to which expectation $\widetilde{\mathcal{C}}$ belongs.
    Let

$$\Pi^{(1)}(x) = \sum_{(u,v)} J(u, v)\,\mathbb{E}_0\left(\mathbb{1}_{\{0 \Longleftrightarrow u\}}\,\mathbb{P}_1\left(E'\left(v, x; \widetilde{\mathcal{C}}_0^{(u,v)}(0)\right)\right)\right). \tag{6.2.25}$$

Inserting (6.2.24) into (6.2.20), and using (6.2.25), we obtain

$$\tau(x) = \delta_{0,x} + \Pi^{(0)}(x) - \Pi^{(1)}(x) + \sum_{(u,v)}\left(\delta_{0,u} + \Pi^{(0)}(u) - \Pi^{(1)}(u)\right)J(u, v)\tau(x - v)$$
$$+ \sum_{(u,v)} J(u, v)\sum_{(u',v')} J(u', v')$$
$$\times \mathbb{E}_0\left(\mathbb{1}_{\{0 \Longleftrightarrow u\}}\,\mathbb{E}_1\left(\mathbb{1}_{E'(v,u';\widetilde{\mathcal{C}}_0^{(u,v)}(0))}\,\mathbb{P}_2\left(v' \xleftarrow{\widetilde{\mathcal{C}}_1^{(u',v')}(v)} x\right)\right)\right). \tag{6.2.26}$$

This proves (6.2.2) for $M = 1$, with $R_1(x)$ given by the last two line of (6.2.26).

Iteration of Inclusion–Exclusion and Expansion for General M. We now repeat this procedure recursively, rewriting

$$\mathbb{P}_2\left(v' \xleftarrow{\widetilde{\mathscr{C}}_1^{(u',v')}(v)} x\right)$$

using (6.2.24), and so on. This leads to (6.2.2), with $\Pi^{(0)}$ and $\Pi^{(1)}$ given by (6.2.7) and (6.2.25), and, for $N \geq 2$,

$$\Pi^{(N)}(x) = \sum_{(u_0,v_0)} \cdots \sum_{(u_{N-1},v_{N-1})} \left[\prod_{i=0}^{N-1} J(u_i,v_i)\right] \mathbb{E}_0 \, \mathbb{1}_{\{0 \Leftrightarrow u_0\}}$$
$$\times \mathbb{E}_1 \, \mathbb{1}_{E'(v_0,u_1;\widetilde{\mathscr{C}}_0)} \cdots \mathbb{E}_{N-1} \, \mathbb{1}_{E'(v_{N-2},u_{N-1};\widetilde{\mathscr{C}}_{N-2})}$$
$$\times \mathbb{E}_N \, \mathbb{1}_{E'(v_{N-1},x;\widetilde{\mathscr{C}}_{N-1})}, \qquad (6.2.27)$$

and $R_0(x)$ and $R_1(x)$ defined in (6.2.20) and (6.2.26), and, for $M \geq 2$,

$$R_M(x) = (-1)^{M+1} \sum_{(u_0,v_0)} \cdots \sum_{(u_M,v_M)} \left[\prod_{i=0}^{M} J(u_i,v_i)\right] \mathbb{E}_0 \, \mathbb{1}_{\{0 \Leftrightarrow u_0\}}$$
$$\mathbb{E}_1 \, \mathbb{1}_{E'(v_0,u_1;\widetilde{\mathscr{C}}_0)} \cdots \mathbb{E}_{M-1} \, \mathbb{1}_{E'(v_{M-2},u_{M-1};\widetilde{\mathscr{C}}_{M-2})}$$
$$\times \mathbb{E}_M \left[\mathbb{1}_{E'(v_{M-1},u_M;\widetilde{\mathscr{C}}_{M-1})} \mathbb{P}_{M+1}\left(v_M \xleftarrow{\widetilde{\mathscr{C}}_M} x\right) \right], \qquad (6.2.28)$$

where we have used the abbreviation $\widetilde{\mathscr{C}}_j = \widetilde{\mathscr{C}}_j^{(u_j,v_j)}(v_{j-1})$, with $v_{-1} = 0$.

Exercise 6.4 (Iteration for $M = 2$). Perform the iteration one more step, and thus prove (6.2.2) for $M = 2$ by identifying $R_2(x)$ and $\Pi^{(2)}(x)$.

Exercise 6.5 (Solution of iteration). Use induction to prove (6.2.2) for general $M \geq 2$ and identify $R_M(x)$ and $\Pi^{(N)}(x)$ as in (6.2.28) and (6.2.27).

6.3   Full Expansion: How to Deal with the Error Term

In this section, we explain how to obtain the full lace expansion as in (5.3.1), or its Fourier version in (5.3.2). The essential difference between (6.2.2) and (5.3.1) is the error term $R_M(x)$ that appears in (6.2.2), while it is absent in (5.3.1). We obtain (5.3.1) from (6.2.2) by letting $M \to \infty$. Here we explain how one might prove bounds on this error term that would allow us to take this limit.

Since

$$\mathbb{P}_{M+1}\left(v_M \xleftarrow{\widetilde{\mathscr{C}}_M} x\right) \leq \tau(x - v_M), \qquad (6.3.1)$$

it follows from (6.2.27)–(6.2.28) that

$$|R_M(x)| \le \sum_{u_M, v_M \in \mathbb{Z}^d} \Pi^{(M)}(u_M) J(u_M, v_M) \tau(x - v_M)$$

$$\le \sum_{u_M, v_M \in \mathbb{Z}^d} \Pi^{(M)}(u_M) J(u_M, v_M) = 2dp\,\widehat{\Pi}^{(M)}(0). \tag{6.3.2}$$

Equation (6.3.2) is crucial in order to show that $|R_M(x)| \to 0$ as $M \to \infty$. If we assume that $\widehat{\Pi}^{(M)}(0) \to 0$ as $M \to \infty$, then $|R_M(x)| \to 0$ as well, and the alternating series

$$\Pi(x) = \sum_{N=0}^{\infty} (-1)^N \Pi^{(N)}(x) \tag{6.3.3}$$

converges, so that (6.1.2) follows. Further, summing the first bound in (6.3.2) over x gives

$$\sum_{x \in \mathbb{Z}^d} |R_M(x)| \le 2dp\,\widehat{\Pi}^{(M)}(0)\chi(p). \tag{6.3.4}$$

When $p < p_c$, we know that $\chi(p) < \infty$. As a result, (6.2.4) follows for every $p < p_c$ when $\widehat{\Pi}^{(M)}(0) \to 0$ as $M \to \infty$. Of course, the fact that $|R_M(x)| \to 0$ as $M \to \infty$ is one of the key steps in the analysis of the inclusion–exclusion lace expansion, as it also allows us to give the implicit relation for p_c given by (6.1.6)! The key ingredient to this is to show that $\widehat{\Pi}^{(M)}(0)$ is exponentially small in M uniformly in $p < p_c$, where the base of the exponential is closely related to the triangle diagram.

Chapter 7
Diagrammatic Estimates for the Lace Expansion

In this chapter, we prove bounds on $\Pi^{(N)}$. These bounds are summarized in Lem. 7.1 for $N = 0$, Lem. 7.2 for $N = 1$, and Prop. 7.4 for $N \geq 2$. We refer to the methods of this section as *diagrammatic estimates*, as we use Feynman diagrams to provide a convenient representation for upper bounds on $\Pi^{(N)}$. This chapter follows Borgs et al. [60, Sect. 4] closely and is in parts a verbatim copy of it. In turn, this approach is essentially identical to what is done by Hara and Slade in [132, Sect. 2.2], apart from some notational differences. The overview in Sect. 7.1 is novel, and we expand on the explanations. In Sect. 7.2, we give a warm up by bounding $\Pi^{(0)}$ and $\Pi^{(1)}$. In Sects. 7.3–7.5, we provide bounds on $\Pi^{(N)}$ in terms of *diagrams*, and bound these in terms of triangle diagrams and related objects. These bounds are performed in three key steps. In the first step in Sect. 7.3, we provide bounding events on the lace-expansion events $E'(v, y; A)$ tailored for an application of the BK inequality (1.3.4). In Sect. 7.4, we apply the BK inequality to bound the lace-expansion coefficients in terms of diagrams, and conveniently organize the large arising sums. Finally, in Sect. 7.5, we reduce the necessary bounds to bounds in terms of five simple diagrams like the triangle diagram. We close this chapter in Sect. 7.6 by giving a summary of what we have achieved and an outlook on how to proceed.

7.1 Overview of the Bounds

Recall 6.1.3, which implies that

$$\tau_p(x) = \big((1 + \Pi_p) \star V_p\big)(x) , \tag{7.1.1}$$

where we define $V_p(x)$ in terms of its Fourier transform $\widehat{V}_p(k)$ as

$$\widehat{V}_p(k) = \frac{1}{1 - 2dp\widehat{D}(k)[1 + \widehat{\Pi}_p(k)]} . \tag{7.1.2}$$

© Springer International Publishing Switzerland 2017
M. Heydenreich and R. van der Hofstad, *Progress in High-Dimensional Percolation and Random Graphs*, CRM Short Courses, DOI 10.1007/978-3-319-62473-0_7

In this chapter, we answer the following question:

What bounds do we need on Π_p for $\hat{\tau}_p(k)$ to be a small perturbation of $\widehat{C}_{\lambda_p}(k)$ for an appropriate λ_p, where

$$\hat{\tau}_p(k) = \frac{1 + \widehat{\Pi}_p(k)}{1 - 2dp\widehat{D}(k)[1 + \widehat{\Pi}_p(k)]} \ ? \tag{7.1.3}$$

In anticipation of the fact that Π_p is a small perturbation, we assume that Π_p is small in high dimensions. Therefore, it is reasonable to assume that the asymptotics of $\tau_p(x)$ are mainly determined by that of $V_p(x)$. Let us assume this for the time being; we return to this issue at the end of this section.

If $2dp\widehat{D}(k)[1 + \widehat{\Pi}_p(k)]$ were the Fourier transform of a nonnegative and summable function $Q_p(x)$, then we could interpret $V_p(x)$ as the Green's function associated with the random walk with transition probabilities given by

$$D_p(x) = \frac{Q_p(x)}{\sum_{y \in \mathbb{Z}^d} Q_p(y)} \ . \tag{7.1.4}$$

Then,

$$V_p(x) = G_{\mu_p}(x) \ , \tag{7.1.5}$$

where

$$\widehat{G}_\mu(k) = \frac{1}{1 - \mu\widehat{D}_p(k)} \ , \tag{7.1.6}$$

and

$$\mu_p = \sum_{y \in \mathbb{Z}^d} Q_p(y) = \widehat{Q}_p(0) = 2dp\widehat{D}(0)[1 + \widehat{\Pi}_p(0)] = 2dp[1 + \widehat{\Pi}_p(0)] \ . \tag{7.1.7}$$

Therefore, the critical value $p_c = p_c(\mathbb{Z}^d)$ should be determined by

$$\mu_{p_c} = 2dp_c[1 + \widehat{\Pi}_{p_c}(0)] = 1 \ . \tag{7.1.8}$$

Thus, for $p = p_c$,

$$\widehat{V}_{p_c}(k) = \widehat{G}_1(k) = \frac{1}{1 - \widehat{D}_{p_c}(k)} \ , \tag{7.1.9}$$

which is the critical Green's function with random walk transition probabilities $x \mapsto D_{p_c}(x)$. Taylor expansion shows that

$$\widehat{V}_{p_c}(k) \sim \frac{A}{|k|^2} \tag{7.1.10}$$

whenever the random walk transition probabilities have finite second moment, i.e., when

$$\sum_{x \in \mathbb{Z}^d} |x|^2 D_{p_c}(x) < \infty. \tag{7.1.11}$$

We know that $\widehat{D}_{p_c}(k) = 2dp_c \widehat{D}(k)[1 + \widehat{\Pi}_{p_c}(k)]$, so that

$$\sum_{x \in \mathbb{Z}^d} |x|^2 D_{p_c}(x) = 2dp_c \sum_{x \in \mathbb{Z}^d} |x|^2 D(x) \sum_{x \in \mathbb{Z}^d} \Pi_{p_c}(x)$$

$$+ 2dp_c \sum_{x \in \mathbb{Z}^d} D(x) \sum_{x \in \mathbb{Z}^d} |x|^2 \Pi_{p_c}(x)$$

$$= 2dp_c[1 + \widehat{\Pi}_{p_c}(0)] + 2dp_c \sum_{x \in \mathbb{Z}^d} |x|^2 \Pi_{p_c}(x) , \tag{7.1.12}$$

where we use that, for symmetric functions $x \mapsto f(x)$ and $x \mapsto g(x)$,

$$\sum_{x \in \mathbb{Z}^d} |x|^2 (f \star g)(x) = \sum_{x \in \mathbb{Z}^d} |x|^2 f(x) \sum_{y \in \mathbb{Z}^d} g(y) + \sum_{x \in \mathbb{Z}^d} f(x) \sum_{y \in \mathbb{Z}^d} |y|^2 g(y). \tag{7.1.13}$$

Exercise 7.1 (Second moment of convolution). Prove (7.1.13).

As a result, this leads us to the requirement that

$$\widehat{\Pi}_{p_c}(0) < \infty , \quad \sum_{x \in \mathbb{Z}^d} |x|^2 \Pi_{p_c}(x) < \infty . \tag{7.1.14}$$

It turns out to be convenient to work in Fourier space instead. The first bound in (7.1.14) is already formulated in Fourier language. Of course, the second statement in (7.1.14) should be equivalent to the statement that, for small k,

$$\sum_{x \in \mathbb{Z}^d} [1 - \cos(k \cdot x)] \Pi_{p_c}(x) = \widehat{\Pi}_{p_c}(0) - \widehat{\Pi}_{p_c}(k) = O(|k|^2) , \tag{7.1.15}$$

where we have used the spatial symmetry of $x \mapsto \Pi_p(x)$. Since $1 - \widehat{D}(k) = O(|k|^2)$, this can also be interpreted as saying that

$$\widehat{\Pi}_{p_c}(0) < \infty , \quad \widehat{\Pi}_{p_c}(0) - \widehat{\Pi}_{p_c}(k) = O\big(1 - \widehat{D}(k)\big) . \tag{7.1.16}$$

The main aim in this chapter is to prove slightly stronger bounds, proving that $\widehat{\Pi}_p(k)$ is in fact a *small* perturbation in high dimensions:

Our main aim in the analysis of the lace expansion in this and the next chapter is to show that there exists a constant C_Π independent of the dimension d such that, for all d sufficiently large and uniformly in $p < p_c$,

$$\left|\widehat{\Pi}_p(k)\right| \leq C_\Pi/d, \quad \left|\widehat{\Pi}_p(0) - \widehat{\Pi}_p(k)\right| \leq (C_\Pi/d)[1 - \widehat{D}(k)]. \tag{7.1.17}$$

The extra factors of $1/d$ are crucial in order to show that $\widehat{\Pi}_p(k)$ indeed is a small perturbation. This suggests what we need to prove for the lace-expansion coefficients. These bounds are the main aim of this section. However, there are a few problems that we have conveniently glossed over:

(a) The first is that $x \mapsto D_{p_c}(x) = 2dp_c(D \star (\delta_{0,\cdot} + \Pi_{p_c}))(x)$ is *not* a random walk step distribution. The point is that the function $x \mapsto \Pi_{p_c}(x)$ can have both signs, so that also $D_{p_c}(x) = 2dp_c(D \star (\delta_{0,\cdot} + \Pi_{p_c}))(x)$ may have negative values. It *is* true that $x \mapsto D_{p_c}(x)$ sums up to one (recall (7.1.8)). Thus, probabilistic arguments do not apply, and we have to resort to more analytical tools.

(b) The second problem is possibly even more daunting. Indeed, the lace-expansion coefficients are complicated objects. For example, by the BK inequality,

$$\Pi_p^{(0)}(x) = \mathbb{P}_p(0 \Longleftrightarrow x) \leq \mathbb{P}_p(0 \longleftrightarrow x)^2 = \tau_p(x)^2.$$

Since $\tau_{p_c}(x)$ should be small for x large, at first glance this looks quite promising. Unfortunately, this bound is in terms of the object we are trying to bound in the first place! Thus, at a second glance, this method seems doomed...

Fortunately, there is a clever way around this, which is through a *bootstrap* or *forbidden-region* argument that we discuss in detail in Sect. 8.1, and its application is performed in all detail in Chap. 8. Intuitively, the bootstrap argument argues that *if* the two-point function satisfies some nice bound of the form that we are trying to prove, then even *better* bounds hold. *If* these bounds can be formulated in terms of *continuous functions* (of p), and *if* the bounds are trivially satisfied when $p = 0$, then they

have to be true for all $p \in [0, p_c)$.

Indeed, these assumptions imply that a certain forbidden region arises, and thus, the functions should avoid that forbidden region. In particular, when the region is chosen appropriately, this implies that the functions at hand cannot blow up when p approached p_c.

We apply these arguments to functions such as

$$p \mapsto 2dp \quad \text{or} \quad p \mapsto \sup_{k \in (-\pi, \pi]^d} \left[1 - \widehat{D}(k)\right] \hat{\tau}_p(k).$$

Of course, the fact that all the stated assumptions are true is highly nontrivial and requires a precise proof, which is called the *bootstrap analysis* of the lace expansion, and is performed in the following chapter.

In the next section, as a warm up, we start by giving bounds on $\Pi^{(0)}$ and $\Pi^{(1)}$, as these give a quite clear picture of how such bounds can be formulated and proved in general. In Sect. 7.4, we introduce notation to state and derive the general bounds.

7.2 A Warm Up: Bounds on $\Pi^{(0)}$ and $\Pi^{(1)}$

In this section, we prove upper bounds on $\Pi^{(0)}$ and $\Pi^{(1)}$ to explain the philosophy behind the proofs. Before stating the bounds, we introduce some notation.

The upper bounds we prove are in terms of various quantities related to the triangle diagram. We recall the definition of $\triangle_p(x)$ from (4.2.1) and define

$$
\triangle_p = \max_{x \in \mathbb{Z}^d} \triangle_p(x) = \max_{x \in \mathbb{Z}^d} \sum_{y,z \in \mathbb{Z}^d} \tau_p(y) \tau_p(z-y) \tau_p(x-z)
$$

$$
= \sum_{y,z \in \mathbb{Z}^d} \tau_p(y) \tau_p(z-y) \tau_p(z) , \tag{7.2.1}
$$

where the last equality is established in Exerc. 5.2. We further define a related quantity

$$
\tilde{\triangle}_p(x) = \sum_{y,z,u \in \mathbb{Z}^d} \tau_p(y) \tau_p(z-y) 2dp D(u) \tau_p(x-z-u)
$$

$$
= (\tau_p \star \tau_p \star \tilde{\tau}_p)(x) , \tag{7.2.2}
$$

where we let

$$
\tilde{\tau}_p(x) = 2dp(D \star \tau_p)(x) . \tag{7.2.3}
$$

Finally, let

$$
\tilde{\triangle}_p = \max_{x \in \mathbb{Z}^d} \tilde{\triangle}_p(x) . \tag{7.2.4}
$$

The extra convolution with D present in $\tilde{\triangle}_p$ makes that $\tilde{\triangle}_p = O(1/d)$ (in contrast to $\triangle_p$, which is always larger than 1), and this is essential for showing that the lace-expansion coefficients are small in high dimensions.

Further, we need two diagrams that allow us to formulate the bounds on $\widehat{\Pi}_{p_c}^{(0)}(0) - \widehat{\Pi}_{p_c}^{(0)}(k)$ and $\widehat{\Pi}_{p_c}^{(1)}(0) - \widehat{\Pi}_{p_c}^{(1)}(k)$. For this, and for $k \in (-\pi, \pi]^d$ and $y \in \mathbb{Z}^d$, we let

$$
W_p(y; k) = \sum_{x \in \mathbb{Z}^d} [1 - \cos(k \cdot x)] \tilde{\tau}_p(x) \tau_p(x+y) , \tag{7.2.5}
$$

$$
W_p(k) = \max_{y \in \mathbb{Z}^d} W_p(y; k) . \tag{7.2.6}
$$

Bounds on $\widehat{\Pi}^{(0)}$. We crucially use the BK inequality in order to prove the following bounds on $\widehat{\Pi}^{(0)}$:

Lemma 7.1 (Bounds on $\Pi^{(0)}$). For $N = 0$, and all $k \in (-\pi, \pi]^d$,

$$\left|\widehat{\Pi}^{(0)}(k)\right| \le \tilde{\Delta}_p ,\tag{7.2.7}$$

and

$$\widehat{\Pi}^{(0)}(0) - \widehat{\Pi}^{(0)}(k) \le W_p(0; k) .\tag{7.2.8}$$

Proof By (6.2.7) and the BK inequality (1.3.4),

$$\Pi^{(0)}(x) = \mathbb{P}(0 \longleftrightarrow x) - \delta_{0,x} \le \tau_p(x)^2 - \delta_{0,x} .\tag{7.2.9}$$

For $x \ne 0$, the event $\{0 \longleftrightarrow x\}$ is the union over neighbors e of the origin of the events $\{\{0, e\} \text{ occupied}\} \circ \{e \longleftrightarrow x\}$. Thus, by the BK inequality,

$$\tau_p(x) \le 2dp(D \star \tau_p)(x) = \tilde{\tau}_p(x) \quad (x \ne 0) .\tag{7.2.10}$$

Therefore, using further that $\tau_p(x) \le (\tau_p \star \tau_p)(x)$,

$$\left|\widehat{\Pi}_p(k)\right| = \left|\sum_{x \in \mathbb{Z}^d : x \ne 0} \cos(k \cdot x) \Pi^{(0)}(x)\right|$$
$$\le \sum_{x \in \mathbb{Z}^d : x \ne 0} \Pi^{(0)}(x) \le \sum_{x \in \mathbb{Z}^d} \tau_p(x) \tilde{\tau}_p(x) \le \tilde{\Delta}_p(0) \le \tilde{\Delta}_p .\tag{7.2.11}$$

Similarly,

$$\sum_{x \in \mathbb{Z}^d} [1 - \cos(k \cdot x)] \Pi^{(0)}(x) \le \sum_{x \in \mathbb{Z}^d : x \ne 0} [1 - \cos(k \cdot x)] \tau_p(x)^2$$
$$\le \sum_{x \in \mathbb{Z}^d : x \ne 0} [1 - \cos(k \cdot x)] \tau_p(x) \tilde{\tau}_p(x)$$
$$= W_p(0; k) .\tag{7.2.12}$$

This proves (7.2.7)–(7.2.8). These proofs in particular highlight the importance of the BK inequality in obtaining bounds on $\Pi^{(0)}$! $\qquad\qquad\square$

Bounds on $\widehat{\Pi}^{(1)}$. We continue to bound $\Pi^{(1)}$, which is more involved due to the occurrence of the difficult event $E'(v, x; \widetilde{\mathscr{C}}_0^{(u,v)}(0))$ in (6.2.25). The main result is as follows:

Lemma 7.2 (Bounds on $\Pi^{(1)}$). For $N = 1$,

$$\sum_{x \in \mathbb{Z}^d} \Pi^{(1)}(x) \le \tilde{\Delta}_p(\Delta_p)^2 ,\tag{7.2.13}$$

and

$$\sum_{x \in \mathbb{Z}^d} [1 - \cos(k \cdot x)] \Pi^{(1)}(x) \leq 3(\Delta_p)^2 W_p(k) + 6\tilde{\Delta}_p \Delta_p W_p(k) . \tag{7.2.14}$$

Alternatively,

$$\sum_{x \in \mathbb{Z}^d} [1 - \cos(k \cdot x)] \Pi^{(1)}(x) \leq W_p(0; k) + 16\tilde{\Delta}_p \Delta_p W_p(k) . \tag{7.2.15}$$

In the proof of Lem. 7.2, we make essential use of the following trigonometric inequality:

Lemma 7.3 (Split of cosines [107]). *Let $J \geq 1$ and $t_i \in \mathbb{R}$ for $i = 1, \ldots, J$. Then*

$$1 - \cos\left(\sum_{i=1}^{J} t_i\right) \leq J \sum_{i=1}^{J} [1 - \cos(t_i)] . \tag{7.2.16}$$

Proof Abbreviate $t = \sum_{i=1}^{J} t_i$. By taking the real part of the telescoping sum identity

$$1 - e^{it} = \sum_{i=1}^{J} [1 - e^{it_i}] \prod_{j=1}^{i-1} e^{it_j} , \tag{7.2.17}$$

we obtain

$$1 - \cos(t) \leq \sum_{i=1}^{J} [1 - \cos(t_i)] + \sum_{i=2}^{J} \sin(t_i) \sin\left(\sum_{j=1}^{i-1} t_j\right) . \tag{7.2.18}$$

In the following, we use $|\sin(x+y)| \leq |\sin(x)| + |\sin(y)|$, $|ab| \leq (a^2 + b^2)/2$ and $1 - \cos^2(a) \leq 2[1 - \cos(a)]$ to conclude from (7.2.18) that

$$\begin{aligned}
1 - \cos(t) &\leq \sum_{i=1}^{J} [1 - \cos(t_i)] + \sum_{i=2}^{J} \sum_{j=1}^{i-1} |\sin(t_i)||\sin(t_j)| \\
&\leq \sum_{i=1}^{J} [1 - \cos(t_i)] + \frac{1}{2} \sum_{i=2}^{J} \sum_{j=1}^{i-1} [\sin^2(t_i) + \sin^2(t_j)] \\
&= \sum_{i=1}^{J} [1 - \cos(t_i)] + \frac{J-1}{2} \sum_{i=1}^{J} \sin^2(t_i) \\
&\leq J \sum_{i=1}^{J} [1 - \cos(t_i)] .
\end{aligned} \tag{7.2.19}$$

$\square$

Now we are ready to complete the proof of Lem. 7.2:

Proof of Lem. 7.2 except (7.2.15). Recall that

$$\Pi^{(1)}(x) = \sum_{(u,v)} J(u,v) \, \mathbb{E}_0 \left(\mathbb{1}_{\{0 \Longleftrightarrow u\}} \, \mathbb{P}_1 \left(E'\left(v, x; \widetilde{\mathcal{C}}_0^{(u,v)}(0)\right)\right)\right). \tag{7.2.20}$$

We start by giving a bound on $\mathbb{P}_1\left(E'(v,x;\widetilde{\mathcal{C}}_0^{(u,v)}(0))\right)$. For this, we note that

$$E'(v,x;\widetilde{\mathcal{C}}_0) \subseteq \bigcup_{z_1 \in \widetilde{\mathcal{C}}_0} \bigcup_{t_1 \in \mathbb{Z}^d} \{v \leftrightarrow t_1\} \circ \{t_1 \leftrightarrow z_1\} \circ \{t_1 \leftrightarrow x\} \circ \{z_1 \leftrightarrow x\}. \tag{7.2.21}$$

Therefore,

$$\mathbb{P}_1\left(E'\left(v,x;\widetilde{\mathcal{C}}_0^{(u,v)}(0)\right)\right)$$
$$\leq \sum_{z_1} \mathbb{1}_{\{z_1 \in \widetilde{\mathcal{C}}_0\}} \sum_{t_1 \in \mathbb{Z}^d} \mathbb{P}_1(\{v \leftrightarrow t_1\} \circ \{t_1 \leftrightarrow z_1\} \circ \{t_1 \leftrightarrow x\} \circ \{z_1 \leftrightarrow x\}). \tag{7.2.22}$$

Substituting this bound into (7.2.20) leads to

$$\Pi^{(1)}(x) \leq \sum_{(u,v)} J(u,v) \sum_{z_1} \mathbb{E}_0\left(\mathbb{1}_{\{0 \Longleftrightarrow u, z_1 \in \widetilde{\mathcal{C}}_0\}}\right)$$
$$\times \sum_{t_1 \in \mathbb{Z}^d} \mathbb{P}_1(\{v \leftrightarrow t_1\} \circ \{t_1 \leftrightarrow z_1\} \circ \{t_1 \leftrightarrow x\} \circ \{z_1 \leftrightarrow x\}). \tag{7.2.23}$$

Using that

$$\mathbb{P}_0\left(0 \Longleftrightarrow u, z_1 \in \widetilde{\mathcal{C}}_0\right)$$
$$\leq \mathbb{P}_0(0 \Longleftrightarrow u, 0 \leftrightarrow z_1)$$
$$\leq \sum_{w_0 \in \mathbb{Z}^d} \mathbb{P}_0(\{0 \leftrightarrow u\} \circ \{0 \leftrightarrow w_0\} \circ \{w_0 \leftrightarrow u\} \circ \{w_0 \leftrightarrow z_1\}), \tag{7.2.24}$$

we arrive at the bound

$$\Pi^{(1)}(x) \leq \sum_{(u,v)} J(u,v)$$
$$\times \sum_{w_0,z_1 \in \mathbb{Z}^d} \mathbb{P}_0(\{0 \leftrightarrow u\} \circ \{0 \leftrightarrow w_0\} \circ \{w_0 \leftrightarrow u\} \circ \{w_0 \leftrightarrow z_1\})$$
$$\times \sum_{t_1 \in \mathbb{Z}^d} \mathbb{P}_1(\{v \leftrightarrow t_1\} \circ \{t_1 \leftrightarrow z_1\} \circ \{t_1 \leftrightarrow x\} \circ \{z_1 \leftrightarrow x\}). \tag{7.2.25}$$

Using the BK inequality, this leads to

$$\Pi^{(1)}(x) \leq \sum_{u,w_0 \in \mathbb{Z}^d} \left[\tau_p(u)\tau_p(w_0)\tau_p(u-w_0) \right] \sum_{z_1 \in \mathbb{Z}^d} \tau_p(z_1 - w_0)$$
$$\times \sum_{v,t_1 \in \mathbb{Z}^d} J(u,v)\tau_p(t_1 - v)\left[\tau_p(z_1 - t_1)\tau_p(x - t_1)\tau_p(x - z_1) \right]$$
$$= \sum_{u,w_0 \in \mathbb{Z}^d} \left[\tau_p(u)\tau_p(w_0)\tau_p(u-w_0) \right] \sum_{z_1 \in \mathbb{Z}^d} \tau_p(z_1 - w_0)$$
$$\times \sum_{t_1 \in \mathbb{Z}^d} \tilde{\tau}_p(t_1 - u)\left[\tau_p(z_1 - t_1)\tau_p(x - t_1)\tau_p(x - z_1) \right]. \tag{7.2.26}$$

We further sum the above over $x \in \mathbb{Z}^d$, and let $t_1' = t_1 - z_1, x' = x - z_1$ to obtain

$$\sum_{x \in \mathbb{Z}^d} \Pi^{(1)}(x) \leq \sum_{u,w_0 \in \mathbb{Z}^d} \left[\tau_p(u)\tau_p(w_0)\tau_p(u-w_0) \right] \sum_{z_1 \in \mathbb{Z}^d} \tau_p(z_1 - w_0)$$
$$\times \sum_{x',t_1' \in \mathbb{Z}^d} \tilde{\tau}_p(t_1' + z_1 - u)\left[\tau_p(t_1')\tau_p(x' - t_1')\tau_p(x') \right]$$
$$= \sum_{u,w_0 \in \mathbb{Z}^d} \left[\tau_p(u)\tau_p(w_0)\tau_p(u-w_0) \right]$$
$$\times \sum_{t_1',x' \in \mathbb{Z}^d} (\tau_p \star \tilde{\tau}_p)(t_1' + w_0 - u)\left[\tau_p(t_1')\tau_p(x' - t_1')\tau_p(x') \right], \tag{7.2.27}$$

where we use that

$$\sum_{z_1 \in \mathbb{Z}^d} \tau_p(z_1 - w_0)\tilde{\tau}_p(t_1' + z_1 - u) = \sum_{z_1 \in \mathbb{Z}^d} \tau_p(w_0 - z_1)\tilde{\tau}_p(t_1' + z_1 - u)$$
$$= (\tau_p \star \tilde{\tau}_p)(t_1' + w_0 - u). \tag{7.2.28}$$

Uniformly in t_1', w_0, u,

$$(\tau_p \star \tilde{\tau}_p)(t_1' + w_0 - u) \leq \max_{x \in \mathbb{Z}^d}(\tau_p \star \tilde{\tau}_p)(x) \leq \tilde{\Delta}_p. \tag{7.2.29}$$

After applying this bound, we use that

$$\sum_{t_1',x' \in \mathbb{Z}^d} \tau_p(t_1')\tau_p(x' - t_1')\tau_p(x') = \sum_{u,w_0 \in \mathbb{Z}^d} \tau_p(u)\tau_p(w_0)\tau_p(u - w_0) = \Delta_p, \tag{7.2.30}$$

to arrive at (7.2.13).

For (7.2.14), we write $x = u + (t_1 - u) + (x - t_1)$, and use Lem. 7.3 to bound

$$[1 - \cos(k \cdot x)] \leq 3[1 - \cos(k \cdot u)] + 3[1 - \cos(k \cdot (t_1 - u))]$$
$$+ 3[1 - \cos(k \cdot (x - t_1))] . \tag{7.2.31}$$

This gives rise to three terms that we bound one by one. The term with $[1 - \cos(k \cdot (t_1 - u))]$ is the easiest and can be bounded in an identical way as for $\sum_{x \in \mathbb{Z}^d} \Pi^{(1)}(x)$. Indeed, following these computations, it gives rise to an extra factor $[1 - \cos(k \cdot (t_1' + z_1 - u))]$. This can be bounded by replacing (7.2.28)–(7.2.29) by

$$\max_{t_1', w_0, u} \sum_{z_1 \in \mathbb{Z}^d} [1 - \cos(k \cdot (t_1' + z_1 - u))] \tau_p(z_1 - w_0) \tilde{\tau}_p(t_1' + z_1 - u) = W_p(k) . \tag{7.2.32}$$

We proceed with the contribution due to $3[1 - \cos(k \cdot u)]$, which we bound as

$$\sum_{x,u,w_0 \in \mathbb{Z}^d} [3[1 - \cos(k \cdot u)] \tau_p(u) \tau_p(w_0)] \sum_{z_1 \in \mathbb{Z}^d} \tau_p(u - w_0) \tau_p(z_1 - w_0)$$

$$\times \sum_{t_1 \in \mathbb{Z}^d} \tilde{\tau}_p(t_1 - u) [\tau_p(z_1 - t_1) \tau_p(x - t_1) \tau_p(x - z_1)]$$

$$\leq \sum_{u,w_0' \in \mathbb{Z}^d} [3[1 - \cos(k \cdot u)] \tau_p(u) \tau_p(u + w_0')] \sum_{z_1' \in \mathbb{Z}^d} \tau_p(w_0') \tau_p(z_1' - w_0')$$

$$\times \sum_{x',t_1' \in \mathbb{Z}^d} \tilde{\tau}_p(t_1') [\tau_p(z_1' - t_1') \tau_p(x' - t_1') \tau_p(x' - z_1')] , \tag{7.2.33}$$

where $w_0' = w_0 - u, t_1' = t_1 - u, z_1' = z_1 - u, x' = x - u$. Now, uniformly in w_0',

$$\sum_{u,w_0' \in \mathbb{Z}^d} [3[1 - \cos(k \cdot u)] \tau_p(u) \tau_p(u + w_0')] \leq 3W_p(k) , \tag{7.2.34}$$

so that the contribution due to $3[1 - \cos(k \cdot u)]$ is bounded by

$$3W_p(k) \sum_{z_1'', w_0'', t_1' \in \mathbb{Z}^d} \tau_p(w_0'' + t_1') \tau_p(z_1'' - w_0'') \tilde{\tau}_p(t_1')$$

$$\times \sum_{x'' \in \mathbb{Z}^d} [\tau_p(z_1'') \tau_p(x'') \tau_p(x'' - z_1'')] , \tag{7.2.35}$$

where $x'' = x' - t_1', z_1'' = z_1' - t_1'$ and $w_0'' = w_0' - t_1'$. Uniformly in z_1'',

$$\sum_{w_0'', t_1' \in \mathbb{Z}^d} \tilde{\tau}_p(t_1') \tau_p(w_0'' - t_1') \tau_p(z_1'' - w_0'') = \tilde{\Delta}_p(z_1'') \leq \tilde{\Delta}_p , \tag{7.2.36}$$

and

$$\sum_{z_1'',x''\in\mathbb{Z}^d} \left[\tau_p(z_1'')\tau_p(x'')\tau_p(x''-z_1'')\right] \le \Delta_p , \tag{7.2.37}$$

so that that the contribution due to $3[1-\cos(k\cdot u)]$ is bounded by $3W_p(k)\tilde{\Delta}_p\Delta_p$. It is not hard to see that by symmetry, the contribution due to $3\left[1-\cos(k\cdot(x-t_1))\right]$ is the same:

Exercise 7.2 (Contribution due to $1-\cos(k\cdot(x-t_1))$). Prove that the contribution due to $3[1-\cos(k\cdot(x-t_1))]$ can also be bounded by $3W_p(k)\tilde{\Delta}_p\Delta_p$. This can be done by either using symmetry, or adapting the steps of the above proof.

The proof of the improved inequality (7.2.15) is deferred to Sect. 7.5.3. This is an essential improvement, as it turns out that the right-hand side of (7.2.15) contains a factor $1/d$, while the right-hand side of (7.2.14) does not.   □

Conclusion of Bounds on $\widehat{\Pi}^{(0)}$ an $\widehat{\Pi}^{(1)}$. Lemmas 7.1 and 7.2 bring us substantially closer to achieving our main goal in bounding $\widehat{\Pi}_p(k)$ as formulated in (7.1.17), provided that we can prove that the bounding diagrams that we encounter in Lems. 7.1 and 7.2 can indeed be bounded by C_Π/d and $(C_\Pi/d)[1-\widehat{D}(k)]$. How this can be achieved is explained in detail in the next chapter.

In the proof of Lem. 7.2, we see that the bounds, even for $N=1$, rapidly become quite involved. The estimates are all quite standard, but the organization of the arising sums is a challenge. This becomes significantly more involved when dealing with $\Pi^{(N)}$ for $N \ge 2$. In the following section, we explain how such sums can be performed in an organized way. We also explain how these bounds can be represented by pictures.

7.3   Bounds on the Lace Expansion: Bounding Events

In the coming three sections, we show how $\Pi^{(N)}$ of (6.2.27) can be bounded in terms of Feynman diagrams. Given increasing events E, F, recall the notation $E \circ F$ to denote the event that E and F occur disjointly, as introduced in Sect. 1.3.

The analysis consists of three key steps. In the first step, performed in this section, we bound the difficult events $E'(v, x; A)$ appearing in (6.2.27) by simpler connection events. In the second step, performed in Sect. 7.4, we bound the probability of these connection events by products of two-point functions using the BK inequality and efficiently summarize the arising products of two-point functions in terms of simple diagrams. In the final step, performed in Sect. 7.5, we bound these diagrams in terms of triangles and related diagrams.

Let $\mathbb{P}^{(N)}$ denote the product measure on $N+1$ independent copies of percolation on $\mathbb{Z}^d$. By Fubini's Theorem and (6.2.27),

$$\Pi^{(N)}(x) = \sum_{(u_0,v_0)} \cdots \sum_{(u_{N-1},v_{N-1})} \left[\prod_{i=0}^{N-1} J(u_i, v_i) \right]$$

$$\times \mathbb{P}^{(N)}\left(\{0 \leftrightarrow u_0\}_0 \cap \left(\bigcap_{i=1}^{N-1} E'(v_{i-1}, u_i; \widetilde{\mathscr{C}}_{i-1})_i \right) \cap E'(v_{N-1}, x; \widetilde{\mathscr{C}}_{N-1})_N \right), \quad (7.3.1)$$

where, for an event F, we write F_i to denote that F occurs on the ith copy of the percolation configurations. To estimate $\Pi^{(N)}(x)$, it is convenient to define the events (for $N \geq 1$)

$$F_0(0, u_0, w_0, z_1) = \{0 \leftrightarrow u_0\} \circ \{0 \leftrightarrow w_0\} \circ \{w_0 \leftrightarrow u_0\}$$
$$\circ \{w_0 \leftrightarrow z_1\}, \quad (7.3.2)$$

$$F'(v_{i-1}, t_i, z_i, u_i, w_i, z_{i+1}) = \{v_{i-1} \leftrightarrow t_i\} \circ \{t_i \leftrightarrow z_i\} \circ \{t_i \leftrightarrow w_i\}$$
$$\circ \{z_i \leftrightarrow u_i\} \circ \{w_i \leftrightarrow u_i\} \circ \{w_i \leftrightarrow z_{i+1}\}, \quad (7.3.3)$$

$$F''(v_{i-1}, t_i, z_i, u_i, w_i, z_{i+1}) = \{v_{i-1} \leftrightarrow w_i\} \circ \{w_i \leftrightarrow t_i\} \circ \{t_i \leftrightarrow z_i\}$$
$$\circ \{t_i \leftrightarrow u_i\} \circ \{z_i \leftrightarrow u_i\} \circ \{w_i \leftrightarrow z_{i+1}\}, \quad (7.3.4)$$

$$F_N(v_{N-1}, t_N, z_N, x) = \{v_{N-1} \leftrightarrow t_N\} \circ \{t_N \leftrightarrow z_N\} \circ \{t_N \leftrightarrow x\} \circ \{z_N \leftrightarrow x\}, \quad (7.3.5)$$

and to write

$$F(v_{i-1}, t_i, z_i, u_i, w_i, z_{i+1}) = F'(v_{i-1}, t_i, z_i, u_i, w_i, z_{i+1})$$
$$\cup F''(v_{i-1}, t_i, z_i, u_i, w_i, z_{i+1}). \quad (7.3.6)$$

The events F_0, F', F'', F_N are depicted in Fig. 7.1. Note that

$$F_N(v, t, z, y) = F_0(y, z, t, v). \quad (7.3.7)$$

By the definition of E' in (6.2.11), and as in (7.2.21),

$$E'(v_{N-1}, y; \widetilde{\mathscr{C}}_{N-1})_N \subseteq \bigcup_{z_N \in \widetilde{\mathscr{C}}_{N-1}} \bigcup_{t_N \in \mathbb{Z}^d} F_N(v_{N-1}, t_N, z_N, y)_N. \quad (7.3.8)$$

Indeed, viewing the connection from v_{N-1} to y as a string of sausages beginning at v_{N-1} and ending at y, for the event E' to occur, there must be a vertex $z_N \in \widetilde{\mathscr{C}}_{N-1}$ that lies on the last sausage, on a path from v_{N-1} to y. (In fact, both "sides" of the sausage must contain a vertex in $\widetilde{\mathscr{C}}_{N-1}$, but we do not need or use this here.) This leads to (7.3.8), with t_N representing the other endpoint of the sausage that terminates at y.

Assume, for the moment, that $N \geq 2$. The condition in (7.3.8) that $z_N \in \widetilde{\mathscr{C}}_{N-1}$ is a condition on the graph $N-1$ that must be satisfied in conjunction with the event $E'(v_{N-2}, u_{N-1}; \widetilde{\mathscr{C}}_{N-2})_{N-1}$. It is not difficult to see that for $i \in \{1, \ldots, N-1\}$,

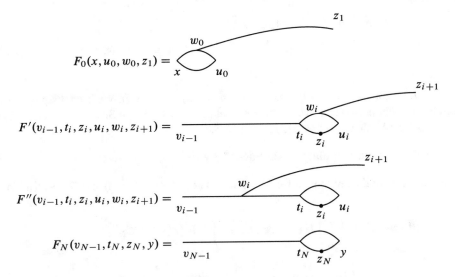

Fig. 7.1 Diagrammatic representations of the events appearing in (7.3.8) and (7.3.9). Lines indicate disjoint connections.

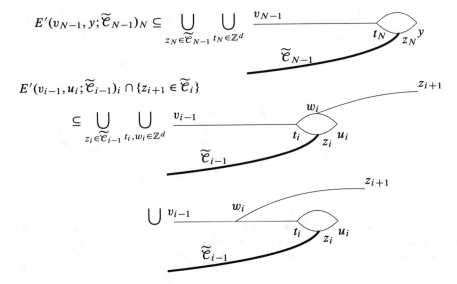

Fig. 7.2 Diagrammatic representations of the inclusions in (7.3.8) and (7.3.9).

$$E'(v_{i-1}, u_i; \widetilde{\mathscr{C}}_{i-1})_i \cap \{z_{i+1} \in \widetilde{\mathscr{C}}_i\}$$

$$\subseteq \bigcup_{z_i \in \widetilde{\mathscr{C}}_{i-1}} \bigcup_{t_i, w_i \in \mathbb{Z}^d} F(v_{i-1}, t_i, z_i, u_i, w_i, z_{i+1})_i \,. \tag{7.3.9}$$

See Fig. 7.2 for a depiction of the inclusions in (7.3.8) and (7.3.9). We leave its proof as an exercise. Further details are given in the original paper by Hara and Slade [132, Lem. 2.5] or in the book by Madras and Slade [211, Lem. 5.5.8].

Exercise 7.3 (Bounding diagrams). Prove (7.3.9).

With an appropriate treatment for $\{x \leftrightarrow u_0\}_0 \cap \{z_1 \in \widetilde{\mathscr{C}}_0\}$, (7.3.8) and (7.3.9) lead to

$$\{0 \leftrightarrow u_0\}_0 \cap \left(\bigcap_{i=1}^{N-1} E'(v_{i-1}, u_i; \widetilde{\mathscr{C}}_{i-1})_i \right) \cap E'(v_{N-1}, x; \widetilde{\mathscr{C}}_{N-1})_N$$

$$\subseteq \bigcup_{\vec{t}, \vec{w}, \vec{z}} \left(F_0(0, u_0, w_0, z_1)_0 \cap \left(\bigcap_{i=1}^{N-1} F(v_{i-1}, t_i, z_i, u_i, w_i, z_{i+1})_i \right) \right.$$

$$\left. \cap F_N(v_{N-1}, t_N, z_N, x)_N \right), \tag{7.3.10}$$

where $\vec{t} = (t_1, \ldots, t_N)$, $\vec{w} = (w_0, \ldots, w_{N-1})$, and $\vec{z} = (z_1, \ldots, z_N)$. Therefore,

$$\Pi^{(N)}(x) \leq \sum \left[\prod_{i=0}^{N-1} J(u_i, v_i) \right] \mathbb{P}\big(F_0(0, u_0, w_0, z_1)\big)$$

$$\times \prod_{i=1}^{N-1} \mathbb{P}\big(F(v_{i-1}, t_i, z_i, u_i, w_i, z_{i+1})\big)$$

$$\times \mathbb{P}\big(F_N(v_{N-1}, t_N, z_N, x)\big), \tag{7.3.11}$$

where the summation is over $z_1, \ldots, z_N, t_1, \ldots, t_N, w_0, \ldots, w_{N-1}, u_0, \ldots, u_{N-1}, v_0, \ldots, v_{N-1}$. The probability in (7.3.11) factors because the events $F_0, \ldots, F_N$ are events on different percolation models, and these different percolation models are *independent*. In the next section, we explain how such a huge sum can be conveniently organized.

7.4 Bounds on the Lace Expansion: Diagrammatic Estimates

Each probability in (7.3.11) can be estimated using the BK inequality. The result is that each of the connections $\{a \leftrightarrow b\}$ present in the events F_0, F and F_N is replaced by a two-point function $\tau_p(a, b)$. This results in a large sum of products of two-point functions. We call such a large product a *diagram*.

To organize the arising large sum of products of two-point functions, we recall (7.2.3) and let

$$\tilde{\tau}_p(x, y) = (J \star \tau_p)(x, y) = 2dp(D \star \tau_p)(y - x) ,$$ (7.4.1)

and define the simple diagrams

$$A_3(s, u, v) = \tau_p(s, v)\tau_p(s, u)\tau_p(u, v) ,$$ (7.4.2)

$$B_1(s, t, u, v) = \tilde{\tau}_p(t, v)\tau_p(s, u) ,$$ (7.4.3)

$$B_2(u, v, s, t) = \tau_p(u, v)\tau_p(u, t)\tau_p(v, s)\tau_p(s, t)$$
$$+ \sum_{a \in \mathbb{Z}^d} \tau_p(s, a)\tau_p(a, u)\tau_p(a, t)\delta_{v, s}\tau_p(u, t) .$$ (7.4.4)

The two terms in B_2 arise from the two events F' and F'' in (7.3.6). We write them as $B_2^{(1)}$ and $B_2^{(2)}$, respectively. The above quantities are represented diagrammatically in Fig. 7.3. In the diagrams, a line joining a and b represents $\tau_p(a, b)$. In addition, small bars are used to distinguish a line that represents $\tilde{\tau}_p$, as in B_1.

Application of the BK inequality yields

$$\mathbb{P}\big(F_0(0, u_0, w_0, z_1)\big) \leq A_3(0, u_0, w_0)\tau_p(w_0, z_1) ,$$ (7.4.5)

$$\sum_{v_{N-1}} J(u_{N-1}, v_{N-1}) \mathbb{P}\big(F_N(v_{N-1}, t_N, z_N, x)\big)$$
$$\leq \frac{B_1(w_{N-1}, u_{N-1}, z_N, t_N)}{\tau_p(w_{N-1}, z_N)} A_3(x, t_N, z_N) .$$ (7.4.6)

For F' and F'', application of the BK inequality yields

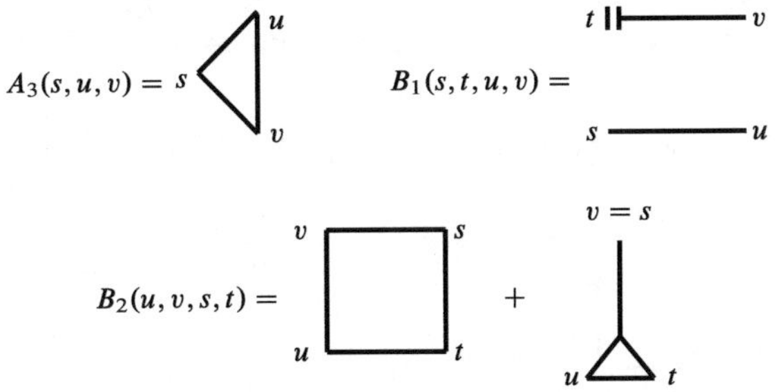

Fig. 7.3 Diagrammatic representations of $A_3(s, u, v)$, $B_1(s, t, u, v)$ and $B_2(u, v, s, t)$.

$$\sum_{v_{i-1}} J(u_{i-1}, v_{i-1}) \, \mathbb{P}\left(F'(v_{i-1}, t_i, z_i, u_i, w_i, z_{i+1})\right)$$

$$\leq \frac{B_1(w_{i-1}, u_{i-1}, z_i, t_i)}{\tau_p(w_{i-1}, z_i)} B_2^{(1)}(z_i, t_i, w_i, u_i) \tau_p(w_i, z_{i+1}) , \qquad (7.4.7)$$

$$\sum_{v_{i-1}, t_i} J(u_{i-1}, v_{i-1}) \, \mathbb{P}\left(F''(v_{i-1}, t_i, z_i, u_i, w_i, z_{i+1})\right)$$

$$\leq \frac{B_1(w_{i-1}, u_{i-1}, z_i, w_i)}{\tau_p(w_{i-1}, z_i)} B_2^{(2)}(z_i, w_i, w_i, u_i) \tau_p(w_i, z_{i+1}) . \qquad (7.4.8)$$

Since the second and the third arguments of $B_2^{(2)}$ are equal by virtue of the Kronecker delta in (7.4.4), we can combine (7.4.7)–(7.4.8) to obtain

$$\sum_{v_{i-1}, t_i} J(u_{i-1}, v_{i-1}) \, \mathbb{P}\left(F(v_{i-1}, t_i, z_i, u_i, w_i, z_{i+1})\right)$$

$$\leq \sum_{t_i} \frac{B_1(w_{i-1}, u_{i-1}, z_i, t_i)}{\tau_p(w_{i-1}, z_i)} B_2(z_i, t_i, w_i, u_i) \tau_p(w_i, z_{i+1}) . \qquad (7.4.9)$$

Upon substitution of the bounds on the probabilities in (7.4.5), (7.4.6), and (7.4.9) into (7.3.11), the ratios of two-point functions form a telescoping product that disappears. After relabelling the summation indices, (7.3.11) becomes

$$\Pi^{(N)}(x) \leq \sum_{\vec{u}, \vec{w}, \vec{z}, \vec{t}} A_3(0, u_0, w_0) \prod_{i=1}^{N-1} [B_1(w_{i-1}, u_{i-i}, z_i, t_i) B_2(z_i, t_i, w_i, u_i)]$$

$$\times B_1(w_{N-1}, u_{N-1}, z_N, t_N) A_3(z_N, t_N, x) . \qquad (7.4.10)$$

The bound (7.4.10) is valid for all $N \geq 1$, and the summation is over all the vertices $u_0, \ldots, u_{N-1}, w_0, \ldots, w_{N-1}, z_1, \ldots, z_N, t_1, \ldots, t_N$ (with x fixed). For $N = 1, 2$, the right side is represented diagrammatically in Fig. 7.4. In the diagrams, unlabelled vertices are summed over $\mathbb{Z}^d$.

Exercise 7.4 (Verification of (7.4.10) for $N = 1$). Verify that (7.4.10) for $N = 1$ agrees with (7.2.26).

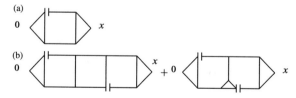

Fig. 7.4 The diagrams bounding $\Pi^{(1)}(x)$ in the first line and $\Pi^{(2)}(x)$ in the second line.

The sum in (7.4.10) still looks quite daunting. In the next section, we explain how its sum over x can be bounded in terms of triangle diagrams. Furthermore, we also bound the sum over x when involving a factor $[1 - \cos(k \cdot x)]$.

7.5 Bounds on the Lace Expansion: Reduction to Simple Diagrams

In this section, we bound the sum over x of the right-hand side of (7.4.10) in terms of products of triangle diagrams, implying that $\widehat{\Pi}_p^{(N)}(0)$ is exponentially small in N whenever these triangles are small. We further bound the sum $\sum_x [1 - \cos(k \cdot x)] \Pi_p^{(N)}(x)$ in terms of three simple diagrams that depend on k.

Recall that $B_2^{(2)}$ denotes the second term of (7.4.4). For $k \in (-\pi, \pi]^d$ and $a_1, a_2 \in \mathbb{Z}^d$, we also define

$$H_p(a_1, a_2; k) = \sum_{u,v,s,t} [1 - \cos(k \cdot (t - u))] B_1(0, a_1, u, s)$$
$$\times B_2^{(2)}(u, s, s, t) B_1(s, t, v, v + a_2), \qquad (7.5.1)$$

and

$$H_p(k) = \max_{a_1, a_2 \in \mathbb{Z}^d} H_p(a_1, a_2; k). \qquad (7.5.2)$$

The remainder of this section is devoted to the proof of the following proposition that provides bounds on the lace-expansion coefficients for $N \geq 2$:

Proposition 7.4 (Diagrammatic estimates for $N \geq 1$). *For $N \geq 1$,*

$$\sum_{x \in \mathbb{Z}^d} \Pi^{(N)}(x) \leq \Delta_p (2\tilde{\Delta}_p \Delta_p)^N, \qquad (7.5.3)$$

and

$$\sum_{x \in \mathbb{Z}^d} [1 - \cos(k \cdot x)] \Pi^{(N)}(x)$$
$$\leq (2N + 1) \big[\Delta_p W_p(k) \big(2\tilde{\Delta}_p + [1 + 2dp] N \Delta_p \big)(2\tilde{\Delta}_p \Delta_p)^{N-1}$$
$$+ (N - 1)\big(\tilde{\Delta}_p^2 W_p(k) + H_p(k) \big)(\Delta_p)^2 (2\tilde{\Delta}_p \Delta_p)^{N-2} \big]. \qquad (7.5.4)$$

The bound for $N = 1$ is similar to that in (7.2.14), so we restrict to $N \geq 2$. We prove the two inequalities in (7.5.3) and (7.5.4) separately, the proof of (7.5.3) is in Sect. 7.5.1, that of (7.5.4) in Sect. 7.5.2.

Before we start with the proof, let us take a closer look at the structure of the bounds in Prop. 7.4. Mind that the triangle diagram Δ_p, as defined in (7.2.1), is always larger than 1, since 1 appears as a summand for the "trivial" contribution $y = z = 0$. On the other hand, $\tilde{\Delta}_p$ requires at least one displacement (since one of the τ terms is replaced by $\tilde{\tau}$) and therefore can be smaller than 1. Indeed, in order for the upper bound in (7.5.3) to be convergent, it is quite handy that the $\tilde{\Delta}_p$ and Δ_p appear together, since we need that $2\tilde{\Delta}_p \Delta_p < 1$ in order for the geometric sum (over N) to converge. Similarly, in (7.5.4), it should be noted that every summand contains either a $W_p(k)$ or a $H_p(k)$ term, since only these two diagrams carry the k-dependence, which we certainly want to keep in the bounds.

7.5.1   Proof of (7.5.3)

For $N \geq 1$, let

$$
\Psi^{(N)}(w_N, u_N)
$$
$$
= \sum_{\vec{u}, \vec{w}, \vec{z}, \vec{t}} A_3(0, u_0, w_0) \prod_{i=1}^{N} [B_1(w_{i-1}, u_{i-1}, z_i, t_i) B_2(z_i, t_i, w_i, u_i)] , \qquad (7.5.5)
$$

where we recall that the summation is over all $u_0, \ldots, u_{N-1}, w_0, \ldots, w_{N-1}, z_1, \ldots, z_N,$ $t_1, \ldots, t_N$ (now with w_N, u_N fixed). For convenience, we define $\Psi^{(0)}(u_0, w_0) = A_3(0, u_0, w_0)$, so that, for $N \geq 1$,

$$
\Psi^{(N)}(w_N, u_N) = \sum_{u_{N-1}, w_{N-1}, z_N, t_N} \Psi^{(N-1)}(w_{N-1}, u_{N-1}) B_1(w_{N-1}, u_{N-1}, z_N, t_N) \\
\times B_2(z_N, t_N, w_N, u_N) . \qquad (7.5.6)
$$

Since

$$
\sum_x A_3(z_N, t_N, x) = \sum_x B_2^{(1)}(z_N, t_N, x, x)
$$
$$
\leq \sum_{w_N, u_N} B_2(z_N, t_N, w_N, u_N) , \qquad (7.5.7)
$$

for any z_N, t_N, it follows from (7.4.10) that

$$
\sum_x \Pi^{(N)}(x) \leq \sum_{w_N, u_N} \Psi^{(N)}(w_N, u_N), \qquad (7.5.8)
$$

and bounds on $\Pi^{(N)}$ can be obtained from bounds on $\Psi^{(N)}$. We prove bounds on $\Psi^{(N)}$, and hence on $\Pi^{(N)}$, by induction on N.

The induction hypothesis is that

$$
\sum_{w_N, u_N} \Psi^{(N)}(w_N, u_N) \leq \Delta_p (2\tilde{\Delta}_p \Delta_p)^N . \qquad (7.5.9)
$$

For $N = 0$, (7.5.9) is true since

$$
\sum_{w_0, u_0} A_3(0, w_0, u_0) \leq \Delta_p . \qquad (7.5.10)
$$

For the induction step, we use (7.5.6) to get

$$\sum_{w_N,u_N} \Psi^{(N)}(w_N,u_N) \leq \left(\sum_{u_{N-1},w_{N-1}} \Psi^{(N-1)}(u_{N-1},w_{N-1}) \right)$$

$$\times \left(\max_{u_{N-1},w_{N-1}} \sum_{z_N,t_N,w_N,u_N} B_1(w_{N-1},u_{N-1},z_N,t_N) B_2(z_N,t_N,w_N,u_N) \right). \quad (7.5.11)$$

If we assume that (7.5.9) is valid for $N-1$, then it follows for N once we prove that

$$\max_{u_{N-1},w_{N-1}} \sum_{z_N,t_N,w_N,u_N} B_1(w_{N-1},u_{N-1},z_N,t_N) B_2(z_N,t_N,w_N,u_N)$$
$$\leq 2\tilde{\Delta}_p \Delta_p. \quad (7.5.12)$$

It remains to prove (7.5.12). There are two terms, due to the two terms in (7.4.4), and we bound each term separately. The first term is bounded as

$$\max_{u_{N-1},w_{N-1}} \sum_{z_N,t_N,w_N,u_N} \tilde{\tau}_p(t_N - u_{N-1})\tau_p(z_N - w_{N-1})\tau_p(t_N - z_N)$$

$$\times \tau_p(w_N - t_N)\tau_p(u_N - w_N)\tau_p(u_N - z_N)$$

$$= \max_{u_{N-1},w_{N-1}} \sum_{z_N,t_N} \tilde{\tau}_p(t_N - u_{N-1})\tau_p(z_N - w_{N-1})\tau_p(t_N - z_N)$$

$$\times \left(\sum_{w_N,u_N} \tau_p(w_N - t_N)\tau_p(u_N - w_N)\tau_p(u_N - z_N) \right)$$

$$= \tilde{\Delta}_p \Delta_p. \quad (7.5.13)$$

The second term is bounded similarly, making use of translation invariance, by

$$\max_{u_{N-1},w_{N-1}} \sum_{z_N,t_N,w_N,u_N,a} \tilde{\tau}_p(t_N - u_{N-1})\tau_p(z_N - w_{N-1})\delta_{t_N,w_N}$$

$$\times \tau_p(t_N - a)\tau_p(u_N - z_N)\tau_p(z_N - a)\tau_p(u_N - a)$$

$$= \max_{u_{N-1},w_{N-1}} \sum_{a,z_N,u_N} \left((\tilde{\tau}_p \star \tau_p)(u_{N-1} - a)\tau_p(z_N - w_{N-1}) \right)$$
$$\times \left(\tau_p(u_N - z_N)\tau_p(z_N - a)\tau_p(u_N - a) \right)$$

$$= \max_{u_{N-1},w_{N-1}} \sum_{a',z_N,u'_N} \left((\tilde{\tau}_p \star \tau_p)(u_{N-1} - z_N - a')\tau_p(z_N - w_{N-1}) \right)$$
$$\times \left(\tau_p(u'_N)\tau_p(-a')\tau_p(u'_N - a') \right)$$

$$\leq \left(\max_{a',u_{N-1},w_{N-1}} \tilde{\Delta}_p(u_{N-1} - w_{N-1} - a') \right)$$

$$\times \left(\sum_{a',u'_N} \tau_p(u'_N)\tau_p(a')\tau_p(u'_N - a') \right)$$

$$= \tilde{\Delta}_p \Delta_p, \quad (7.5.14)$$

where $a' = a - z_N$, $u'_N = u_N - z_N$. This completes the proof of (7.5.12) and hence of (7.5.3). $\qquad \square$

7.5.2 Proof of (7.5.4)

Next, we estimate $\sum_x [1 - \cos(k \cdot x)] \Pi^{(N)}(x)$. In a term in (7.4.10), there is a sequence of $2N + 1$ two-point functions along the "top" of the diagram, such that the sum of the displacements of these two-point functions is exactly equal to x. For example, in Fig. 7.4(a), there are three displacements along the top of the diagram, and in Fig. 7.4(b), there are five in the first diagram and four in the second. We regard the second diagram as also having five displacements, with the understanding that the third is constrained to vanish. With a similar general convention, each of the 2^{N-1} diagrams bounding $\Pi^{(N)}$ has $2N + 1$ displacements along the top of the diagram. We denote these displacements by $d_1, \ldots, d_{2N+1}$, so that $x = \sum_{j=1}^{2N+1} d_j$. We distribute the factor $1 - \cos(k \cdot x)$ among the displacements d_j, as we explain now.

We apply Lem. 7.3 with $t = k \cdot x = \sum_{j=1}^{2N+1} k \cdot d_j$ to obtain a sum of $2N + 1$ diagrams like the ones for $\Pi^{(N)}(x)$, except now in the jth term, the jth line in the top of the diagram represents $[1 - \cos(k \cdot d_j)] \tau_p(d_j)$ rather than $\tau_p(d_j)$, or $[1 - \cos(k \cdot d_j)] \tilde{\tau}_p(d_j)$ rather than $\tilde{\tau}_p(d_j)$.

We distinguish three cases: (a) The displacement d_j is in a line of A_3, (b) the displacement d_j is in a line of B_1, and (c) the displacement d_j is in a line of B_2.

Case (a): The Displacement is in a Line of A_3. We consider the case where the weight $[1 - \cos(k \cdot d_j)]$ falls on the last of the factors A_3 in (7.4.10). This contribution is equal to

$$\sum_{u,v} \psi^{(N-1)}(u, v) \sum_{w,x,y} B_1(u, v, w, y) \tau_p(y - w) \left[1 - \cos\left(k \cdot (x - y)\right)\right] \tau_p(x - y)$$
$$\times \tau_p(x - w) . \qquad (7.5.15)$$

Applying (7.2.10) to $\tau_p(x - y)$, we have

$$\max_{u,v} \sum_{w,x,y} B_1(u, v, w, y) \tau_p(y - w) \left[1 - \cos\left(k \cdot (x - y)\right)\right] \tau_p(x - y) \tau_p(x - w)$$
$$\leq \tilde{\Delta}_p W_p(k) . \qquad (7.5.16)$$

It then follows from (7.5.9) that (7.5.15) is bounded above by $\Delta_p (2 \tilde{\Delta}_p \Delta_p)^{N-1} \tilde{\Delta}_p W_p(k)$. By symmetry, the same bound applies when the weight falls into the first factor of A_3, i.e., when we have a factor $[1 - \cos(k \cdot d_1)]$. Thus, case (a) leads to an upper bound

$$2 \Delta_p (2 \tilde{\Delta}_p \Delta_p)^{N-1} \tilde{\Delta}_p W_p(k) . \qquad (7.5.17)$$

Case (b): The Displacement is in a Line of B_1. Suppose that the factor $[1 - \cos(k \cdot d_j)]$ falls on the ith factor B_1 in (7.4.10). Depending on i, it falls either on $\tilde{\tau}_p$ or on τ_p in (7.4.3). We write the right-hand side of (7.4.10) with the extra factor as

$$\sum_x \sum_{s,t,u,v} \psi^{(i-1)}(s, t) \widetilde{B}_1(s, t, u, v) \bar{\psi}^{(N-i)}(u - x, v - x) . \qquad (7.5.18)$$

In (7.5.18),

$$\widetilde{B}_1(s,t,u,v) = \begin{cases} \tau_p(u-s)\big[1 - \cos\big(k \cdot (v-t)\big)\big]\tilde{\tau}_p(v-t) & \text{if } i \text{ is odd ;} \\ \big[1 - \cos\big(k \cdot (u-s)\big)\big]\tau_p(u-s)\tilde{\tau}_p(v-t) & \text{if } i \text{ is even ,} \end{cases} \tag{7.5.19}$$

and $\bar{\Psi}^{(N-i)}$ denotes a small variant of $\Psi^{(N-i)}$, defined inductively by $\bar{\Psi}^{(0)} = \Psi^{(0)}$ and

$$\bar{\Psi}^{(i)}(x,y) = \sum_{s,t,u,v} B_2(x,y,s,t)B_1(s,t,u,v)\bar{\Psi}^{(i-1)}(u,v) . \tag{7.5.20}$$

It can be verified that $\bar{\Psi}^{(N-i)}$ also obeys (7.5.9):

Exercise 7.5 (Bound on $\bar{\Psi}^{(N)}$). Prove that $\bar{\Psi}^{(N)}$ also obeys (7.5.9).

The distinction between even and odd values of i in (7.5.19) stems from the fact that we distribute the displacement along the top line of the diagram, and after every B_2 term, the two factors τ_p and $\tilde{\tau}_p$ of B_1 interchange top and bottom position; see also Fig. 7.4(b).

If i is odd, then we let $a_1 = t - s$, $a_2 = v - u$, and $x' = u - x$. With this notation, the contribution to (7.5.18) due to (7.5.19) is bounded above by

$$\left(\sum_{s,a_1} \Psi^{(i-1)}(s, s + a_1)\right)\left(\sum_{x',a_2} \bar{\Psi}^{(N-i)}(x', x' + a_2)\right)$$

$$\times \left(\max_{s,a_1,a_2} \sum_u \widetilde{B}_1(s, s + a_1, u, u + a_2)\right)$$

$$= \left(\sum_{s,t} \Psi^{(i-1)}(s, t)\right)\left(\sum_{x,y} \bar{\Psi}^{(N-i)}(x, y)\right)W_p(k)$$

$$\leq \Delta_p(2\tilde{\Delta}_p\Delta_p)^{i-1}\Delta_p(2\tilde{\Delta}_p\Delta_p)^{N-i}W_p(k)$$

$$= \Delta_p(2\tilde{\Delta}_p\Delta_p)^{N-1}\Delta_p W_p(k) , \tag{7.5.21}$$

where we have used (7.5.9) and the identical bound on $\bar{\Psi}^{(N)}$ following from Exerc. 7.5. If i in (7.5.19) is even, then we proceed as in (7.5.21) to obtain

$$\left(\sum_{s,a_1} \Psi^{(i-1)}(s, s + a_1)\right)\left(\sum_{x',a_2} \bar{\Psi}^{(N-i)}(x', x' + a_2)\right)$$

$$\times \left(\max_{s,a_1,a_2} \sum_u \widetilde{B}_1'(s, s + a_1, u, u + a_2)\right)$$

$$= \left(\sum_{s,t} \Psi^{(i-1)}(s, t)\right)\left(\sum_{x,y} \bar{\Psi}^{(N-i)}(x, y)\right)W_p'(k)$$

$$\leq \Delta_p(2\tilde{\Delta}_p\Delta_p)^{i-1}\Delta_p(2\tilde{\Delta}_p\Delta_p)^{N-i}W_p'(k)$$

$$= \Delta_p(2\tilde{\Delta}_p\Delta_p)^{N-1}\Delta_p W_p'(k) , \tag{7.5.22}$$

where
$$W_p'(k) = \max_{s,a_1,a_2} \sum_u \widetilde{B}_1(s, s + a_1, u, u + a_2) . \tag{7.5.23}$$

To bound $W_p'(k)$, we use (7.2.10) for $\tau_p(u - s)$, write

$$\tilde{\tau}_p(v - t) = \sum_y 2dpD(y)\tau_p(v - t - y) ,$$

then use $\sum_y 2dpD(y) = 2dp$ to obtain

$$W_p'(k) \leq 2dpW_p(k) . \tag{7.5.24}$$

Since there are N choices of factors B_1, case (b) leads to an overall upper bound

$$N[1 + 2dp]\Delta_p(2\tilde{\Delta}_p\Delta_p)^{N-1}\Delta_pW_p(k) . \tag{7.5.25}$$

Case (c): The Displacement is in a Line of B_2. It is sufficient to estimate

$$\max_{\substack{a,b,u,v \\ s,t,w,y,x}} \sum \Psi^{(i-1)}(a, b)\bar{\Psi}^{(N-i-1)}(w - x, y - x)[1 - \cos(k \cdot d)]$$
$$\times B_1(a, b, u, v)B_2(u, v, s, t)B_1(s, t, w, y) , \tag{7.5.26}$$

where the maximum is over the choices $d = s - v$ or $d = t - u$. We separately consider the contributions due to $B_2^{(1)}$ and $B_2^{(2)}$ of (7.4.4), beginning with $B_2^{(2)}$.

Recall the definition of $H(a_1, a_2; k)$ in (7.5.1). The contribution to (7.5.26) due to $B_2^{(2)}$ can be rewritten, using $x' = w - x$, $a_2 = y - w$, $a_1 = b - a$, as

$$\sum_{a,a_1,a_2,x'} \Psi^{(i-1)}(a, a + a_1)\bar{\Psi}^{(N-i-1)}(x', x' + a_2)H(a_1, a_2; k)$$
$$\leq H_p(k)\left(\sum_{x,y} \Psi^{(i-1)}(x, y)\right)\left(\sum_{x,y} \bar{\Psi}^{(N-i-1)}(x, y)\right)$$
$$\leq H_p(k)(\Delta_p)^2(2\tilde{\Delta}_p\Delta_p)^{N-2} . \tag{7.5.27}$$

Since there are $N - 1$ factors B_2 to choose, this contribution to case (c) contributes at most

$$(N - 1)H_p(k)(\Delta_p)^2(2\tilde{\Delta}_p\Delta_p)^{N-2} . \tag{7.5.28}$$

It is not difficult to check that the contribution to case (c) due to $B_2^{(1)}$ is at most

$$(N - 1)(\tilde{\Delta}_p^2 W_p(k))(\Delta_p)^2(2\tilde{\Delta}_p\Delta_p)^{N-2} . \tag{7.5.29}$$

The desired estimate (7.5.4) then follows from Lem. 7.3, (7.5.17) and (7.5.25), and (7.5.28)–(7.5.29).

This completes the proof of Prop. 7.4.   □

7.5.3 Proof of (7.2.15)

We close the bounds in this section by improving the bound on $\Pi_p^{(1)}$ in (7.2.15). In terms of the definitions in (7.4.2) and (7.4.3), (7.2.26) is equivalent to

$$\Pi_p^{(1)}(x) \le \sum_{u,w_0,t_1,z_1} A_3(0,u,w_0)B_1(u,w_0,t_1,z_1)A_3(t_1,z_1,x) . \tag{7.5.30}$$

We define $A_3'(u,v,x)$ by

$$A_3'(u,v,x) = A_3(u,v,x) - \delta_{u,x}\delta_{v,x} . \tag{7.5.31}$$

Then,

$$\sum_x [1 - \cos(k \cdot x)]\Pi_p^{(1)}(x)$$
$$\le \sum_x [1 - \cos(k \cdot x)]B_1(0,0,x,x)$$
$$+ \sum_{\substack{x,u,w_0 \\ t_1,z_1}} [1 - \cos(k \cdot x)]A_3'(0,u,w_0)B_1(u,w_0,t_1,z_1)$$
$$\times A_3(t_1,z_1,x)$$
$$+ \sum_{x,t_1,z_1} [1 - \cos(k \cdot x)]B_1(0,0,u,v)A_3'(t_1,z_1,x) . \tag{7.5.32}$$

The first term equals $W_p(0;k)$. The second and third terms are bounded above by $3 \cdot 3\tilde{\Delta}_p\Delta_p W_p(k)$ and $2 \cdot 2\tilde{\Delta}_p W_p(k) \le 4\tilde{\Delta}_p\Delta_p W_p(k)$, respectively, using (7.2.16) (with $J = 3$ and $J = 2$) and the methods of Sect. 7.5.2.

7.6 Outlook on the Remainder of the Argument

In this chapter, we have provided bounds on the complicated functions $\widehat{\Pi}_p^{(N)}(k)$ and $\widehat{\Pi}_p^{(N)}(0) - \widehat{\Pi}_p^{(N)}(k)$ in terms of the five simple diagrams $\tilde{\Delta}_p$, Δ_p, $W_p(0;k)$, $W_p(k)$ and $H_p(k)$. While being relatively simple, these diagrams still involve the two-point function $\tau_p(x)$ or its Fourier transform $\hat{\tau}_p(k)$. *Should* we have bounds on the two-point function τ_p, *then* these simple diagrams could easily be bounded. The next chapter explains how this can be done in detail.

Chapter 8
Bootstrap Analysis of the Lace Expansion

In this chapter, we complete the proof of the infrared bound in Thm. 5.1. For this, we analyze the lace-expansion recurrence relation in Fourier space and reduce its asymptotics to necessary bounds on the lace-expansion coefficients that have been obtained in the previous chapter. This chapter is organized as follows. In Sect. 8.1, we start by giving an overview of the bootstrap argument. In Sect. 8.2, we introduce the three bootstrap functions $f_1(p)$, $f_2(p)$, and $f_3(p)$ and explain their merits. In Sect. 8.3, we derive bounds on the simple diagrams that were introduced in Chap. 7 to bound the lace-expansion coefficients. In Sect. 8.4, we show that the bootstrap argument can be successfully carried out. Finally, in Sect. 8.5, we summarize some consequences of the completed bootstrap argument that are interesting in their own right.

8.1 Overview of the Bootstrap Argument

In this chapter, we answer the following question:

How to prove asymptotics for $\hat{\tau}_p(k)$ when

$$\hat{\tau}_p(k) = \frac{1 + \widehat{\Pi}_p(k)}{1 - 2dp\widehat{D}(k)[1 + \widehat{\Pi}_p(k)]} \tag{8.1.1}$$

and $\widehat{\Pi}_p(k)$ is bounded in terms of $\hat{\tau}_p(k)$ itself?

By Lems. 7.1 and 7.2, as well as Prop. 7.4, we see that in order to prove our main aim formulated in (7.1.17), it suffices to prove the following diagrammatic bounds:

© Springer International Publishing Switzerland 2017
M. Heydenreich and R. van der Hofstad, *Progress in High-Dimensional Percolation and Random Graphs*, CRM Short Courses, DOI 10.1007/978-3-319-62473-0_8

Prove that there exists a constant C independent of the dimension d such that, for all d sufficiently large and uniformly in $p < p_c$,

$$\tilde{\Delta}_p \leq C/d \, , \quad \Delta_p \leq 1 + C/d \, , \quad W_p(0;k) \leq (C/d)[1 - \widehat{D}(k)] \tag{8.1.2}$$

and

$$W_p(k) \leq C[1 - \widehat{D}(k)] \, , \quad H_p(k) \leq C[1 - \widehat{D}(k)] \, . \tag{8.1.3}$$

Indeed, the diagrams appearing on the right-hand sides of (8.1.2) and (8.1.3) are themselves formulated in terms of $\tau_p(x)$ or $\hat{\tau}_p(k)$. The downside seems that we have set forth to bound $\hat{\tau}_p(k)$ and are now faced with even more complicated objects than $\hat{\tau}_p(k)$! However, on the bright side, these diagrams are much simpler than the original objects of study $\widehat{\Pi}_p(k)$.

The proof makes use of a clever *bootstrap* or *forbidden-region* argument that has been used successfully in order to analyze the lace expansion. It appeared first in the study of self-avoiding walk by Slade in [243] and for percolation and lattice trees and lattice animals in the works by Hara and Slade [132, 133]. Earlier work on the lace expansion for self-avoiding walk by Brydges and Spencer [66] made use of an induction argument in terms of a finite memory instead. We start by discussing this bootstrap argument.

At the heart of the proof is the following lemma:

Lemma 8.1 (The bootstrap/forbidden-region argument). *Let f be a continuous function on the interval $[0, p_c)$, let $0 < a < b < \infty$, and assume that $f(0) \leq a$. Suppose for each $p \in (0, p_c)$ that if $f(p) \leq b$, then in fact $f(p) \leq a$ holds. Then, it is true that $f(p) \leq a$ for all $p \in [0, p_c)$.*

Proof. Suppose there exists $p \in (0, p_c)$ with $f(p) > a$. Since we have assumed that for no p there is $f(p) \in [a, b)$, we conclude that $f(p) > b$. The intermediate value theorem for continuous functions implies the existence of $p' \in (0, p_c)$ with $f(p') \in (a, b)$, which yields a contradiction. □

The bootstrap argument in Lem. 8.1 is often used in analyzing the lace expansion, see, e.g., the Madras and Slade book [211, Sect. 6.1]. We now explain why it is so useful. In Chap. 7, we have analyzed the lace-expansion coefficients and have proved that they satisfy bounds in terms of the two-point function $\hat{\tau}_p(k)$. However, our whole aim is to *prove* bounds on the two-point function. In particular, we are on our way to proving Thm. 5.1, which proves the infrared bound on the two-point function $\hat{\tau}_p(k)$ in k-space. Therefore, the bounds presented in Prop. 7.4 might, at first glance, look completely useless since we have no initial control over the two-point function at our disposal. However, *if* we were to have a bound on the two-point function, and possibly some related objects, *then* the bounds in Prop. 7.4 might be used to obtain excellent bounds on the lace-expansion coefficients. In turn, such bounds could then lead to improved bounds on $\hat{\tau}_p(k)$ through (8.1.1).

The bootstrap argument allows us to do precisely that. We introduce a bootstrap function $p \mapsto f(p)$ that provides us with initial bounds on the two-point function, as well as several related objects. These objects are chosen in such a way that they allow us to analyze the quantities appearing in the bounds on the lace-expansion coefficients in Prop. 7.4. Thus, the initial bound $f(p) \leq b$ allows us to give sharp bounds on $\widehat{\Pi}_p(k)$. We can then feed these sharp bounds into the asymptotic identity for $\hat{\tau}_p(k)$ in (8.1.1) and obtain an *improved* bound on the two-point function that reads that $f(p) \leq a$ in fact does hold. When the bootstrap function that we have chosen is indeed continuous, then we conclude that $f(p) \leq a$ must hold for *all* $p \in [0, p_c)$. We have then turned the initial weak bounds into strong bounds.

One can think of the bootstrap lemma as a form of induction in the continuum. Let $(a_n)_{n \geq 0}$ be a sequence for which a_n can be bounded in terms of $(a_k)_{k=0}^{n-1}$, and for which a_0 is well understood. From the recursive properties of the sequence $(a_n)_{n \geq 0}$, an induction step can sometimes be used to prove that bounds on $(a_k)_{k=0}^{n-1}$ imply that the same bound on a_n also holds. Then, in fact, the assumed bound in the induction hypothesis holds for *every* n.

In the bootstrap argument, induction cannot be used, since the function $p \mapsto f(p)$ maps the continuum set $[0, p_c)$ to $[0, \infty)$. However, the pattern in Lem. 8.1 is similar. The initialization of the induction is replaced by the verification that $f(0) \leq a$, while the induction step is replaced by the verification that, for some $b > a$, the weak bound $f(p) \leq b$ implies the stronger bound $f(p) \leq a$. The problem of dealing with the continuous set $[0, p_c)$ is resolved by verifying that $p \mapsto f(p)$ is continuous on $[0, p_c)$.

We conclude this section by giving a simple example where the bootstrap argument can be used to prove an inequality. Let $u: [0, \infty) \to [0, \infty)$ be a function that satisfies $u(0) = 1$ and, for all $t \geq 0$,

$$\partial_t u(t) = \sqrt{u(t)} . \tag{8.1.4}$$

Of course, we can *solve* this equation to yield $u(t) = (t/2 + 1)^2$, but that is not our point. Suppose that we wish to prove a good upper bound on $u(t)$ for every $t \geq 0$. We could define

$$f(t) = \sup_{s \in [0,t]} \frac{u(s)}{(s/2 + 1)^2} . \tag{8.1.5}$$

We take $b = 1.1$ and $a = \sqrt{1.1}$. Then, assume that t is such that $f(t) \leq b$. Take $s \leq t$. Improving the bootstrap bound follows if we can show that $f(s) \leq a$ for every $s \in [0, t]$. Integrating (8.1.4) from 0 to s and using that $u(0) = 1$ give rise to

$$u(s) - 1 = \int_0^s \sqrt{u(r)} \, dr . \tag{8.1.6}$$

Now, in the integral we can apply our assumption that $u(r) \leq b(r/2 + 1)^2$ that is provided by the weak bootstrap assumption, and this leads us to

$$u(s) - 1 \leq \int_0^s \sqrt{b(r/2 + 1)^2} \, dr = \sqrt{b} \int_0^s (r/2 + 1) \, dr . \tag{8.1.7}$$

Noting that $\sqrt{b} = a$ and performing the integral, we arrive at

$$u(s) - 1 \leq a(s^2/4 + s) , \tag{8.1.8}$$

so that, using $a \geq 1$,

$$u(s) \leq a(s^2/4 + s) + 1 \leq a(s^2/4 + s + 1) = a(s/2 + 1)^2 . \tag{8.1.9}$$

Since this is true for every $s \leq t$, we thus obtain that $f(t) = \sup_{s \in [0,t]} u(s)/(s/2 + 1)^2 \leq a$, as required.

Of course, the above method is cumbersome for the relatively simple problem at hand. However, when we would have more involved inequalities, such as

$$\partial_t u(t) \leq \sqrt{u(t)} + \varepsilon(u(t))^{1/4} , \quad \text{or} \quad \partial_t u(t) \leq \sqrt{u(t - \varepsilon)} , \tag{8.1.10}$$

then such results could possibly even be interesting. Here, the main point is that we see how weak bounds can imply stronger bounds, and this fact implies that the strong bounds always hold. Naturally, in order to *apply* the bootstrap argument, we need to choose the bootstrap functions in such a way that they allow us to bound the lace-expansion coefficients. In the next section, we explain how this can be done.

Exercise 8.1 (Example of bootstrap 1). Formulate a bound on the solution $u(t)$ to $\partial_t u(t) \leq \sqrt{u(t)} + \varepsilon(u(t))^{1/4}$ with $u(0) = 1$ and prove it using an appropriate bootstrap argument.

Exercise 8.2 (Example of bootstrap 2). Formulate a bound on the solution $u(t)$ to $\partial_t u(t) \leq \sqrt{u(t - \varepsilon)}$ with $u(0) = 1$ and prove it using an appropriate bootstrap argument.

8.2   The Bootstrap Functions

In this section, we introduce the bootstrap functions that we are going to apply. We start with some notation. To this end, we introduce the quantity

$$\lambda_p := 1 - \frac{1}{\chi(p)} \in [0, 1] . \tag{8.2.1}$$

Clearly, $\lambda_p = 1$ precisely when $p = p_c(\mathbb{Z}^d)$, and $\lambda_p \nearrow 1$ when $p \nearrow p_c$ by Thm. 3.2(3). Recall that

$$\widehat{C}_\lambda(k) = \frac{1}{1 - \lambda \widehat{D}(k)} . \tag{8.2.2}$$

Then λ_p satisfies the equality

$$\hat{\tau}_p(0) = \widehat{C}_{\lambda_p}(0) . \tag{8.2.3}$$

Exercise 8.3 (Relation λ_p and $\chi(p)$). Prove (8.2.3).

The proof of Thm. 5.1 is motivated by the intuition that $\hat{\tau}_p(k)$ and $\widehat{C}_{\lambda_p}(k)$ are comparable in size and, moreover, the discretized second derivative

$$\Delta_k \hat{\tau}_p(l) := \hat{\tau}_p(l - k) + \hat{\tau}_p(l + k) - 2\hat{\tau}_p(l) \tag{8.2.4}$$

is bounded by

$$\widehat{U}_{\lambda_p}(k, l) := 200 \, \widehat{C}_1(k)^{-1}$$
$$\times \left\{ \widehat{C}_{\lambda_p}(l - k)\widehat{C}_{\lambda_p}(l) + \widehat{C}_{\lambda_p}(l)\widehat{C}_{\lambda_p}(l + k) + \widehat{C}_{\lambda_p}(l - k)\widehat{C}_{\lambda_p}(l + k) \right\}. \tag{8.2.5}$$

More precisely, we show that the function $f \colon [0, p_c) \to \mathbb{R}$, defined by

$$f(p) := \max\{f_1(p), f_2(p), f_3(p)\}, \tag{8.2.6}$$

with

$$f_1(p) := 2dp, \quad f_2(p) := \sup_{k \in (-\pi, \pi]^d} \frac{\hat{\tau}_p(k)}{\widehat{C}_{\lambda_p}(k)}, \tag{8.2.7}$$

and

$$f_3(p) := \sup_{k, l \in (-\pi, \pi]^d} \frac{|\Delta_k \hat{\tau}_p(l)|}{\widehat{U}_{\lambda_p}(k, l)}, \tag{8.2.8}$$

satisfies a good bound, given that d is sufficiently large. We aim to perform a bootstrap argument on the function $p \mapsto f(p)$ in (8.2.6), for $p \in [0, p_c)$. The precise form of $f(p)$ in (8.2.6) may look somewhat daunting, so we start by explaining why the functions f_1, f_2, and f_3 are useful:

Intuitive Explanation of $f(p)$. The function $f(p)$ can be understood as bounding three percolation quantities in terms of their random walk equivalents. Indeed, the function $f_1(p)$ allows us to bound factors of p, appearing, for example, in (7.2.2)–(7.2.3). The critical value for the random walk Green's function is 1, and $f_1(p) \le K$ tells us that the percolation threshold is at most a factor K off the value $1/(2d)$.

The function $f_2(p)$ is what we are after. Indeed, $f_2(p) \le K$ implies that

$$\hat{\tau}_p(k) \le K\widehat{C}_{\lambda_p}(k) \le \frac{2K}{1 + \lambda_p}\widehat{C}_1(k) = \frac{K}{1 + \lambda_p} \frac{1}{1 - \widehat{D}(k)}, \tag{8.2.9}$$

where we use that obviously $\lambda_p \le 1$ and we rely on the inequality

$$\widehat{C}_\lambda(k) \le \frac{2}{1 + \lambda}\widehat{C}_1(k), \tag{8.2.10}$$

which follows directly from $|\widehat{D}(k)| \le 1$ and (8.2.2). Thus, the bound on $f_2(p)$ implies the infrared bound in Thm. (5.1) with $A(d) = 2K$. Again, the bound $f_2(p) \le K$ can be interpreted as saying that the percolation two-point function is bounded by a constant times the random walk Green's function.

The function $f_3(p)$ is for two reasons the most difficult to explain. First, since it is unclear why we wish a bound on $\Delta_k \hat{\tau}_p(l)$, and secondly why such a bound should involve $\widehat{U}_{\lambda_p}(k, l)$ in (8.2.5). We start by explaining how the bound $f_3(p) \le K$ can help us. We recall from Lem. 7.1 that

$$\sum_{x \in \mathbb{Z}^d} [1 - \cos(k \cdot x)] \Pi_p^{(0)}(x) \le W_p(0; k) , \tag{8.2.11}$$

where

$$W_p(0; k) = \sum_{x \in \mathbb{Z}^d} [1 - \cos(k \cdot x)] \tau_p(x) \tilde{\tau}_p(x) = (\tau_{p,k} \star \tilde{\tau}_p)(0) , \tag{8.2.12}$$

with

$$\tau_{p,k}(x) = [1 - \cos(k \cdot x)] \tau_p(x) . \tag{8.2.13}$$

By the Fourier inversion formula, we can rewrite

$$W_p(0; k) = \int_{(-\pi,\pi]^d} \hat{\tau}_{p,k}(l) \hat{\tilde{\tau}}_p(l) \frac{dl}{(2\pi)^d} . \tag{8.2.14}$$

The whole point is that

$$\hat{\tau}_{p,k}(l) = -\tfrac{1}{2} \Delta_k \hat{\tau}_p(l) , \tag{8.2.15}$$

so that a bound on the discrete derivative of $\hat{\tau}_p$ helps in bounding $\widehat{\Pi}_p(0) - \widehat{\Pi}_p(k)$. We next explain this in more detail.

Exercise 8.4 (Discrete second derivative $\hat{\tau}_p$). Prove (8.2.15).

Using that $\hat{\tilde{\tau}}_p(l) = 2dp \widehat{D}(l) \hat{\tau}_p(l)$, we obtain

$$\sum_{x \in \mathbb{Z}^d} [1 - \cos(k \cdot x)] \Pi_p^{(0)}(x) \le dp \int_{(-\pi,\pi]^d} 2|\Delta_k \hat{\tau}_p(l)| |\widehat{D}(l)| \hat{\tau}_p(l) \frac{dl}{(2\pi)^d} . \tag{8.2.16}$$

Now, we can use our bounds on f_1, f_2, and f_3 and $|\widehat{D}| \le 1$ in order to bound the resulting integral as

$$\sum_{x \in \mathbb{Z}^d} [1 - \cos(k \cdot x)] \Pi_p^{(0)}(x) \le K^3 \int_{(-\pi,\pi]^d} \widehat{U}_{\lambda_p}(k, l) \widehat{C}_{\lambda_p}(l) \frac{dl}{(2\pi)^d} . \tag{8.2.17}$$

This is a random walk integral, which is small when the dimension is large. See also Exerc. 5.4. As a result,

$$\widehat{\Pi}_p^{(0)}(0) - \widehat{\Pi}_p^{(0)}(k) = \sum_{x \in \mathbb{Z}^d} [1 - \cos(k \cdot x)] \Pi_p^{(0)}(x) \le C c_{2,3}^{(\mathrm{RW})} K^3 [1 - \widehat{D}(k)] . \quad (8.2.18)$$

With a little more effort, we can improve this bound to $(c_K/d)[1 - \widehat{D}(k)]$ for some constant c_K independent of d. This explains how the bound $f_3(p)$ can be used to obtain the required bounds on the lace-expansion coefficients as formulated in (7.1.16).

We next explain that $\widehat{U}_{\lambda_p}(k,l)$ can be interpreted as an upper bound on $\Delta_k \widehat{C}_{\lambda_p}(l)$, so as to show that $|\Delta_k \hat{\tau}_p(l)|/\widehat{U}_{\lambda_p}(k,l)$ can be interpreted as a bound of a percolation quantity in terms of a random walk quantity. Using $\Delta_k \widehat{C}_{\lambda_p}(l)$ directly could not work, since $\Delta_k \widehat{C}_{\lambda_p}(l)$ might be equal to zero at different places than $\Delta_k \hat{\tau}_p(l)$. The fact that $\widehat{U}_{\lambda_p}(k,l)$ can be seen as (a multiple of) an upper bound on $\Delta_k \widehat{C}_{\lambda_p}(l)$ is a consequence of the following lemma, that is also crucial for the improvement of the bound on f_3:

Lemma 8.2 (Slade [246]). *Suppose that $a(x) = a(-x)$ for all $x \in \mathbb{Z}^d$, and let*

$$\hat{A}(k) = \frac{1}{1 - \hat{a}(k)} . \quad (8.2.19)$$

Then, for all $k, l \in [-\pi, \pi)^d$,

$$\left| \Delta_k \hat{A}(l) \right| \le \left(\hat{A}(l-k) + \hat{A}(l+k) \right) \hat{A}(l) \left(\widehat{|a|}(0) - \widehat{|a|}(k) \right)$$
$$+ 8 \hat{A}(l-k) \, \hat{A}(l) \, \hat{A}(l+k) \left(\widehat{|a|}(0) - \widehat{|a|}(l) \right) \left(\widehat{|a|}(0) - \widehat{|a|}(k) \right) , \quad (8.2.20)$$

where $\widehat{|a|}$ denotes the Fourier transform of the absolute value of a. In particular,

$$\left| \Delta_k \widehat{C}_\lambda(l) \right| \le [1 - \widehat{D}(k)][\widehat{C}_\lambda(l) \, \widehat{C}_\lambda(l+k) + \widehat{C}_\lambda(l) \, \widehat{C}_\lambda(l-k)$$
$$+ 8 \widehat{C}_\lambda(l-k) \, \widehat{C}_\lambda(l+k)] . \quad (8.2.21)$$

The proof of Lem. 8.2 uses several bounds on trigonometric quantities and can be found in Slade's lecture notes [246, Lem. 5.7]. The main ingredients can also be found in [60, Lem. 5.3 and (5.32)–(5.33)].

Proof. Let $\hat{a}_\pm = \hat{a}(l \pm k)$ and write $\hat{a} = \hat{a}(l)$. We define

$$\hat{a}^{\cos}(l,k) = \sum_x a(x) \cos(l \cdot x) \cos(k \cdot x) = \tfrac{1}{2}[\hat{a}_- + \hat{a}_+] , \quad (8.2.22)$$

$$\hat{a}^{\sin}(l,k) = \sum_x a(x) \sin(l \cdot x) \sin(k \cdot x) = \tfrac{1}{2}[\hat{a}_- - \hat{a}_+] . \quad (8.2.23)$$

Then,

$$\tfrac{1}{2} \Delta_k \hat{a}(l) = \hat{a}(l) - \hat{a}^{\cos}(l,k) . \quad (8.2.24)$$

Direct computation using (8.2.4) gives

$$
\begin{aligned}
-\tfrac{1}{2}\Delta_k \hat{A}(l) &= \tfrac{1}{2}\hat{A}(l)\hat{A}(l+k)\hat{A}(l-k)[[2\hat{a} - \hat{a}_+ - \hat{a}_-] \\
&\qquad\qquad + [2\hat{a}_+\hat{a}_- - \hat{a}\hat{a}_- - \hat{a}\hat{a}_+]] \\
&= \hat{A}(l)\hat{A}(l+k)\hat{A}(l-k)[[\hat{a}(l) - \hat{a}^{\cos}(l,k)] \\
&\qquad\qquad + [\hat{a}_+\hat{a}_- - \hat{a}(l)\hat{a}^{\cos}(l,k)]] ,
\end{aligned}
\tag{8.2.25}
$$

using (8.2.22) in the last step. By definition,

$$
\hat{a}_-\hat{a}_+ = \left(\tfrac{1}{2}[\hat{a}_+ + \hat{a}_-]\right)^2 - \left(\tfrac{1}{2}[\hat{a}_- - \hat{a}_+]\right)^2 = \hat{a}^{\cos}(l,k)^2 - \hat{a}^{\sin}(l,k)^2 .
\tag{8.2.26}
$$

Substitution in (8.2.25) gives that for all $k, l \in (-\pi, \pi]^d$,

$$
\begin{aligned}
-\tfrac{1}{2}\Delta_k \hat{A}(l) &= \hat{A}(l+k)\hat{A}(l-k)\hat{A}(l) \\
&\qquad \times \left[[\hat{a}(l) - \hat{a}^{\cos}(l,k)] + [\hat{a}^{\cos}(l,k)^2 - \hat{a}(l)\hat{a}^{\cos}(l,k)]\right] \\
&\qquad - \hat{A}(l-k)\hat{A}(l)\hat{A}(l+k)\hat{a}^{\sin}(l,k)^2 \\
&= \hat{A}(l+k)\hat{A}(l-k)\hat{A}(l)[\hat{a}(l) - \hat{a}^{\cos}(l,k)][1 - \hat{a}^{\cos}(l,k)] \\
&\qquad - \hat{A}(l-k)\hat{A}(l)\hat{A}(l+k)\hat{a}^{\sin}(l,k)^2 \\
&= \tfrac{1}{2}[\hat{A}(l-k) + \hat{A}(l+k)]\hat{A}(l)[\hat{a}(l) - \hat{a}^{\cos}(l,k)] \\
&\qquad - \hat{A}(l-k)\hat{A}(l)\hat{A}(l+k)\hat{a}^{\sin}(l,k)^2 ,
\end{aligned}
\tag{8.2.27}
$$

since $1 - \hat{a}^{\cos}(l,k) = \tfrac{1}{2}[1 - \hat{a}(l-k)] + \tfrac{1}{2}[1 - \hat{a}(l+k)]$, so that

$$
\hat{A}(l+k)\hat{A}(l-k)[1 - \hat{a}^{\cos}(l,k)] = \tfrac{1}{2}[\hat{A}(l-k) + \hat{A}(l+k)] .
\tag{8.2.28}
$$

Now we use that $a(x) = a(-x)$ to write

$$
\begin{aligned}
\tfrac{1}{2}|\Delta_k \hat{a}(l)| = |\hat{a}(l) - \hat{a}^{\cos}(l,k)| &= \left| \sum_x [1 - \cos(k \cdot x)]\cos(l \cdot x)a(x) \right| \\
&\le \sum_x [1 - \cos(k \cdot x)]|a(x)| = \widehat{|a|}(0) - \widehat{|a|}(k) .
\end{aligned}
\tag{8.2.29}
$$

Also, by the Cauchy–Schwarz inequality and the elementary estimate $1 - \cos^2 t = (1 + \cos t)(1 - \cos t) \le 2[1 - \cos t]$,

$$\hat{a}^{\sin}(k,l)^2 \leq \sum_x \sin^2(k \cdot x)|g(x)| \sum_y \sin^2(l \cdot y)|a(y)|$$

$$= \sum_x [1 - \cos^2(k \cdot x)]|a(x)| \sum_y [1 - \cos^2(l \cdot y)]|a(y)|$$

$$\leq 4 \sum_x [1 - \cos(k \cdot x)]|a(x)| \sum_y [1 - \cos(l \cdot y)]|a(y)|$$

$$= 4\left(\widehat{|a|}(0) - \widehat{|a|}(l)\right)\left(\widehat{|a|}(0) - \widehat{|a|}(k)\right). \tag{8.2.30}$$

We obtain (8.2.20) by combining (8.2.27) with (8.2.29) and (8.2.30). The bound in (8.2.21) follows immediately, by using that $\widehat{|a|}(l) = \lambda \widehat{D}(l)$ with $\lambda \leq 1$. $\square$

By Lem. 8.2, we can interpret $f_3(p)$ as giving a bound on $|\Delta_k \hat{\tau}_p(l)|/|\Delta_k \widehat{C}_{\lambda_p}(l)|$. Since $\Delta_k \widehat{C}_{\lambda_p}(l)$ is a somewhat more involved object that can have zeros, it is technically more useful to make use of a bound on $|\Delta_k \hat{\tau}_p(l)|/\widehat{U}_{\lambda_p}(k,l)$ instead. It can also be viewed as a double derivative. For k small, the discrete double-derivative $\Delta_k \widehat{C}_{\lambda_p}(l)$ extracts a small factor $1 - \widehat{D}(k)$ at the expense of changing the single Green's function into two Green's functions. Thus, we can think of the double derivative as adding an extra two-point function.

As it turns out, the fact that $f(p) \leq K$ allows us to use Prop. 7.4 to prove general bounds on the lace-expansion coefficients. These bounds provide the necessary conditions to be able to improve the bootstrap. See Prop. 8.3 in the next section. We explore the consequences of the bootstrap bounds in detail now.

8.3 Consequences of the Bootstrap Bound

In this section, we investigate the consequences of the bound assumed in the bootstrap argument. The main result is the following proposition that proves strong bounds on the lace-expansion coefficients when $f(p) \leq K$.

Proposition 8.3 (Consequences of the bootstrap bound). *Let* $M = 0, 1, 2, \ldots$ *If* $p < p_c$ *and* $f(p)$ *of (8.2.6) obeys* $f(p) \leq K$, *then there are positive constants* c_K *and* $d_0 = d_0(K)$ *such that for* $d \geq d_0$,

$$\sum_{x \in \mathbb{Z}^d} |\Pi_M(x)| \leq c_K/d, \tag{8.3.1}$$

and

$$\sum_{x \in \mathbb{Z}^d} [1 - \cos(k \cdot x)]|\Pi_M(x)| \leq (c_K/d)[1 - \widehat{D}(k)], \tag{8.3.2}$$

and for M *sufficiently large (depending on* K *and* p),

$$\sum_{x \in \mathbb{Z}^d} |R_M(x)| \leq c_K(c_K/d)^M \chi(p). \tag{8.3.3}$$

Consequently, $\hat{\tau}_p(k)$ can be written as

$$\hat{\tau}_p(k) = \frac{1 + \widehat{\Pi}_p(k)}{1 - 2dp\widehat{D}(k)[1 + \widehat{\Pi}_p(k)]} \ , \quad with \ \widehat{\Pi}_p(k) = \sum_{N=0}^{\infty}(-1)^N \widehat{\Pi}^{(N)}(k) \ . \quad (8.3.4)$$

In the remainder of this section, we prove Prop. 8.3. The main ingredient is the following lemma:

Lemma 8.4 (Bounds on the lace expansion). *Let $N = 0, 1, 2, \ldots, p < p_c$, and assume that $d \geq d_0 > 6$. For each $K > 0$, there is a constant $\bar{c}_K$ such that if $f(p)$ of (8.2.6) obeys $f(p) \leq K$, then*

$$\sum_{x \in \mathbb{Z}^d} \Pi^{(N)}(x) \leq (\bar{c}_K/d)^{N \vee 1} \quad (8.3.5)$$

and

$$\sum_{x \in \mathbb{Z}^d} [1 - \cos(k \cdot x)]\Pi^{(N)}(x) \leq [1 - \widehat{D}(k)](\bar{c}_K/d)^{(N-1) \vee 1} \ . \quad (8.3.6)$$

Before proving Lem. 8.4, we show that it implies Prop. 8.3:
Proof of Prop. 8.3 subject to Lem. 8.4. The bounds (8.3.1)–(8.3.2) are immediate consequences of Lem. 8.4. The constant c_K can be taken to be equal to $4\bar{c}_K$, where the factor 4 stems from summing the geometric series.

For the remainder term $R_M(x)$, we conclude from (6.3.2) and $f_1(p) \leq K$ that

$$|R_M(x)| \leq K \sum_{u,v} \Pi^{(M)}(u) D(v - u)\tau_p(x - v) \ , \quad (8.3.7)$$

and hence, $\sum_x |R_M(x)| \leq K\widehat{\Pi}^{(M)}(0)\chi(p)$. The claim now follows from (8.3.5). The lace-expansion equation in (8.3.4) follows, since $\hat{R}_M(k) \to 0$ uniformly in k and $\widehat{\Pi}_p(k)$ in (8.3.4) is well defined.   □

Lemma 8.4 follows from Prop. 7.4 combined with the following three lemmas. For these three lemmas, we recall the quantities defined in (7.2.1)–(7.2.6) and (7.5.1)–(7.5.2), and we also define

$$\tilde{\tilde{\Delta}}_p = \int_{(-\pi,\pi]^d} \widehat{D}(k)^2 \hat{\tau}_p(k)^3 \frac{dk}{(2\pi)^d} = (D * D * \tau_p * \tau_p * \tau_p)(0) \ . \quad (8.3.8)$$

Lemma 8.5 (Triangle bounds from the bootstrap assumptions). *Fix $p \in (0, p_c)$, assume that $f(p)$ of (8.2.6) obeys $f(p) \leq K$, and assume that $d \geq d_0 > 6$. There is a constant c'_K, independent of p, such that*

$$\tilde{\tilde{\Delta}}_p \leq c'_K/d \ , \quad \tilde{\Delta}_p \leq c'_K/d, \quad \Delta_p \leq 1 + c'_K/d \ . \quad (8.3.9)$$

The bound on $\tilde{\tilde{\Delta}}_p$ also applies if $\hat{\tau}_p(k)^3$ in (8.3.8) is replaced by $\hat{\tau}_p(k)$ or $\hat{\tau}_p(k)^2$.

Proof. We begin with $\tilde{\tilde{A}}_p$. We use $f_2(p) \leq K$ and Prop. 5.5 to obtain

$$\tilde{\tilde{A}}_p \leq \int_{(-\pi,\pi]^d} \widehat{D}(k)^2 K^3 \widehat{C}_{\lambda_p}(k)^3 \frac{dk}{(2\pi)^d} \leq K^3 c_{2,3}^{(RW)}/d \leq c_K'/d \qquad (8.3.10)$$

whenever $c_K' \geq K^3 c_{2,3}^{(RW)}$. The conclusion concerning replacement of $\hat{\tau}_p(k)^3$ by $\hat{\tau}_p(k)$ or $\hat{\tau}_p(k)^2$ can be obtained by going to x-space and using $\tau_p(x) \leq (\tau_p \star \tau_p)(x) \leq (\tau_p \star \tau_p \star \tau_p)(x)$.

For $\tilde{A}_p$, we extract the term in (7.2.2) due to $y = z = 0$ and $u = x$, which is $2dpD(x) \leq K/(2d)$ using $f_1(p) \leq K$. This gives

$$\tilde{A}_p(x) \leq K/(2d) + \sum_{u,y,z:(y,z-y,x+z-u)\neq(0,0,0)} \tau_p(y)\tau_p(z-y)KD(u)\tau_p(x+z-u). \qquad (8.3.11)$$

Therefore, by (7.2.10),

$$\tilde{A}_p \leq K/(2d) + 3K^2 \max_x \sum_{y,z\in\mathbb{Z}^d} \tau_p(y)(D \star \tau_p)(z-y)(D \star \tau_p)(x+z), \qquad (8.3.12)$$

where the factor 3 comes from the 3 factors τ_p whose argument can differ from 0. In terms of the Fourier transform, this gives

$$\tilde{A}_p \leq K/(2d) + 3K^2 \max_x \int_{(-\pi,\pi]^d} \widehat{D}(k)^2 \hat{\tau}_p(k)^3 e^{-ik\cdot x} \frac{dk}{(2\pi)^d}$$

$$\leq K/(2d) + 3K^2 \tilde{\tilde{A}}_p. \qquad (8.3.13)$$

Our bound on $\tilde{\tilde{A}}_p$ then gives the desired estimate for $\tilde{A}_p$.

The bound on A_p is a consequence of $A_p \leq 1 + 3\tilde{A}_p$. Here, the term 1 is due to the contribution to (7.2.1) with $y = z - y = x - z = 0$, so that $x = y = z = 0$. If at least one of $y, z - y, x - z$ is nonzero, then we can use (7.2.10) for the corresponding two-point function. $\qquad\square$

Lemma 8.6 (Bounds on the Fourier diagrams). *Fix $p \in (0, p_c)$, assume that $f(p)$ of (8.2.6) obeys $f(p) \leq K$, and that $d \geq d_0 > 6$. There is a constant c_K', independent of p, such that*
$$W_p(0;k) \leq (c_K'/d)[1 - \widehat{D}(k)], \quad W_p(k) \leq c_K'[1 - \widehat{D}(k)]. \qquad (8.3.14)$$

Proof. For the bound on $W_p(0;k)$, we use (7.2.10) to obtain

$$\tilde{\tau}_p(x) = 2dpD(x) + \sum_{v:v\neq x} 2dpD(v)\tau(x-v)$$

$$\leq 2dpD(x) + [2dp]^2(D \star D \star \tau_p)(x). \qquad (8.3.15)$$

We insert (8.3.15) into the definition (7.2.5) of $W_p(0;k)$ to get

$$W_p(0; k) \leq 2dp \sum_x [1 - \cos(k \cdot x)] D(x) \tau_p(x)$$

$$+ [2dp]^2 \sum_x [1 - \cos(k \cdot x)] \tau_p(x) (D \star D \star \tau_p)(x) . \quad (8.3.16)$$

We begin with the first term in (8.3.16), which receives no contribution from $x = 0$. Using (7.2.10) and (8.3.15) again, we obtain

$$2dp \sum_{x \neq 0} [1 - \cos(k \cdot x)] D(x) \tau_p(x)$$

$$\leq [2dp]^2 \sum_x [1 - \cos(k \cdot x)] D(x)^2$$

$$+ [2dp]^2 \sum_x [1 - \cos(k \cdot x)] D(x) \sum_{v \neq x} D(v) \tau_p(x - v)$$

$$\leq [2dp]^2 \sum_x [1 - \cos(k \cdot x)] D(x)^2$$

$$+ [2dp]^3 \sum_x [1 - \cos(k \cdot x)] D(x) (D \star D)(x)$$

$$+ [2dp]^3 \sum_x [1 - \cos(k \cdot x)] D(x) (D \star D \star \tau_p)(x) . \quad (8.3.17)$$

The first term on the right side is bounded by $(K^2/(2d))[1 - \widehat{D}(k)]$. The second term can be bounded similarly, using $\max_x (D \star D)(x) \leq 1/(2d)$. For the last term in (8.3.17), we use the Fourier inversion theorem, together with the fact that the Fourier transform of $[1 - \cos(k \cdot x)] D(x)$ is $\widehat{D}(l) - \widehat{D}^{\cos}(k, l)$ by (8.2.22), to obtain

$$\sum_x [1 - \cos(k \cdot x)] D(x) (D \star D \star \tau_p)(x)$$

$$= \int_{(-\pi, \pi]^d} [\widehat{D}(l) - \widehat{D}^{\cos}(k, l)] \widehat{D}(l)^2 \hat{\tau}_p(l) \frac{dl}{(2\pi)^d} . \quad (8.3.18)$$

It follows as in (8.2.29) that for all $k, l \in (-\pi, \pi]^d$,

$$\left| \widehat{D}(l) - \widehat{D}^{\cos}(k, l) \right| = \left| \sum_x [1 - \cos(k \cdot x)] \cos(l \cdot x) D(x) \right| \leq 1 - \widehat{D}(k) , \quad (8.3.19)$$

Applying the bound on $\tilde{\widehat{A}}_p$ (with $\hat{\tau}_p(k)^3$ replaced by $\hat{\tau}_p(k)$) and using Prop. 5.5, this is bounded by

$$[1 - \widehat{D}(k)] \int_{(-\pi, \pi]^d} \widehat{D}(l)^2 \hat{\tau}_p(l) \frac{dl}{(2\pi)^d} \leq (K c_{2,1}^{(RW)}/d)[1 - \widehat{D}(k)] . \quad (8.3.20)$$

This completes the bound on the first term of (8.3.16).

For the second term in (8.3.16), we again use the Fourier inversion theorem to obtain

$$\sum_x [1 - \cos(k \cdot x)]\tau_p(x)(D \star D \star \tau_p)(x)$$

$$= \int_{(-\pi,\pi]^d} \left[\hat{\tau}_p(l) - \frac{1}{2}(\hat{\tau}_p(l+k) + \hat{\tau}_p(l-k))\right] \widehat{D}(l)^2 \hat{\tau}_p(l) \frac{dl}{(2\pi)^d} . \quad (8.3.21)$$

By the assumed bounds on $f_2(p)$ and $f_3(p)$, this is at most

$$200K^2[1 - \widehat{D}(k)]$$

$$\times \int_{(-\pi,\pi]^d} \widehat{D}(l)^2 \widehat{C}_{\lambda_p}(l)[\widehat{C}_{\lambda_p}(l-k)\widehat{C}_{\lambda_p}(l) + \widehat{C}_{\lambda_p}(l)\widehat{C}_{\lambda_p}(l+k)$$

$$+ \widehat{C}_{\lambda_p}(l-k)\widehat{C}_{\lambda_p}(l+k)]\frac{dl}{(2\pi)^d} . \quad (8.3.22)$$

All the arising integrals are bounded by $c_{2,3}^{(RW)}/d$ by Exerc. 5.4. This completes the bound on the second term of (8.3.16), and thus the proof that $W_p(0;k) \le (c'_K/d)[1 - \widehat{D}(k)]$.

Finally, we estimate $W_p(k)$. Note that no factor $1/d$ appears in the desired bound. By (7.2.5)–(7.2.6),

$$W_p(k) = 2dp \max_{y \in \mathbb{Z}^d} \sum_{x,v \in \mathbb{Z}^d} [1 - \cos(k \cdot x)]D(v)\tau_p(x-v)\tau_p(x+y) . \quad (8.3.23)$$

Let

$$D_k(x) = [1 - \cos(k \cdot x)]D(x) , \quad \tau_{p,k}(x) = [1 - \cos(k \cdot x)]\tau_p(x) . \quad (8.3.24)$$

Applying Lem. 7.3 with $t = k \cdot v + k \cdot (x - v)$ and $J = 2$, we obtain

$$W_p(k) \le 4dp \max_{y \in \mathbb{Z}^d} \sum_{x,v \in \mathbb{Z}^d} [1 - \cos(k \cdot v)]D(v)\tau_p(x-v)\tau_p(y-x)$$

$$+ 4dp \max_{y \in \mathbb{Z}^d} \sum_{x,v \in \mathbb{Z}^d} D(v)[1 - \cos(k \cdot (x-v))]\tau_p(x-v)\tau_p(y-x)$$

$$\le 2K \max_{y \in \mathbb{Z}^d}(D_k \star \tau_p \star \tau_p)(y) + 2K \max_{y \in \mathbb{Z}^d}(D \star \tau_{p,k} \star \tau_p)(y) . \quad (8.3.25)$$

For the first term,

$$(D_k \star \tau_p \star \tau_p)(y) = \int_{(-\pi,\pi]^d} e^{-il \cdot y} \widehat{D}_k(l)\hat{\tau}_p(l)^2 \frac{dl}{(2\pi)^d}$$

$$\le K^2 \int_{(-\pi,\pi]^d} |\widehat{D}_k(l)||\widehat{C}_{\lambda_p}(l)|^2 \frac{dl}{(2\pi)^d} . \quad (8.3.26)$$

By (8.3.19) combined with Prop. 5.5,

$$\max_{y \in \mathbb{Z}^d} (D_k \star \tau_p \star \tau_p)(y) \leq [1 - \widehat{D}(k)]K^2 \int_{(-\pi,\pi]^d} \widehat{C}_{\lambda_p}(l)^2 \frac{dl}{(2\pi)^d}$$

$$\leq K^2 c_{0,2}^{(RW)}[1 - \widehat{D}(k)] . \tag{8.3.27}$$

The remaining term to estimate in (8.3.25) is

$$\max_{y \in \mathbb{Z}^d} (D \star \tau_{p,k} \star \tau_p)(y) = \max_{y \in \mathbb{Z}^d} \int_{(-\pi,\pi]^d} e^{-il \cdot y} \widehat{D}(l) \widehat{\tau}_p(l) \widehat{\tau}_{p,k}(l) \frac{dl}{(2\pi)^d} . \tag{8.3.28}$$

Since

$$\widehat{\tau}_{p,k}(l) = \widehat{\tau}_p(l) - \tfrac{1}{2}\big(\widehat{\tau}_p(l + k) + \widehat{\tau}_p(l - k)\big) , \tag{8.3.29}$$

we can use the bounds on $f_2(p)$ and $f_3(p)$ to see that (8.3.28) is at most

$$200K^2[1 - \widehat{D}(k)]$$

$$\times \int_{(-\pi,\pi]^d} |\widehat{D}|\widehat{C}_{\lambda_p}(l)[\widehat{C}_{\lambda_p}(l - k)\widehat{C}_{\lambda_p}(l) + \widehat{C}_{\lambda_p}(l)\widehat{C}_{\lambda_p}(l + k)$$

$$+ \widehat{C}_{\lambda_p}(l - k)\widehat{C}_{\lambda_p}(l + k)]\frac{dl}{(2\pi)^d} . \tag{8.3.30}$$

The above sums can all be bounded using the methods employed for the previous terms, Exerc. 5.4 and Prop. 5.5. □

Lemma 8.7 (Bound on $H_p(k)$). *Fix* $p \in (0, p_c)$, *assume that* $f(p)$ *of (8.2.6) obeys* $f(p) \leq K$, *and assume that* $d \geq d_0 > 6$. *Then, there is a constant* c_K', *independent of* p, *such that*

$$H_p(k) \leq (c_K'/d)[1 - \widehat{D}(k)] . \tag{8.3.31}$$

Proof. Recall the definition of $H_p(a_1, a_2; k)$ in (7.5.1). In terms of the Fourier transform, recalling (8.3.24),

$$H(a_1, a_2; k) = \int_{(-\pi,\pi]^{3d}} e^{-il_1 \cdot a_1} e^{-il_2 \cdot a_2} \widehat{D}(l_1) \widehat{\tau}_p(l_1)^2 \widehat{D}(l_2) \widehat{\tau}_p(l_2)^2 \widehat{\tau}_{p,k}(l_3)$$

$$\times \widehat{\tau}_p(l_1 - l_2)\widehat{\tau}_p(l_2 - l_3)\widehat{\tau}_p(l_1 - l_3) \frac{dl_1 \, dl_2 \, dl_3}{(2\pi)^{3d}} . \tag{8.3.32}$$

We use $f(p) \leq K$ to replace $\widehat{\tau}_p(k)$ by $K\widehat{C}_{\lambda_p}(k)$ and (recalling (8.3.29)) $\widehat{\tau}_{p,k}(l_3)$ by

$$200K[1 - \widehat{D}(k)][\widehat{C}_{\lambda_p}(l_3 - k)\widehat{C}_{\lambda_p}(l_3) + \widehat{C}_{\lambda_p}(l_3)\widehat{C}_{\lambda_p}(l_3 + k)$$

$$+ \widehat{C}_{\lambda_p}(l_3 - k)\widehat{C}_{\lambda_p}(l_3 + k)] . \tag{8.3.33}$$

This gives an upper bound for (8.3.32) consisting of a sum of three terms.

The last of these terms can be bounded by

$$200K^8[1 - \widehat{D}(k)]$$

$$\times \int_{(-\pi,\pi]^{3d}} |\widehat{D}(l_1)| \widehat{C}_{\lambda_p}(l_1)^2 |\widehat{D}(l_2)| \widehat{C}_{\lambda_p}(l_2)^2 \widehat{C}_{\lambda_p}(l_3 - k) \widehat{C}_{\lambda_p}(l_3 + k)$$

$$\times \widehat{C}_{\lambda_p}(l_1 - l_2) \widehat{C}_{\lambda_p}(l_2 - l_3) \widehat{C}_{\lambda_p}(l_1 - l_3) \frac{dl_1\, dl_2\, dl_3}{(2\pi)^{3d}} \,. \quad (8.3.34)$$

Using Hölder's inequality with $p = 3$ and $q = 3/2$, (8.3.34) is bounded above by $200K^8[1 - \widehat{D}(k)]$ times

$$\left(\int_{(-\pi,\pi]^{3d}} |\widehat{D}(l_1)|^{3/2} \widehat{C}_{\lambda_p}(l_1)^3 |\widehat{D}(l_2)|^{3/2} \widehat{C}_{\lambda_p}(l_2)^3 \right.$$
$$\left. \times \widehat{C}_{\lambda_p}(l_3 + k)^{3/2} \widehat{C}_{\lambda_p}(l_1 - l_3)^{3/2} \frac{dl_1\, dl_2\, dl_3}{(2\pi)^{3d}} \right)^{2/3}$$

$$\times \left(\int_{(-\pi,\pi]^{3d}} \widehat{C}_{\lambda_p}(l_1 - l_2)^3 \widehat{C}_{\lambda_p}(l_2 - l_3)^3 \widehat{C}_{\lambda_p}(l_3 - k)^3 \frac{dl_1\, dl_2\, dl_3}{(2\pi)^{3d}} \right)^{1/3} \,. \quad (8.3.35)$$

For $\alpha \geq 0$, let

$$S_p^{(\alpha)} = \int_{(-\pi,\pi]^d} |\widehat{D}(l)|^\alpha \widehat{C}_{\lambda_p}(l)^3 \frac{dl}{(2\pi)^d} \,. \quad (8.3.36)$$

The Cauchy–Schwarz inequality implies that for all k and l_1,

$$\int_{(-\pi,\pi]^d} \widehat{C}_{\lambda_p}(l_3 + k)^{3/2} \widehat{C}_{\lambda_p}(l_1 - l_3)^{3/2} \frac{dl_3}{(2\pi)^d} \leq S_p^{(0)} \,. \quad (8.3.37)$$

Therefore, (8.3.35) is bounded above by

$$\left(S_p^{(0)} \right)^{5/3} \left(S_p^{(3/2)} \right)^{4/3} \,. \quad (8.3.38)$$

To complete the proof, we note that by Hölder's inequality,

$$S_p^{(3/2)} \leq \left(S_p^{(2)} \right)^{3/4} \left(S_p^{(0)} \right)^{1/4} \,. \quad (8.3.39)$$

Thus, (8.3.38) is bounded above by $[1 - \widehat{D}(k)] S_p^{(2)} (S_p^{(0)})^2$. These factors can be bounded using (5.4.2). This gives a bound of the desired form. Routine bounds can be used to deal with the other two terms in a similar fashion.

Exercise 8.5 (Bounding the remaining cases). Shows that the remaining two contributions from (8.3.33) in (8.3.32) are also bounded by $(c_K'/d)[1 - \widehat{D}(k)]$. $\qquad \square$

Proof of Lem. 8.4. This is an immediate consequence of Prop. 7.4 and Lems. 8.5–8.7. The bound (7.2.14) is used for (8.3.6) when $N = 1$ (as (7.5.4) is not sufficient since it misses a factor $1/d$). $\qquad \square$

8.4 The Bootstrap Argument Completed

In this section, we prove that the bootstrap argument can be applied successfully. This is formulated in the following proposition:

Proposition 8.8 (Successful application of the bootstrap). *There exist constants* const *and* $d_0 > 6$ *with* const *independent of the dimension, such that the bound*

$$f(p) \leq 1 + \text{const}/d \tag{8.4.1}$$

holds uniformly for $p < p_c$ *and* $d \geq d_0$.

In the remainder of this section, we prove Prop. 8.8 and, by doing so, complete the proof of Thm. 5.1 by completing the bootstrap argument. This is achieved by proving that the function f defined in (8.2.6) obeys the hypotheses of Lem. 8.1. We therefore have to show that

(1) $f(0) \leq 3$;
(2) f is continuous on $[0, p_c)$; and
(3) $f(p) \leq 4$ implies $f(p) \leq 3$ for $p \in (0, p_c)$.

The latter is referred to as the *improvement of the bounds*. We verify these conditions one by one.

(1) Verification of f(0) ≤ 3. Let us first check that $f(0) \leq 3$. Clearly, $f_1(0) = 0$. Note that $\widehat{\Pi}_0(k) \equiv 0$, where $\widehat{\Pi}_0(k)$ is $\widehat{\Pi}_p(k)$ with $p = 0$. This leads to $\hat{\tau}_0(k) \equiv 1$ and $\lambda_0 = 0$, hence $f_2(0) = 1$ and $f_3(0) = 0$.

(2) Verification of Continuity of p ↦ f(p) for p ∈ [0, p_c). Next, we prove continuity of f. Continuity of f on $[0, p_c)$ is equivalent to proving continuity on $[0, p_c - \varepsilon]$ for every $\varepsilon > 0$. To prove the continuity on $[0, p_c - \varepsilon]$, we rely on the continuity of the supremum of a family of equicontinuous functions, cf. Lem. 4.3.

Lemma 8.9 (Continuity). *The function* f *defined in* (8.2.6) *is continuous on* $[0, p_c)$.

Proof. It is sufficient to show that f_1, f_2, and f_3 are continuous. The continuity of f_1 is obvious. We show that f_2 is continuous on the closed interval $[0, p_c - \varepsilon]$ for any $\varepsilon > 0$ by taking derivatives with respect to p and bounding it uniformly in k on $[0, p_c - \varepsilon]$. To this end, we consider the derivative

$$\frac{d}{dp} \frac{\hat{\tau}_p(k)}{\widehat{C}_{\lambda_p}(k)} = \frac{1}{\widehat{C}_{\lambda_p}(k)^2} \left[\widehat{C}_{\lambda_p}(k) \frac{d\hat{\tau}_p(k)}{dp} - \hat{\tau}_p(k) \frac{d\widehat{C}_\lambda(k)}{d\lambda} \bigg|_{\lambda = \lambda_p} \frac{d\lambda_p}{dp} \right]. \tag{8.4.2}$$

We proceed by showing that each of the terms on the right-hand side is uniformly bounded in k and $p \in [0, p_c - \varepsilon]$, and hence, the derivative is bounded. First, we recall the definition of λ_p in (8.2.1) to see that

$$\frac{1}{2} \leq \frac{1}{1 - \lambda_p \widehat{D}(k)} = \widehat{C}_{\lambda_p}(k) \leq \widehat{C}_{\lambda_p}(0) = \chi(p). \tag{8.4.3}$$

Additionally, $\chi(p) \leq \chi(p_c - \varepsilon)$, and the latter is finite by the fact that $p_c = p_T$ (recall Thm. 3.1). For every $k \in [-\pi, \pi)^d$, the two-point function is bounded from above by

$$\hat{\tau}_p(k) \leq \hat{\tau}_p(0) = \chi(p) \leq \chi(p_c - \varepsilon) , \qquad (8.4.4)$$

For the derivative of the two-point function, we bound

$$\left| \frac{d}{dp} \hat{\tau}_p(k) \right| = \left| \sum_{x \in \mathbb{Z}^d} e^{ik \cdot x} \frac{d}{dp} \tau_p(x) \right| \leq \sum_{x \in \mathbb{Z}^d} \frac{d}{dp} \tau_p(x)$$

$$= \frac{d}{dp} \sum_{x \in \mathbb{Z}^d} \tau_p(x) = \chi'(p) , \qquad (8.4.5)$$

where the exchange in the order of sum and derivative is validated by the fact that both $\sum_{x \in \mathbb{Z}^d} e^{ik \cdot x} \tau_p(x)$ and $\sum_{x \in \mathbb{Z}^d} \tau_p(x)$ are uniformly convergent series of functions. By the mean-field bound $\chi'(p) \leq 2dp\chi(p)^2$ (4.2.4), (8.4.5) is bounded above by $2d\chi(p_c - \varepsilon)^2$.

Moreover, we obtain from (8.2.2) that $|d\hat{C}_\lambda(k)/d\lambda| \leq \hat{C}_\lambda(k)^2$, and, for $\lambda = \lambda_p$, this is in turn bounded by $\chi(p_c - \varepsilon)^2$, cf. (8.4.3). Finally, $|d\lambda_p/dp| = \chi'(p)/\chi(p)^2 \leq 2d$ by (8.2.1) and (4.2.4). Uniform boundedness of the derivative $(d/dp)(\hat{\tau}_p(k)/\hat{C}_{\lambda_p}(k))$ implies that $p \mapsto \hat{\tau}_p(k)/\hat{C}_{\lambda_p}(k)$ is an equicontinuous family of functions indexed by $k \in [-\pi, \pi)^d$. Therefore, by Lem. 4.3, we conclude that $p \mapsto f_2(p)$ is continuous on $[0, p_c - \varepsilon]$ for every $\varepsilon > 0$ and hence on $[0, p_c)$.

We treat f_3 in exactly the same way as f_2 and leave the proof of continuity as an exercise: □

Exercise 8.6 (Continuity of f_3). Verify that $p \mapsto f_3(p)$ is continuous for $p \in [0, p_c - \varepsilon]$ for every $\varepsilon > 0$.

(3) Improvement of the Bounds. The following proposition covers the remaining hypotheses of Lem. 8.1 and thus proves the final ingredient needed for the proof of Prop. 8.8:

Proposition 8.10 .(Improvement of the bounds). *If the assumptions of Prop. 8.8 are satisfied for some sufficiently large d_0, and if $f(p) \leq 4$ for all $p \in (0, p_c)$, then there exists a constant such that $f(p) \leq 1 + \text{const}/d$ for all $p \in (0, p_c)$ and $d \geq d_0$. In particular, $f(p) \leq 3$ if d_0 is large enough.*

Proof of Prop. 8.10. Fix $p \in (0, p_c)$ arbitrarily and assume that $f(p) \leq 4$. Our strategy is to show that $f_i(p)$ for $i = 1, 2, 3$ is smaller than $(1 + \text{const}/d)$ uniformly in p and thus, by taking $d \geq d_0$ with d_0 large, $f(p) \leq 3$.

We next improve the bounds on f_1, f_2, and f_3 one at a time. We start with f_1 and f_2:

(3.1)–(3.2) Improvement of the Bounds for f_1 and f_2 The main result is in the following lemma:

Lemma 8.11 (Improvement of the bounds for f_1 and f_2). *If the assumptions of Prop. 8.8 are satisfied for some sufficiently large d_0, and if $f(p) \leq 4$ for all $p \in (0, p_c)$, then there exists a constant $c > 0$ such that $f_1(p) \leq 1 + c/d$ and $f_2(p) \leq 1 + c/d$ for all p and $d \geq d_0$.*

Proof. The bound on f_1 is easy. First note that

$$\lambda_p = 1 - \hat{\tau}_p(0)^{-1} = 1 - \frac{1 - 2dp[1 + \widehat{\Pi}_p(0)]}{1 + \widehat{\Pi}_p(0)} \leq 1 . \tag{8.4.6}$$

By (8.2.1) and c_4 as in Prop. 8.3 (with $K = 4$),

$$f_1(p) = 2dp = \lambda_p + \frac{\widehat{\Pi}_p(0)}{1 + \widehat{\Pi}_p(0)} \leq \lambda_p + 2|\widehat{\Pi}_p(0)| \leq 1 + 2c_4/d . \tag{8.4.7}$$

The bound on f_2 is slightly more involved. We write $\hat{\tau}_p(k) = \widehat{N}(k)/\widehat{F}(k)$, with

$$\widehat{N}(k) = \frac{1 + \widehat{\Pi}_p(k)}{1 + \widehat{\Pi}_p(0)}, \qquad \widehat{F}(k) = \frac{1 - 2dp\widehat{D}(k)[1 + \widehat{\Pi}_p(k)]}{1 + \widehat{\Pi}_p(0)} . \tag{8.4.8}$$

Recall from (8.2.2) that $\widehat{C}_{\lambda_p}(k) = [1 - \lambda_p \widehat{D}(k)]^{-1}$ and, by (6.1.3) and (8.2.1),

$$\lambda_p = 1 - \frac{1 - 2dp[1 + \widehat{\Pi}_p(0)]}{1 + \widehat{\Pi}_p(0)} = \frac{\widehat{\Pi}_p(0) + 2dp[1 + \widehat{\Pi}_p(0)]}{1 + \widehat{\Pi}_p(0)} . \tag{8.4.9}$$

This yields

$$\begin{aligned}
\frac{\hat{\tau}_p(k)}{\widehat{C}_{\lambda_p}(k)} &= \widehat{N}(k)\frac{[1 - \lambda_p\widehat{D}(k)]}{\widehat{F}(k)} \\
&= \widehat{N}(k) + \widehat{N}(k)\frac{[1 - \lambda_p\widehat{D}(k) - \widehat{F}(k)]}{\widehat{F}(k)} \\
&= \widehat{N}(k) + \hat{\tau}_p(k)\left[1 - \lambda_p\widehat{D}(k) - \widehat{F}(k)\right] ,
\end{aligned} \tag{8.4.10}$$

where

$$\begin{aligned}
&1 - \lambda_p\widehat{D}(k) - \widehat{F}(k) \\
&= \frac{[1+\widehat{\Pi}_p(0)] - \{\widehat{\Pi}_p(0) + 2dp[1+\widehat{\Pi}_p(0)]\}\widehat{D}(k) - 1 + 2dp[1+\widehat{\Pi}_p(k)]\widehat{D}(k)}{1 + \widehat{\Pi}_p(0)} \\
&= \frac{[1 - \widehat{D}(k)]\widehat{\Pi}_p(0) + 2dp[\widehat{\Pi}_p(k) - \widehat{\Pi}_p(0)]\widehat{D}(k)}{1 + \widehat{\Pi}_p(0)} .
\end{aligned} \tag{8.4.11}$$

By taking $c_4/d \leq \frac{1}{2}$, we obtain the bound

$$\frac{1 + \ell c_4/d}{1 - c_4/d} \leq 1 + (2\ell + 2) c_4/d , \quad \ell = 0, 1, 2, \dots , \tag{8.4.12}$$

which we frequently use below. For example, together with $d \geq d_0 > 6$, it enables us to bound

$$|\hat{N}(k)| = \left| \frac{1 + \hat{\Pi}_p(k)}{1 + \hat{\Pi}_p(0)} \right| \leq \frac{1 + |\hat{\Pi}_p(k)|}{1 - |\hat{\Pi}_p(0)|} \leq 1 + 4 c_4/d .$$

Together with (8.4.11) and $f_1(p) = 2dp \leq 4$, we obtain in the same fashion that

$$\begin{aligned}
\left| 1 - \lambda_p \hat{D}(k) - \hat{F}(k) \right| &\leq \frac{[1 - \hat{D}(k)]|\hat{\Pi}_p(0)| + 2dp|\hat{\Pi}_p(k) - \hat{\Pi}_p(0)|}{1 - |\hat{\Pi}_p(0)|} \\
&\leq \frac{5(c_4/d)[1 - \hat{D}(k)]}{1 - c_4/d} \leq 12 (c_4/d) \hat{C}_{\lambda_p}(k)^{-1}.
\end{aligned}$$

By our assumption $\hat{\tau}_p(k) \leq 4\hat{C}_{\lambda_p}(k)$ (which follows from $f(p) \leq 4$) and the above inequalities, we can bound (8.4.10) from above by

$$\begin{aligned}
\left| \frac{\hat{\tau}_p(k)}{\hat{C}_{\lambda_p}(k)} \right| &\leq 1 + 4 c_4/d + 12 \cdot 4 (c_4/d) \left| \hat{C}_{\lambda_p}(k) \hat{C}_{\lambda_p}(k)^{-1} \right| \\
&= 1 + 52 c_4/d
\end{aligned} \tag{8.4.13}$$

for every $k \in (-\pi, \pi]^d$. This improves the bound on f_2. $\qquad\square$

(3.3) Improvement of the Bound for f_3. The improvement of f_3, which closes the cycle of bounds that we need to prove, is technically the most challenging.

In the improvement of f_3, we make crucial use of Lem. 8.2. We further rely on the following bounds on discrete derivatives:

$$\begin{aligned}
|\partial_k^\pm \hat{a}(l)| \\
&\leq \sum_x |\mathrm{Re}(e^{il \cdot x}[e^{\pm ik \cdot x} - 1])a(x)| \\
&\leq \sum_x [[1 - \cos(k \cdot x)] + |\sin(k \cdot x)||\sin(l \cdot x)|]|a(x)| \\
&\leq \sum_x [1 - \cos(k \cdot x)]|a(x)| \\
&\quad + \left\{ 4 \sum_x [1 - \cos(k \cdot x)]|a(x)| \sum_y [1 - \cos(l \cdot y)]|a(y)| \right\}^{1/2} , \tag{8.4.14}
\end{aligned}$$

using the same technique as in (8.2.30) for the third inequality. The improvement of f_3 is formulated in the following lemma:

Lemma 8.12 (Improvement of the bounds for f_3). *If the assumptions of Prop. 8.8 are satisfied for some sufficiently large d_0, and if $f(p) \leq 4$ for all $p \in (0, p_c)$, then there exists a constant $c > 0$ such that $f_3(p) \leq 1 + c/d$ for all $p \in (0, p_c)$ and $d \geq d_0$.*

Proof. Keeping track of the precise constants appearing in the proof is fairly messy, and not needed for the result. We therefore relay on the Landau symbol $O(\cdot)$ in the sequel.

We start the proof by writing

$$\hat{\tau}_p(k) = \frac{\hat{b}(k)}{1 - \hat{a}(k)},$$

where

$$\hat{b}(k) = 1 + \widehat{\Pi}_p(k), \qquad \hat{a}(k) = 2dp\widehat{D}(k)[1 + \widehat{\Pi}_p(k)] . \tag{8.4.15}$$

A straightforward calculation (see also [76, (4.18)]) shows that

$$\Delta_k \hat{\tau}_p(l) = \frac{\Delta_k \hat{b}(l)}{1 - \hat{a}(l)} + \sum_{\sigma \in \{1, -1\}} \frac{\big(\hat{a}(l + \sigma k) - \hat{a}(l)\big)\big(\hat{b}(l + \sigma k) - \hat{b}(l)\big)}{\big(1 - \hat{a}(l)\big)\big(1 - \hat{a}(l + \sigma k)\big)}$$

$$+ \hat{b}(l) \, \Delta_k \left[\frac{1}{1 - \hat{a}(l)}\right] . \tag{8.4.16}$$

We now bound all three summands in (8.4.16) and start with the first one:

$$\left|\frac{\Delta_k \hat{b}(l)}{1 - \hat{a}(l)}\right| = \left|\frac{\Delta_k \hat{b}(l)}{\hat{b}(l)}\right| |\hat{\tau}_p(l)| = \left|\frac{\Delta_k \widehat{\Pi}_p(l)}{1 + \widehat{\Pi}_p(l)}\right| |\hat{\tau}_p(l)|$$

$$\leq \left|\Delta_k \widehat{\Pi}_p(l)\right| 6\big(1 + O(1/d)\big)\widehat{C}_{\lambda_p}(l) , \tag{8.4.17}$$

where the last bound uses (8.3.1) and $c_4'/d \leq \frac{1}{2}$ to bound the denominator, and (8.4.13). We apply (8.2.29) with $a(x) = \Pi_p(x)$, combine it with (8.4.17) and (8.3.2), and use $\widehat{C}_{\lambda_p}(l \pm k) \geq \frac{1}{2}$ (cf. (8.4.3)) and the definition of $\widehat{U}_{\lambda_p}(l, k)$ in (8.2.5) to obtain

$$\left|\frac{\Delta_k \hat{b}(l)}{1 - \hat{a}(l)}\right| \leq O(1/d)\,[1 - \widehat{D}(k)]\widehat{C}_{\lambda_p}(l) \leq O(1/d)\,\widehat{U}_{\lambda_p}(l, k) . \tag{8.4.18}$$

The second term in (8.4.16) is bounded as follows. First, since

$$\left|e^{i \cdot x}\big(e^{i \pm k \cdot x} - 1\big)\right| \leq |\sin(k \cdot x)| + 1 - \cos(k \cdot x) , \tag{8.4.19}$$

we obtain

$$
\begin{aligned}
\left|\hat{b}(l \pm k) - \hat{b}(l)\right| &= \left|\widehat{\Pi}_p(l \pm k) - \widehat{\Pi}_p(l)\right| \\
&\leq \sum_x |\sin(k \cdot x)| \left|\Pi_p(x)\right| + \sum_x [1 - \cos(k \cdot x)] \left|\Pi_p(x)\right| .
\end{aligned}
\tag{8.4.20}
$$

The second term on the right-hand side of (8.4.20) is bounded by $O(1/d)\,[1 - \widehat{D}(k)]$; on the first term, we apply the Cauchy–Schwarz inequality and (8.3.2):

$$
\begin{aligned}
\sum_x |\sin(k \cdot x)| \left|\Pi_p(x)\right| &\leq \left(\sum_{x \neq 0} \left|\Pi_p(x)\right|\right)^{1/2} \left(\sum_{x \neq 0} \sin(k \cdot x)^2 \left|\Pi_p(x)\right|\right)^{1/2} \\
&\leq O(1/d)^{1/2} \left(\sum_{x \neq 0} [1 - \cos(k \cdot x)^2] \left|\Pi_p(x)\right|\right)^{1/2} \\
&\leq O(1/d)^{1/2} \left(2 \sum_{x \neq 0} [1 - \cos(k \cdot x)] \left|\Pi_p(x)\right|\right)^{1/2} \\
&\leq O(1/d)\, \widehat{C}_{\lambda_p}(k)^{-1/2} .
\end{aligned}
\tag{8.4.21}
$$

Furthermore,

$$
\hat{a}(l \pm k) - \hat{a}(l) = 2dp\big(\widehat{D}(l \pm k) - \widehat{D}(l)\big)\widehat{\Pi}_p(l) + 2dp\,\widehat{D}(l)\big(\widehat{\Pi}_p(l \pm k) - \widehat{\Pi}_p(l)\big) .
\tag{8.4.22}
$$

In a similar fashion as (8.4.20)–(8.4.21), we bound

$$
\left|\widehat{\Pi}_p(l \pm k) - \widehat{\Pi}_p(l)\right| \leq O(1/d)\, \widehat{C}_{\lambda_p}(k)^{-1/2} ,
$$

and

$$
\begin{aligned}
\left|\widehat{D}(l \pm k) - \widehat{D}(l)\right| &\leq \left(\sum_x D(x)\right)^{1/2} \left(\sum_x [1 - \cos(k \cdot x)] D(x)\right)^{1/2} + \sum_x [1 - \cos(k \cdot x)] D(x) \\
&= 1 \cdot [1 - \widehat{D}(k)]^{1/2} + [1 - \widehat{D}(k)] \\
&\leq 2\widehat{C}_{\lambda_p}(k)^{-1/2} + 2\widehat{C}_{\lambda_p}(k)^{-1} \leq O(1)\, \widehat{C}_{\lambda_p}(k)^{-1/2} ,
\end{aligned}
\tag{8.4.23}
$$

where the last line uses (8.2.10). The combination of (8.4.20)–(8.4.23) and (8.4.7) yields

$$\big(\hat{a}(l \pm k) - \hat{a}(l)\big)\big(\hat{b}(l \pm k) - \hat{b}(l)\big) \leq O(1/d)\,[1 - \widehat{D}(k)] . \tag{8.4.24}$$

On the other hand, by (8.4.13)–(8.4.15) and for all $\sigma \in \{-1, 0, 1\}$,

$$\frac{1}{1 - \hat{a}(l + \sigma k)} = \frac{1}{\hat{b}(l + \sigma k)}\,\hat{\tau}_p(l + \sigma k) \leq (1 + O(1/d))\,\widehat{C}_{\lambda_p}(l + \sigma k) . \tag{8.4.25}$$

Combining (8.4.24) and (8.4.25) yields

$$\frac{\big(\hat{a}(l \pm k) - \hat{a}(l)\big)\big(\hat{b}(l \pm k) - \hat{b}(l)\big)}{\big(1 - \hat{a}(l)\big)\big(1 - \hat{a}(l \pm k)\big)}$$

$$\leq O(1/d)\,[1 - \widehat{D}(k)]\,\widehat{C}_{\lambda_p}(l)\,\widehat{C}_{\lambda_p}(l \pm k)$$

$$\leq O(1/d)\,\widehat{U}_{\lambda_p}(l, k) . \tag{8.4.26}$$

For the third term in (8.4.16), we argue that $|\hat{b}(l)| = 1 + \big|\widehat{\Pi}_p(l)\big| \leq 1 + c_4/d$ by (8.3.1). In order to apply Lem. 8.2 to bound $\Delta_k(1 - \hat{a}(l))^{-1}$, we estimate

$$\hat{A}(l) := \frac{1}{1 - \hat{a}(l)} = \frac{1}{\hat{b}(l)}\,\hat{\tau}_p(l)$$

$$\leq (1 + 2c_4/d)\,(1 + 51c_4/d)\,\widehat{C}_{\lambda_p}(l)$$

$$\leq (1 + O(1/d))\widehat{C}_{\lambda_p}(l) \tag{8.4.27}$$

since $d \geq d_0 > 6$ and (8.4.13), and

$$\widehat{|a|}(0) - \widehat{|a|}(k) = 2dp \sum_x [1 - \cos(k \cdot x)]\big|D(x) + (D \star \Pi_p)(x)\big|$$

$$\leq 2dp[1 - \widehat{D}(k)] + \sum_x [1 - \cos(k \cdot x)]\big|(D \star \Pi_p)(x)\big|$$

$$\leq (2(1 + c_4/d) + 4c_4/d)[1 - \widehat{D}(k)] \leq 5\,[1 - \widehat{D}(k)],$$

where the last line uses again (8.2.10) and, as usual, requires a certain smallness of $1/d$ (here we need $c_4/d \leq \frac{1}{2}$). Plugging these estimates into (8.2.20) yields

$$\Delta_k \frac{1}{1 - \hat{a}(l)} \leq (1 + O(1/d))^3 \cdot 8 \cdot 5^2 \cdot [1 - \widehat{D}(k)]$$

$$\times \Big\{\widehat{C}_{\lambda_p}(l - k)\widehat{C}_{\lambda_p}(l) + \widehat{C}_{\lambda_p}(l)\widehat{C}_{\lambda_p}(l + k)$$

$$+ \widehat{C}_{\lambda_p}(l - k)\widehat{C}_{\lambda_p}(l + k)\Big\} , \tag{8.4.28}$$

so that finally

$$\frac{|\Delta_k \hat{\tau}_p(l)|}{\widehat{U}_{\lambda_p}(k,l)} \le 1 + \text{const}/d \ , \tag{8.4.29}$$

as required.   □

In conclusion, $f_3(p) \le 1 + \text{const}/d$, and thus, we obtain the improved bound $f(p) \le 1 + \text{const}/d$. Thus, Lems. 8.11 and 8.12 complete the proof of Prop. 8.10. □

Proof of Prop. 8.8. Note first that f is continuous on $(0, p_c)$ by Lem. 8.9. Hence, the hypotheses of Lem. 8.1 are satisfied by Prop. 8.10 and the fact that $f(0) = 1$. Therefore, by taking d_0 large enough, we can make $f(p) \le 3$ for all $p < p_c$. Moreover, Prop. 8.10 shows that if $f(p) \le 4$ for some $p \in [0, p_c)$, then in fact $f(p) \le 1 + O(1/d)$. Hence, $f(p) \le 1 + O(1/d)$ uniformly for $p < p_c$. □

8.5   Consequences of the Completed Bootstrap

In this section, we state some consequences of the completed bootstrap argument that are crucial later on. These consequences are formulated in the following corollary:

Corollary 8.13   (Bound p_c and triangle from completed bootstrap). *For percolation, there exist $d_0 > 6$ and $C > 0$ such that for $d \ge d_0 > 6$, and all $p \le p_c$,*

$$\hat{\tau}_p(k) = \frac{1 + \widehat{\Pi}_p(k)}{1 - 2dp\widehat{D}(k)[1 + \widehat{\Pi}_p(k)]} \ , \tag{8.5.1}$$

where

$$\widehat{\Pi}_p(k) = \sum_{N=0}^{\infty} (-1)^N \widehat{\Pi}_p^{(N)}(k) \ . \tag{8.5.2}$$

Further,

$$1 \le 2dp_c \le 1 + C/d \quad and \quad \Delta_p \le 1 + C/d \ . \tag{8.5.3}$$

This corollary proves the triangle condition (4.1.1), from which $\theta(p_c) = 0$ follows. We thus solved Open Problem 1.1 in high dimensions.

Proof. First fix $p < p_c$. The bound on Δ_p follows directly from Lem. 8.5. Lemma 8.4 implies that the sum over N in (8.5.2) exists, and $R_M(x)$ vanishes as $M \to \infty$. As a result, (8.5.1) follows. This completes the proof for $p < p_c$.

To extend the argument to $p = p_c$, we see that the bounds on $\widehat{\Pi}_p^{(N)}(k)$ hold *uniformly in* $p < p_c$. By dominated convergence, $\Pi_p(x)$ converges to $\Pi_{p_c}(x)$ for fixed $x \in \mathbb{Z}^d$ as $p \nearrow p_c$. Π_{p_c} has a well-defined Fourier transform, so also $\widehat{\Pi}_p(k)$ converges to $\widehat{\Pi}_{p_c}(k)$ as $p \nearrow p_c$. For $k \ne 0$, we *define*

$$\hat{\tau}_{p_c}(k) = \frac{1 + \widehat{\Pi}_{p_c}(k)}{1 - 2dp_c\widehat{D}(k)[1 + \widehat{\Pi}_{p_c}(k)]} \ . \tag{8.5.4}$$

For $p = p_c$ and since $\hat{\tau}_{p_c}(0) = \chi(p_c) = \infty$, we obtain that p_c satisfies $1 - 2dp_c[1 + \widehat{\Pi}_{p_c}(0)] = 0$. Then, by (8.3.6) in Lem. 8.4, the right-hand side of (8.5.4) has an *integrable* singularity for $k = 0$ when $d > 2$. Thus, we can use dominated convergence to obtain that

$$
\begin{aligned}
\tau_{p_c}(x) &= \lim_{p \nearrow p_c} \tau_p(x) = \lim_{p \nearrow p_c} \int_{(-\pi,\pi]^d} e^{-ik \cdot x} \hat{\tau}_p(k) \frac{dk}{(2\pi)^d} \\
&= \lim_{p \nearrow p_c} \int_{(-\pi,\pi]^d} e^{-ik \cdot x} \frac{1 + \widehat{\Pi}_p(k)}{1 - 2dp\widehat{D}(k)[1 + \widehat{\Pi}_p(k)]} \frac{dk}{(2\pi)^d} \\
&= \int_{(-\pi,\pi]^d} e^{-ik \cdot x} \lim_{p \nearrow p_c} \frac{1 + \widehat{\Pi}_p(k)}{1 - 2dp\widehat{D}(k)[1 + \widehat{\Pi}_p(k)]} \frac{dk}{(2\pi)^d} \\
&= \int_{(-\pi,\pi]^d} e^{-ik \cdot x} \frac{1 + \widehat{\Pi}_{p_c}(k)}{1 - 2dp_c\widehat{D}(k)[1 + \widehat{\Pi}_{p_c}(k)]} \frac{dk}{(2\pi)^d} .
\end{aligned}
\tag{8.5.5}
$$

In particular, this shows that $\hat{\tau}_{p_c}(k)$ is well defined and $\tau_{p_c}(x)$ is its Fourier inverse. Thus, (8.5.1) extends to $p = p_c$.

Having established the existence of the lace expansion for $p = p_c$, we can now use it to investigate $p_c = p_c(\mathbb{Z}^d)$. The lower bound on $p_c(\mathbb{Z}^d)$ is proved in Exerc. 2.6. For the upper bound, we use (6.1.6) (which is now established rigorously) to obtain

$$
2dp_c = \frac{1}{1 + \widehat{\Pi}_{p_c}(0)} \leq \frac{1}{1 - O(1/d)} = 1 + O(1/d) .
\tag{8.5.6}
$$

$\square$

Exercise 8.7 (Finer asymptotics of $p_c(\mathbb{Z}^d)$). Use (8.5.3) and Lem. 8.4 to prove that

$$
\widehat{\Pi}_{p_c}(0) = -\widehat{\Pi}_{p_c}^{(1)}(0) + O(1/d^2) = -\frac{1}{2d} + O(1/d^2) .
\tag{8.5.7}
$$

In turn, use this in the implicit equation $2dp_c = [1 + \widehat{\Pi}_{p_c}(0)]^{-1}$ for p_c in (6.1.6) to improve the bound on p_c in (8.5.3) to the statement that there exists a constant C such that

$$
1 \leq (2d - 1)p_c \leq 1 + C/d^2 .
\tag{8.5.8}
$$

Exercise 8.8 (Asymptotics of p_c for spread-out percolation). Fix $d > 6$. Adapt the proof of the bound on p_c in (8.5.3) to prove that there exists a constant C such that spread-out percolation satisfies

$$
1 \leq [(2L + 1) - 1]^d p_c \leq 1 + C/L^d .
\tag{8.5.9}
$$

Chapter 9
Proof that $\delta = 2$ and $\beta = 1$ under the Triangle Condition

In this chapter, we use the finiteness of the triangle diagram in order to establish that certain critical exponents take on their mean-field values.

We again rely on the differential inequalities in Lem. 3.5. We complement these two mean-field differential inequalities by a differential inequality involving the triangle, as stated as Lem. 9.1 in Sect. 9.1. We then prove that $\delta = 2$ in Sect. 9.2 and use the inequalities proved for δ to derive that $\beta = 1$ in Sect. 9.3. We complete this chapter by a proof of the differential inequality in Lem. 9.1 in Sect. 9.4.

9.1 A Differential Inequality Involving the Triangle

We investigate the magnetization $M(p, \gamma)$ once more and formulate an extra differential inequality that involves the triangle diagram. We recall from Section 3.4 that

$$M(p, \gamma) = \mathbb{E}_p[1 - (1 - \gamma)^{|\mathscr{C}(0)|}] = \mathbb{P}_{p,\gamma}(0 \leftrightarrow \mathscr{G}) . \tag{9.1.1}$$

Exercise 9.1 (Monotonicity of M). Prove that $\gamma \mapsto M(p, \gamma)$ and $p \mapsto M(p, \gamma)$ are nondecreasing. Prove also that $\gamma \mapsto M(p, \gamma)$ is concave.

The extra differential inequality is crucial to prove that the one-sided bounds on the critical exponents β and δ are actually equalities.

For $p \in [0, p_c]$, we let

$$\triangle_p^{\max} = \sup_{x \in \mathbb{Z}^d} [\triangle_p(x) - \delta_{x,0}] = \sup_{x \in \mathbb{Z}^d} \left[\tau_p^{\star 3}(x) - \delta_{0,x} \right] . \tag{9.1.2}$$

Since at least one of the three τ_p factors needs to have a nonzero displacement in order for $\tau_p^{\star 3}(x) - \delta_{0,x}$ to be nonzero, we can use (7.2.10) and (7.2.2) to bound

$$\triangle_{p_c}^{\max} \leq 3 \sup_{x \in \mathbb{Z}^d} (\tau_{p_c}^{\star 2} \star \tilde{\tau}_{p_c})(x) = 3\tilde{\triangle}_{p_c} . \tag{9.1.3}$$

© Springer International Publishing Switzerland 2017
M. Heydenreich and R. van der Hofstad, *Progress in High-Dimensional Percolation and Random Graphs*, CRM Short Courses, DOI 10.1007/978-3-319-62473-0_9

Thus, by (8.5) and the fact that it now holds for *all* $p \in [0, p_c]$,

$$\triangle_{p_c}^{\max} = O(1/d) . \tag{9.1.4}$$

The differential inequality that we rely on is the following:

Lemma 9.1 (Differential inequalities revisited [59, Lem. 5.5]). *Fix* $p \in [0, p_c]$ *and* $0 < \gamma < 1$. *Then*

$$M(p, \gamma) \geq \kappa(p)(1 - \gamma)M(p, \gamma)^2 \frac{\partial M(p, \gamma)}{\partial \gamma} , \tag{9.1.5}$$

where

$$\kappa(p) = \left[\binom{2d}{2} p^2(1 - p)^{2d-2}\left[(1 - \triangle_p^{\max})^2 - \triangle_p^{\max}\right] - p - \triangle_p^{\max} \right] 2dp . \tag{9.1.6}$$

The differential inequality (9.1.5), which is a variant of an inequality derived by Barsky and Aizenman [28], is used to prove our upper bounds on $M(p_c, \gamma)$. In turn, the upper bound on $M(p_c, \gamma)$ is the input in an *extrapolation argument* to prove the upper bound on $\theta(p)$ for $p > p_c$. This extrapolation inequality bounds $M(p, 0+) = \theta(p)$ in terms of $M(p_c, \gamma)$ for an appropriate $\gamma > 0$. We give a proof of (9.1.5) in Sect. 9.4. While the formula for $\kappa(p)$ may look quite involved, the fact that $2dp_c = 1 + O(1/d)$ implies that $\kappa(p_c) = (1/2e)(1 + O(1/d))$ in high dimensions:

Exercise 9.2 (Asymptotics of κ). Prove that $\kappa(p_c) = (1/2e)\big(1 + O(1/d)\big)$ in high dimensions, and thus, $\kappa(p_c) > 0$.

We start by proving that $\delta = 2$ in Sect. 9.2 and use the inequalities proved for δ to derive that $\beta = 1$ in Sect. 9.3.

9.2   The Cluster Tail Critical Exponent δ

Here, we give the Aizenman–Newman argument [13] to identify the critical exponent $\delta = 2$. The main result is the following theorem:

Theorem 9.2 ($\delta = 2$ when the triangle condition holds). *For percolation, there exist* $d_0 > 6$ *and* $0 < c_\delta < C_\delta < \infty$ *such that for* $d \geq d_0 > 6$ *and all* $n \geq 1$

$$\frac{c_\delta}{\sqrt{n}} \leq \mathbb{P}_{p_c}(|\mathscr{C}(0)| \geq n) \leq \frac{C_\delta}{\sqrt{n}} . \tag{9.2.1}$$

In particular, the critical exponent δ *exists in the bounded-ratio sense and takes on the mean-field value* $\delta = 2$.

In order to prove Thm. 9.2, we start by proving an upper bound on the magnetization in the generating function sense. This provides a matching upper bound on the magnetization to that proved in Cor. 4.5, and also rather directly implies (9.2.1). This upper bound is stated in the following lemma:

Lemma 9.3 (Upper bound on δ assuming the triangle condition [59, (5.38)]). *For percolation , there exists a $d_0 > 6$ such that for $d \geq d_0 > 6$ and all $\gamma \in [0, 1]$,*

$$M(p_c, \gamma) \leq \sqrt{12\gamma}. \tag{9.2.2}$$

In particular, the critical exponent δ exists for the generating function as in (4.3.1) and takes on the mean-field value $\delta = 2$.

Proof. We assume that the triangle condition is satisfied and that the bound on $p_c(\mathbb{Z}^d)$ in (8.5.3) in Cor. 8.13 holds. Under these conditions, $2dp_c = 1 + O(1/d)$, so (9.1.5) implies that

$$M(p_c, \gamma) \geq \frac{1}{2e}[1 - O(1/d)](1 - \gamma)\frac{\partial M(p_c, \gamma)}{\partial \gamma}M(p_c, \gamma)^2 . \tag{9.2.3}$$

For $p = p_c$, we divide this by $M(p_c, \gamma) > 0$ and thus obtain

$$\frac{1}{2}\frac{\partial M^2}{\partial \gamma} \leq \frac{2e}{1 - \gamma}[1 - O(1/d)] . \tag{9.2.4}$$

We further integrate (9.2.4) w.r.t. the second argument of M over the interval $[0, \gamma]$ and use $M(p_c, 0+) = \theta(p_c) = 0$ (which we know from the triangle condition, cf. Corol. 8.13), to see that

$$M^2(p_c, \gamma) \leq \frac{4e\gamma}{1 - \gamma}[1 - O(1/d)] . \tag{9.2.5}$$

For $\gamma \in [0, 1/12]$, this implies (9.2.2), provided d is sufficiently large. Finally, we note that we can remove the restriction $\gamma \in [0, 1/12]$, since trivially, $M(p_c, \gamma) \leq 1 \leq \sqrt{12\gamma}$ if $\gamma \geq 1/12$. $\qquad\square$

Proof of upper bound in Thm. 9.2. For the upper bound, we follow [59, Lem. 6.2(ii)] and use that

$$\mathbb{P}_{p_c}(|\mathcal{C}(0)| \geq n) \leq \frac{e}{e - 1}M(p_c, 1/n) , \tag{9.2.6}$$

since $1 - e^{-1} \leq 1 - (1 - 1/n)^\ell$ whenever $\ell \geq n$. Thus, the upper bound in (9.2.1) follows with $C_\delta = \sqrt{12}e/(e - 1)$. $\qquad\square$

Proof of lower bound in Thm. 9.2. We start by noting that, for all $0 \leq \gamma < \tilde{\gamma} < 1$,

$$\mathbb{P}_{p_c}(|\mathcal{C}(0)| \geq n) \geq M(p_c, \gamma) - \frac{\gamma}{\tilde{\gamma}}e^{\tilde{\gamma}n}M(p_c, \tilde{\gamma}). \tag{9.2.7}$$

Indeed, to prove (9.2.7), we note that $[1 - (1 - \gamma)^\ell] \leq \ell\gamma$. Also, $\ell\tilde{\gamma} \leq e^{\ell\tilde{\gamma}} - 1 = e^{\ell\tilde{\gamma}}(1 - e^{-\ell\tilde{\gamma}})$, which, combined with $e^{-\tilde{\gamma}} \geq 1 - \tilde{\gamma}$, gives $\ell\tilde{\gamma} \leq e^{\ell\tilde{\gamma}}(1 - (1 - \tilde{\gamma})^\ell)$. Therefore,

$$M(p_{\mathrm{c}}, \gamma) = \sum_{\ell \geq 1} (1 - (1 - \gamma)^\ell) \, \mathbb{P}_{p_{\mathrm{c}}}(|\mathcal{C}(0)| = \ell)$$

$$\leq \gamma \sum_{\ell < n} \ell \, \mathbb{P}_{p_{\mathrm{c}}}(|\mathcal{C}(0)| = \ell) + \sum_{\ell \geq n} \mathbb{P}_{p_{\mathrm{c}}}(|\mathcal{C}(0)| = \ell)$$

$$\leq \frac{\gamma}{\tilde{\gamma}} e^{\tilde{\gamma} n} M(p_{\mathrm{c}}, \tilde{\gamma}) + \mathbb{P}_{p_{\mathrm{c}}}(|\mathcal{C}(0)| \geq n) \,. \tag{9.2.8}$$

We apply (9.2.7) with $\tilde{\gamma} = 1/n$. Since

$$M(p_{\mathrm{c}}, 1/n) \leq \sqrt{12/n} \tag{9.2.9}$$

by (9.2.2), (9.2.7) implies that

$$\mathbb{P}_{p_{\mathrm{c}}}(|\mathcal{C}(0)| \geq n) \geq M(p_{\mathrm{c}}, \gamma) - \sqrt{12n}\,\gamma \mathrm{e} \,. \tag{9.2.10}$$

Further, Prop. 3.6, together with (8.5.3) in Cor. 8.13, implies that for $d \geq d_0$ sufficiently large, $M(p_{\mathrm{c}}, \gamma) \geq \frac{1}{3}\sqrt{\gamma}$, and hence

$$\mathbb{P}_{p_{\mathrm{c}}}(|\mathcal{C}(0)| \geq n) \geq \frac{1}{3}\sqrt{\gamma} - \sqrt{12n}\,\gamma \mathrm{e} \geq \frac{1}{3}\sqrt{\gamma}\big(1 - 30\sqrt{\gamma n}\big). \tag{9.2.11}$$

The choice $\gamma = 1/(60^2 n)$ gives the lower bound of (9.2.1) with $c_\delta = \frac{1}{6}$. $\square$

Let us remark that Hara and Slade [139] have provided much stronger results that the ones obtained here. Indeed, they have shown the following result:

Theorem 9.4 ($\delta = 2$ in sharp sense in high dimensions). *There exist $d_0 > 6$ and $0 < C_\delta < \infty$ such that for $d \geq d_0 > 6$,*

$$\mathbb{P}_{p_{\mathrm{c}}}(|\mathcal{C}(0)| = n) = \frac{C_\delta}{\sqrt{8\pi n^3}}(1 + O(n^{-\varepsilon})) \,. \tag{9.2.12}$$

In particular, the critical exponent δ exists in the asymptotic sense and takes on the mean-field value $\delta = 2$.

We return to the Hara and Slade paper [139] in Sect. 15.1 below, where we discuss super-process limits of critical percolation clusters. Unfortunately, Thm. 9.4 only holds in sufficiently high dimensions, *even* for spread-out models.

9.3 The Percolation-Function Critical Exponent β

Here, we give the Barsky–Aizenman argument [28] to identify the critical exponent $\beta = 1$, see also [59, Sect. 5].

Theorem 9.5 ($\beta = 1$ **when the triangle condition holds**). *For percolation, there exist* $d_0 > 6$ *and* $0 < c_\beta < C_\beta < \infty$ *such that for* $d \geq d_0 > 6$ *and all* $p \geq p_c(\mathbb{Z}^d)$,

$$c_\beta(p - p_c) \leq \theta(p) \leq C_\beta(p - p_c) . \tag{9.3.1}$$

In particular, the critical exponent β *exists in the bounded-ratio sense and takes on the mean-field value* $\beta = 1$.

Proof. The lower bound is proved in Thm. 3.2(2), so we focus on the upper bound, which is expected to be true only above the upper critical dimension.

Following Barsky and Aizenman [28], we apply the extrapolation principle used by Aizenman and Barsky [9] (recall Chap. 3), to extend (9.2.2) to (9.3.1). The extrapolation principle was first used by Aizenman and Fernandez in the context of the Ising model [11] (see also the survey by Fernandez, Fröhlich and Sokal [105]). We find it most convenient to reparametrize and use the variable $h = -\log(1 - \gamma)$ rather than γ and define

$$\widetilde{M}(p, h) = M(p, 1 - e^{-h}) \quad \text{for } h \geq 0 .$$

We take $p = p_c + \varepsilon/(2d)$ with $\varepsilon > 0$ small. Assuming that $\varepsilon \leq 1$, the differential inequality (3.4.4) implies that

$$\frac{\partial \widetilde{M}}{\partial p} \leq 2d A \widetilde{M} \frac{\partial \widetilde{M}}{\partial h} , \tag{9.3.2}$$

where $A = \left(1 - p_c - 1/(2d)\right)^{-1} = 1 + O(1/d)$.

Exercise 9.3 (Proof of differential inequality for $\widetilde{M}$). Prove (9.3.2) using (3.4.4).

For fixed $m \in [0, 1]$ and fixed $p \in (0, 1)$, we can solve the equation $\widetilde{M}(p, h) = m$ for $h = h(p)$, so that $\widetilde{M}(p, h(p)) = m$. Differentiation of this identity with respect to p gives

$$\frac{\partial \widetilde{M}}{\partial p} + \frac{\partial \widetilde{M}}{\partial h} \frac{\partial h}{\partial p} \bigg|_{\widetilde{M}=m} = 0 . \tag{9.3.3}$$

Therefore, by (9.3.2),

$$0 \leq -\frac{\partial h}{\partial p} \bigg|_{\widetilde{M}=m} = \frac{\partial \widetilde{M}/\partial p}{\partial \widetilde{M}/\partial h} \leq 2d A m . \tag{9.3.4}$$

The upper and lower bounds of (9.3.4) imply that a contour line $\widetilde{M} = m_1$ in the (p, h)-plane (with p-axis horizontal and h-axis vertical) passing through a point $P_1 = (p_1, h_1)$ is such that $\widetilde{M}(P) \leq m_1$ for all points P in the first quadrant that lie on or below the line of slope 0 through P_1; see Fig. 9.1.

Key Inequality for Extrapolation. The key observation in the extrapolation inequality argument is that if $P_2 = (p_2, h_2)$ is on the line through P_1 with slope $-2d A m_1$, with $p_2 < p_1$, then P_2 lies above the contour line $\widetilde{M} = m_1$, so if we set $m_2 = \widetilde{M}(P_2)$, then $m_2 \geq m_1$.

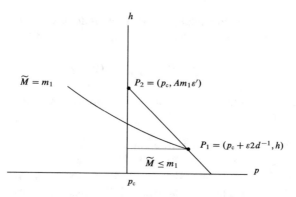

Fig. 9.1 The extrapolation geometry

Indeed, the fact that $m_2 \geq m_1$ is crucially used below. We now complete the argument of the upper bound on β.

Fix h, and fix $\varepsilon > 0$. Let $P_1 = (p_c + \varepsilon/(2d), h)$, and define $m_1 = m_1(\varepsilon) = \widetilde{M}(P_1)$. Let

$$\varepsilon' = \varepsilon + \frac{h}{Am_1} , \tag{9.3.5}$$

further define $P_2 = (p_c, Am_1\varepsilon')$ and $m_2 = \widetilde{M}(P_2)$. The point P_2 lies on the line through P_1 with slope $-2dAm_1$. Therefore, as observed above, $m_1 \leq m_2$. Applying (9.2.2) gives

$$\begin{aligned}
\widetilde{M}(p_c + \varepsilon/(2d), h) = m_1 \leq m_2 &= \widetilde{M}(p_c, Am_1\varepsilon') \\
&\leq \sqrt{12}\left(1 - e^{-Am_1\varepsilon'}\right)^{1/2} \\
&= \sqrt{12}\left(1 - e^{-Am_1\varepsilon} + e^{-Am_1\varepsilon}\left[1 - e^{-h}\right]\right)^{1/2} \\
&\leq \sqrt{12}(Am_1\varepsilon + \gamma)^{1/2} ,
\end{aligned} \tag{9.3.6}$$

with $\gamma = 1 - e^{-h}$. The inequality

$$m_1^2 \leq 12(Am_1\varepsilon + \gamma) \tag{9.3.7}$$

has roots

$$m^{\pm} = 6A\varepsilon \pm \sqrt{12\gamma + (6A\varepsilon)^2}. \tag{9.3.8}$$

The root m^+ is positive and m^- is negative. Thus,

$$\begin{aligned}
M(p_c + \varepsilon/(2d), \gamma) = m_1 \leq m^+ &\leq 6A\varepsilon + \sqrt{12\gamma + (6A\varepsilon)^2} \\
&\leq 12A\varepsilon + \sqrt{12\gamma} ,
\end{aligned} \tag{9.3.9}$$

using $\sqrt{a+b} \leq \sqrt{a} + \sqrt{b}$ in the last step.

We let $\gamma \searrow 0$ in (9.3.9), to obtain (recall Exerc. 3.4)

$$\theta(p_c + \varepsilon/(2d)) = M(p_c + \varepsilon/(2d), 0+) \le 12A\varepsilon . \tag{9.3.10}$$

This completes the proof of (9.3.1), since we can choose A arbitrarily close to 1 by choosing $d \ge d_0$ with d_0 sufficiently large.   □

We close this section with the open problem of proving that $\beta = 1$ in the asymptotic sense. Possibly, a lace-expansion argument might be necessary for this. However, the application of the lace expansion in the *supercritical* regime is highly nontrivial:

> **Open Problem 9.1 ($\beta = 1$ in the asymptotic sense).** Prove that $\beta = 1$ in the asymptotic sense, possibly by using a lace-expansion argument.

9.4 Proof of the Differential Inequality Involving the Triangle

We follow the presentation by Borgs et al. in [59, App. A.2]. Let $\{v \Longleftrightarrow \mathcal{G}\}$ denote the event that there exist $x, y \in \mathcal{G}$, with $x \ne y$, such that there are disjoint connections $v \longleftrightarrow x$ and $v \longleftrightarrow y$. Let $F_{(u,v)}$ denote the event that the bond (u, v) is occupied and pivotal for the connection from 0 to $\mathcal{G}$, with $\{v \Longleftrightarrow \mathcal{G}\}$. Let $F = \bigcup_{(u,v)} F_{(u,v)}$, and note that the union is disjoint. Since $0 \longleftrightarrow \mathcal{G}$ when F occurs,

$$M(p, \gamma) = \mathbb{P}_{p,\gamma}(0 \longleftrightarrow \mathcal{G}) \ge \mathbb{P}_{p,\gamma}(F) = \sum_{(u,v)} \mathbb{P}_{p,\gamma}(F_{(u,v)}) , \tag{9.4.1}$$

and it suffices to prove that $\mathbb{P}_{p,\gamma}(F)$ is bounded below by the right side of (9.1.5).

For $x, y \in \mathbb{Z}^d$, we define a "green-free" analogue of the two-point function by

$$\tau_{p,\gamma}(x, y) = \mathbb{P}_{p,\gamma}(x \longleftrightarrow y, x \nleftrightarrow \mathcal{G}) , \tag{9.4.2}$$

so that

$$\chi(p, \gamma) = \sum_{x \in \mathbb{Z}^d} \tau_{p,\gamma}(0, x) \tag{9.4.3}$$

and $\chi(p, 0) = \chi(p)$. Given a subset $A \subseteq \mathbb{Z}^d$, we define $\tau_{p,\gamma}^A(x, y)$ to be the probability that (i) $x \longleftrightarrow y$ in $\mathbb{Z}^d \setminus A$ and (ii) $x \nleftrightarrow \mathcal{G}$ in $\mathbb{Z}^d \setminus A$, which is to say that $x \nleftrightarrow \mathcal{G}$ after every bond with an endpoint in A is made vacant. We write $\tilde{I}^{\{u,v\}}[E]$ to be the indicator that E occurs after $\{u, v\}$ is made vacant. Then, we have the following generalization of the Cutting-Bond Lemma 6.4:

Lemma 9.6 (Green-free cutting bond lemma). *For every $p \in [0, p_c]$ and $\gamma > 0$,*

$$\mathbb{P}_{p,\gamma}(F_{(u,v)}) = p\, \mathbb{E}_{p,\gamma}\left[\tau_{p,\gamma}^{\tilde{\mathscr{C}}^{(u,v)}(v)}(0, u)\tilde{I}^{\{u,v\}}[v \Longleftrightarrow \mathcal{G}] \right] . \tag{9.4.4}$$

Proof. We first observe that the event $F_{(u,v)}$ is given by

$$F_{(u,v)} = \{0 \leftrightarrow u \text{ in } \mathbb{Z}^d \backslash \widetilde{\mathscr{C}}^{(u,v)}(v)\} \cap \{0 \nleftrightarrow \mathscr{G} \text{ in } \mathbb{Z}^d \backslash \widetilde{\mathscr{C}}^{(u,v)}(v)\}$$
$$\cap \{\{u, v\} \text{ occupied}\}$$
$$\cap \{v \leftrightarrow \mathscr{G} \text{ after } \{u, v\} \text{ is made vacant}\} . \quad (9.4.5)$$

The bond $\{u, v\}$ has an endpoint in $\widetilde{\mathscr{C}}^{(u,v)}(v)$, and hence, the event that $\{u, v\}$ is occupied is independent of the other events above. Therefore,

$$\mathbb{P}_{p,\gamma}(F_{(u,v)}) = p \sum_{A:A \ni v} \mathbb{E}_{p,\gamma}\big[\mathbb{1}_{\{\widetilde{\mathscr{C}}^{(u,v)}(v)=A\}} \tilde{I}^{\{u,v\}}[v \leftrightarrow \mathscr{G}]$$
$$\times \mathbb{1}_{\{(0 \leftrightarrow u \text{ and } 0 \nleftrightarrow \mathscr{G}) \text{ in } \mathbb{Z}^d \backslash A\}}\big] . \quad (9.4.6)$$

The triangle condition in Corol. 8.13 guarantees that $\widetilde{\mathscr{C}}^{(u,v)}(v)$ is finite a.s., so that the sum in (9.4.6) is over finite connected sets $A \subset \mathbb{Z}^d$ containing v.

The two events in the first line depend only on bonds with an endpoint in A (but not on $\{u, v\}$) and vertices in A, while those in the second line depend only on bonds with no endpoint in A (so not on $\{u, v\}$) and on vertices in $\mathbb{Z}^d \backslash A$. Therefore,

$$\mathbb{P}_{p,\gamma}(F_{(u,v)}) = p \sum_{A:A \ni v} \mathbb{E}_{p,\gamma}\big[\mathbb{1}_{\{\widetilde{\mathscr{C}}^{(u,v)}(v)=A\}} \tilde{I}^{\{u,v\}}[v \leftrightarrow \mathscr{G}]\big] \tau_{p,\gamma}^A(0, u) , \quad (9.4.7)$$

which implies the desired result.   □

Proof of Lem. 9.1. We use the identities

$$\tau_{p,\gamma}^{\widetilde{\mathscr{C}}^{(u,v)}(v)}(0, u) = \tau_{p,\gamma}(0, u) - \Big(\tau_{p,\gamma}(0, u) - \tau_{p,\gamma}^{\widetilde{\mathscr{C}}^{(u,v)}(v)}(0, u)\Big) \quad (9.4.8)$$

and

$$\tilde{I}^{\{u,v\}}[v \leftrightarrow \mathscr{G}] = \mathbb{1}_{\{v \leftrightarrow \mathscr{G}\}} - \Big(\mathbb{1}_{\{v \leftrightarrow \mathscr{G}\}} - \tilde{I}^{\{u,v\}}[v \leftrightarrow \mathscr{G}]\Big) . \quad (9.4.9)$$

It follows from Lem. 9.6 and (9.4.3) that

$$\mathbb{P}_{p,\gamma}(F) = 2dp\chi(p, \gamma) \mathbb{P}_{p,\gamma}(0 \leftrightarrow \mathscr{G})$$
$$- p \sum_{(u,v)} \tau_{p,\gamma}(0, u) \mathbb{E}_{p,\gamma}\big[\mathbb{1}_{\{v \leftrightarrow \mathscr{G}\}} - \tilde{I}^{\{u,v\}}[v \leftrightarrow \mathscr{G}]\big]$$
$$- p \sum_{(u,v)} \mathbb{E}_{p,\gamma}\Big[\big(\tau_{p,\gamma}(0, u) - \tau_{p,\gamma}^{\widetilde{\mathscr{C}}^{(u,v)}(v)}(0, u)\big) \tilde{I}^{\{u,v\}}[v \leftrightarrow \mathscr{G}]\Big] . \quad (9.4.10)$$

We write (9.4.10) as $X_1 - X_2 - X_3$, bound X_1 from below, and bound X_2 and X_3 from above.

Lower bound on X_1. We prove that

$$\mathbb{P}_{p,\gamma}(0 \Longleftrightarrow \mathcal{G}) \geq \binom{2d}{2} p^2 (1-p)^{2d-2} M^2(p,\gamma) \left[(1 - \triangle_p^{\max})^2 - \triangle_p^{\max} \right], \qquad (9.4.11)$$

which implies that

$$X_1 \geq 2dp\chi(p,\gamma) \binom{2d}{2} p^2 (1-p)^{2d-2} M^2(p,\gamma) \left[(1 - \triangle_p^{\max})^2 - \triangle_p^{\max} \right]. \qquad (9.4.12)$$

Let $E_{e,f}$ be the event that the bonds $(0,e)$ and $(0,f)$ are occupied, and all other bonds incident on 0 are vacant, and that in the reduced graph $G^- = (\mathbb{Z}_-^d, \mathcal{E}^-)$ obtained by deleting the origin and each of the $2d$ bonds incident on 0 from $\mathbb{Z}^d$ the following three events occur: $e \leftrightarrow \mathcal{G}$, $f \leftrightarrow \mathcal{G}$, and $\mathcal{C}(e) \cap \mathcal{C}(f) = \varnothing$. Let $\mathbb{P}_{p,\gamma}^-$ denote the joint bond/vertex measure of G^-. We note that the event $\{0 \leftrightarrow \mathcal{G}\}$ contains the event $\bigcup_{e,f} E_{e,f}$, where the union is over unordered pairs of neighbors e, f of the origin, and the union is disjoint. Then,

$$\mathbb{P}_{p,\gamma}(0 \Longleftrightarrow \mathcal{G})$$
$$\geq \mathbb{P}_{p,\gamma}\left(\bigcup_{e,f} E_{e,f} \right) = \sum_{e,f} \mathbb{P}_{p,\gamma}(E_{e,f})$$
$$= p^2(1-p)^{2d-2} \sum_{e,f} \mathbb{P}_{p,\gamma}^-(e \leftrightarrow \mathcal{G}, f \leftrightarrow \mathcal{G}, \mathcal{C}(e) \cap \mathcal{C}(f) = \varnothing). \quad (9.4.13)$$

Let $W = W_{e,f}$ denote the event whose probability appears on the right side of (9.4.13). Conditioning on the set $\mathcal{C}(e) = A \subset \mathbb{Z}_-^d$, we see that

$$\mathbb{P}_{p,\gamma}^-(W) = \sum_{A:A \ni e} \mathbb{P}_{p,\gamma}^-(\mathcal{C}(e) = A, e \leftrightarrow \mathcal{G}, f \leftrightarrow \mathcal{G}, \mathcal{C}(e) \cap \mathcal{C}(f) = \varnothing). \qquad (9.4.14)$$

This can be rewritten as

$$\mathbb{P}_{p,\gamma}^-(W) = \sum_{A:A \ni e} \mathbb{P}_{p,\gamma}^-(\mathcal{C}(e) = A, e \leftrightarrow \mathcal{G}, f \leftrightarrow \mathcal{G} \text{ in } \mathbb{Z}_-^d \setminus A), \qquad (9.4.15)$$

where $\{f \leftrightarrow \mathcal{G} \text{ in } \mathbb{Z}_-^d \setminus A\}$ is the event that there exists $x \in \mathcal{G}$ such that $f \leftrightarrow x$ in $\mathbb{Z}_-^d \setminus A$. The intersection of the first two events on the right-hand side of (9.4.15) is independent of the third event, and hence

$$\mathbb{P}_{p,\gamma}^-(W) = \sum_{A:A \ni e} \mathbb{P}_{p,\gamma}^-(\mathcal{C}(e) = A, e \leftrightarrow \mathcal{G}) \mathbb{P}_{p,\gamma}^-(f \leftrightarrow \mathcal{G} \text{ in } \mathbb{Z}_-^d \setminus A). \qquad (9.4.16)$$

Let $M^-(x) = \mathbb{P}_{p,\gamma}^-(x \leftrightarrow \mathcal{G})$, for $x \in \mathbb{Z}_-^d$. Then, by the BK inequality and the fact that the two-point function on G^- is bounded above by the two-point function on G,

$$\mathbb{P}_{p,\gamma}^{-}(f \leftrightarrow \mathcal{G} \text{ in } \mathbb{Z}_{-}^{d} \setminus A) = M^{-}(f) - \mathbb{P}_{p,\gamma}^{-}(f \overset{A}{\leftrightarrow} \mathcal{G})$$
$$\geq M^{-}(f) - \sum_{y \in A} \tau_{p,0}(f, y) M^{-}(y) . \qquad (9.4.17)$$

By definition and the BK inequality,

$$M^{-}(x) = M(p, \gamma) - \mathbb{P}_{p,\gamma}(x \overset{\{0\}}{\longleftrightarrow} \mathcal{G}) \geq M(p, \gamma)\big(1 - \tau_{p,0}(0, x)\big)$$
$$\geq M(p, \gamma)(1 - \Delta_{p}^{\max}) . \qquad (9.4.18)$$

In the above, we have also used $\tau_{p,0}(0, x) \leq \tau_{p}^{\star 3}(x) \leq \Delta_{p}(x)$.

It follows from (9.4.16)–(9.4.18) and the upper bound $M^{-}(x) \leq M(p, \gamma)$ that

$$\mathbb{P}_{p,\gamma}^{-}(W)$$
$$\geq M(p, \gamma) \sum_{A:A \ni e} \mathbb{P}_{p,\gamma}^{-}(\mathscr{C}(e) = A, e \leftrightarrow \mathcal{G}) \left[(1 - \Delta_{p}^{\max}) - \sum_{y \in A} \tau_{p,0}(f, y) \right]$$
$$= M(p, \gamma) \left[M^{-}(e)(1 - \Delta_{p}^{\max}) - \sum_{y \in \mathbb{Z}_{-}^{d}} \tau_{p,0}(f, y) \mathbb{P}_{p,\gamma}^{-}(e \leftrightarrow y, e \leftrightarrow \mathcal{G}) \right] . \qquad (9.4.19)$$

It is not difficult to show, using the BK-inequality and a union bound, that

$$\mathbb{P}_{p,\gamma}^{-}(e \leftrightarrow y, e \leftrightarrow \mathcal{G}) \leq \sum_{w \in \mathbb{Z}_{-}^{d}} \tau_{p,0}(e, w) \tau_{p,0}(w, y) M^{-}(w) , \qquad (9.4.20)$$

and hence, by (9.4.18)–(9.4.19),

$$\mathbb{P}_{p,\gamma}^{-}(W) \geq M(p, \gamma) \left[M^{-}(e)(1 - \Delta_{p}^{\max}) \right.$$
$$\left. - \sum_{y,w \in \mathbb{Z}_{-}^{d}} \tau_{p,0}(f, y) \tau_{p,0}(e, w) \tau_{p,0}(w, y) M^{-}(w) \right]$$
$$\geq M^{2}(p, \gamma) \left[(1 - \Delta_{p}^{\max})^{2} - \Delta_{p}^{\max} \right]. \qquad (9.4.21)$$

This completes the proof of (9.4.11), and hence of (9.4.12).

Upper bound on X_2. This is the easiest term. By definition,

$$X_2 = p \sum_{(u,v)} \tau_{p,\gamma}(0, u) \, \mathbb{E}_{p,\gamma} \left[\mathbb{1}_{\{v \Leftrightarrow \mathcal{G}\}} - \tilde{I}^{\{u,v\}}[v \Leftrightarrow \mathcal{G}] \right] . \qquad (9.4.22)$$

For the difference of indicators to be nonzero, the double connection from v to $\mathcal{G}$ must be realized via the bond $\{u, v\}$, which therefore must be occupied. The difference of indicators is therefore bounded above by the indicator that the events $\{v \leftrightarrow \mathcal{G}\}$, $\{u \leftrightarrow \mathcal{G}\}$, and $\{\{u, v\} \text{ occupied}\}$ occur disjointly. Thus, by the BK inequality,

$$\mathbb{E}_{p,\gamma}\big[\mathbb{1}_{\{v\Leftrightarrow\mathcal{G}\}} - \tilde{I}^{\{u,v\}}[v \Leftrightarrow \mathcal{G}]\big] \le pM^2(p,\gamma)\,, \tag{9.4.23}$$

and hence,

$$X_2 \le 2dp^2 M^2(p,\gamma)\chi(p,\gamma)\,. \tag{9.4.24}$$

Upper bound on X_3. By definition,

$$X_3 = p \sum_{(u,v)} \mathbb{E}_{p,\gamma}\Big[\big(\tau_{p,\gamma}(0,u) - \tau_{p,\gamma}^{\widetilde{\mathcal{C}}^{(u,v)}(v)}(0,u)\big)\tilde{I}^{\{u,v\}}[v \Leftrightarrow \mathcal{G}]\Big]. \tag{9.4.25}$$

The difference of two-point functions is the expectation of

$$\mathbb{1}_{\{0\leftrightarrow u,0\not\leftrightarrow\mathcal{G}\}} - \mathbb{1}_{\{0\leftrightarrow u \text{ in } \mathbb{Z}^d\backslash\widetilde{\mathcal{C}}^{(u,v)}(v),0\not\leftrightarrow\mathcal{G}\}}$$

$$+ \mathbb{1}_{\{0\leftrightarrow u \text{ in } \mathbb{Z}^d\backslash\widetilde{\mathcal{C}}^{(u,v)}(v),0\not\leftrightarrow\mathcal{G}\}} - \mathbb{1}_{\{(0\leftrightarrow u,0\not\leftrightarrow\mathcal{G}) \text{ in } \mathbb{Z}^d\backslash\widetilde{\mathcal{C}}^{(u,v)}(v)\}}$$

$$\le \mathbb{1}_{\{0\xleftarrow{\widetilde{\mathcal{C}}^{(u,v)}(v)}u,0\not\leftrightarrow\mathcal{G}\}}, \tag{9.4.26}$$

since the second line is nonpositive and the first line equals the third line. Since the indicator in (9.4.25) is bounded above by $\mathbb{1}_{\{v\Leftrightarrow\mathcal{G}\}}$, it follows that

$$X_3 \le p \sum_{(u,v)} \mathbb{E}_{p,\gamma}\Big[\mathbb{P}_{p,\gamma}(0\xleftarrow{\widetilde{\mathcal{C}}^{(u,v)}(v)}u,0\not\leftrightarrow\mathcal{G})\mathbb{1}_{\{v\Leftrightarrow\mathcal{G}\}}\Big]. \tag{9.4.27}$$

To investigate $\mathbb{P}_{p,\gamma}(0\xleftarrow{\widetilde{\mathcal{C}}^{(u,v)}(v)}u,0\not\leftrightarrow\mathcal{G})$, we state a more general lemma investigating disjoint occurrence of events while being green-free.

Let E be an event specifying that finitely many pairs of sites are connected, possibly disjointly. In particular, E is increasing. We say that E *occurs and is green-free* if E occurs and the clusters of all the sites for which connections are specified in its definition do not intersect the random set $\mathcal{G}$ of green sites. For such events, Hara and Slade [139, Lem. 4.3] prove the following BK inequality, in which the upper bound retains a green-free condition on one part of the event only:

Lemma 9.7 (Green-free BK inequality [139, Lem. 4.3]). *Let E_1, E_2 be events of the above type. Then, for $\gamma \ge 0$ and $p \in [0,1]$,*

$$\mathbb{P}_{p,\gamma}\big((E_1 \circ E_2) \text{ occurs and is green-free}\big)$$
$$\le \mathbb{P}_{p,\gamma}(E_1 \text{ occurs and is green-free}) \,\mathbb{P}_p(E_2)\,. \tag{9.4.28}$$

Proof. We follow the original proof by Hara and Slade [139] almost verbatim. Given an increasing event F, we write $[F]_n$ to denote that F occurs in Λ_n. Then, we show that

$$\mathbb{P}_{p,\gamma}\big([(E_1 \circ E_2) \text{ occurs and is green-free}]_n\big)$$
$$\le \mathbb{P}_{p,\gamma}([E_1 \text{ occurs and is green-free}]_n) \,\mathbb{P}_p([E_2]_n)\,. \tag{9.4.29}$$

Lemma 9.7 follows by sending $n \to \infty$.

Given a bond-vertex configuration, we let $\mathcal{C}(\mathcal{G})_n$ denote the set of vertices that are connected to the green set in Λ_n. Conditioning on $\mathcal{C}(\mathcal{G})_n$ gives

$$\mathbb{P}_{p,\gamma}([(E_1 \circ E_2) \text{ occurs and is green-free}]_n)$$
$$= \sum_S \mathbb{P}_{p,\gamma}\big(\mathcal{C}(\mathcal{G})_n = S, [(E_1 \circ E_2) \text{ occurs and is green-free}]_n\big), \quad (9.4.30)$$

where the sum is over all subsets $S \subseteq \Lambda_n$. When $\mathcal{C}(\mathcal{G})_n = S$, bonds in the edge boundary of S are vacant and we can replace the event $[(E_1 \circ E_2) \text{ occurs and is green-free}]_n$ by the event $[(E_1 \circ E_2) \text{ occurs in } \mathbb{Z}^d \setminus S]_n$. It is here that we use the specific form of the events E_1 and E_2 of finite intersections of (possibly disjoint occurrence of) connections between pairs of vertices. Thus,

$$\mathbb{P}_{p,\gamma}([(E_1 \circ E_2) \text{ occurs and is green-free}]_n)$$
$$= \sum_S \mathbb{P}_{p,\gamma}(\mathcal{C}(\mathcal{G})_n = S)\, \mathbb{P}_{p,\gamma}\big([(E_1 \circ E_2) \text{ occurs in } \mathbb{Z}^d \setminus S]_n\big). \quad (9.4.31)$$

Now, we apply the usual BK inequality (in the reduced lattice consisting of all the vertices in Λ_n that do not touch S) to obtain

$$\mathbb{P}_{p,\gamma}([(E_1 \circ E_2) \text{ occurs and is green-free}]_n)$$
$$\leq \sum_S \mathbb{P}_{p,\gamma}(\mathcal{C}(\mathcal{G})_n = S)\, \mathbb{P}_{p,\gamma}([E_1 \text{ occurs in } \mathbb{Z}^d \setminus S]_n)$$
$$\times \mathbb{P}_{p,\gamma}([E_2 \text{ occurs in } \mathbb{Z}^d \setminus S]_n). \quad (9.4.32)$$

Since E_2 is increasing, $\mathbb{P}_{p,\gamma}([E_2 \text{ occurs in } \mathbb{Z}^d \setminus S]_n) \leq \mathbb{P}_p([E_2]_n)$, so that

$$\mathbb{P}_{p,\gamma}([(E_1 \circ E_2) \text{ occurs and is green-free}]_n)$$
$$\leq \sum_S \mathbb{P}_{p,\gamma}(\mathcal{C}(\mathcal{G})_n = S)\, \mathbb{P}_{p,\gamma}([E_1 \text{ occurs in } \mathbb{Z}^d \setminus S]_n)\, \mathbb{P}_p([E_2]_n)$$
$$= \mathbb{P}_{p,\gamma}([E_1 \text{ occurs and is green-free}]_n)\, \mathbb{P}_p([E_2]_n), \quad (9.4.33)$$

as required. $\square$

By Lem. 9.7,

$$\mathbb{P}_{p,\gamma}(0 \overset{A}{\leftrightarrow} u, 0 \nleftrightarrow \mathcal{G})$$
$$\leq \sum_{y \in A} \mathbb{P}_{p,\gamma}\big((0 \leftrightarrow y) \circ (y \leftrightarrow u) \text{ occurs and is green-free}\big)$$
$$\leq \sum_{y \in \mathbb{Z}^d} \tau_{p,\gamma}(0, y)\tau_{p,0}(y, u)\mathbb{1}_{\{y \in A\}}. \quad (9.4.34)$$

The important point in (9.4.34) is that the condition $0 \nleftrightarrow \mathcal{G}$ on the left side is retained in the factor $\tau_{p,\gamma}(0, y)$ on the right side (but not in $\tau_{p,0}(y, u)$). With (9.4.27), this gives

$$X_3 \leq p \sum_{(u,v)} \sum_{y \in \mathbb{Z}^d} \tau_{p,\gamma}(0, y) \tau_{p,0}(y, u) \, \mathbb{E}_{p,\gamma} \left[\mathbb{1}_{\{v \Leftrightarrow \mathcal{G}\}} \mathbb{1}_{\{y \in \widetilde{\mathcal{C}}^{(u,v)}(v)\}} \right] . \tag{9.4.35}$$

Since

$$\mathbb{1}_{\{v \Leftrightarrow \mathcal{G}\}} \mathbb{1}_{\{y \in \widetilde{\mathcal{C}}^{(u,v)}(v)\}} \leq \sum_{w \in \mathbb{Z}^d} \mathbb{1}_{\{v \leftrightarrow w\} \circ \{w \leftrightarrow y\} \circ \{w \leftrightarrow \mathcal{G}\} \circ \{v \leftrightarrow \mathcal{G}\}}, \tag{9.4.36}$$

a further application of BK gives

$$X_3 \leq p \sum_{y \in \mathbb{Z}^d} \tau_{p,\gamma}(0, y) \sum_{(u,v)} \tau_{p,0}(y, u) \sum_{w \in \mathbb{Z}^d} \tau_{p,0}(v, w) \tau_{p,0}(y, w) M^2(p, \gamma)$$

$$= p M^2(p, \gamma) \chi(p, \gamma) \sum_{(0,e)} \tilde{\Delta}_p(e) \leq 2dp \Delta_p^{\max} M^2(p, \gamma) \chi(p, \gamma) , \tag{9.4.37}$$

where we have used that $\tilde{\Delta}_p(e) \leq \Delta_p^{\max}$ for every $(0, e)$ in the last step.

The combination of (9.4.12), (9.4.24), and (9.4.37) completes the proof of (9.1.5). $\square$

Part III
Mean-Field Behavior: Recent Results

Chapter 10
The Nonbacktracking Lace Expansion

In the previous chapter, we have concluded the argument that percolation has mean-field behavior in sufficiently high dimensions. The analysis, however, is not very explicit about what dimension suffices in order to make the analysis work. In this chapter, we explain that $d \geq 11$ suffices.

This chapter is organized as follows. In Sect. 10.1, we state the main results for $d \geq 11$. In Sect. 10.2, we introduce nonbacktracking random walk and explain that this yields a better approximation to the percolation two-point function than the random walk Green's function. In Sect. 10.3, we explain the nonbacktracking lace expansion, which makes this perturbation statement precise. In Sect. 10.4, we show how the bootstrap argument in Chap. 8 can be performed in the new setting. The proof in this chapter is *computer assisted*, and we close this chapter in Sect. 10.5 by discussing the numerical aspects of the analysis.

10.1 Mean-Field Behavior for $d \geq 11$

Hara and Slade in [137] wrote that $d \geq 19$ is sufficient to carry out the lace expansion as outlined in Chaps. 6–8, though the actual calculations leading to this number have never been published. This magic number 19 circulated in the community for many years, and few people understood how it came about. We would like to emphasize that the dimension $d = 19$ has no special meaning at all, except for the fact that the numerical arguments, at the time, worked for $d \geq 19$. The reason is that the lace-expansion analysis, as presented in Chaps. 6–8, follows a perturbation approach: The two-point function $\tau_{p_c}(x)$ is compared to the critical random walk Green's function $C_1(x)$. This approach simply fails for percolation when the triangle diagram $\triangle_{p_c}$ is too large. On the other hand, Thm. 4.1 implies that percolation displays mean-field behavior in any dimension for which the triangle diagram $\triangle_{p_c}$ is *finite*. Thus, the main problem in establishing mean-field behavior for all $d > 6$ is as follows:

© Springer International Publishing Switzerland 2017
M. Heydenreich and R. van der Hofstad, *Progress in High-Dimensional Percolation and Random Graphs*, CRM Short Courses, DOI 10.1007/978-3-319-62473-0_10

> We know how to prove that the triangle diagram $\triangle_{p_c}$ is *small* in large dimensions, but we do not know how to prove that it is *finite*.

We now present results obtained by Fitzner and the second author [107, 108] on the minimal dimension above which percolation can be rigorously proved to display mean-field behavior. This not only improves the earlier $d \geq 19$ to $d \geq 11$, but also rigorously justifies that 11 is sufficient for the lace-expansion analysis. A key to this improvement is based on a new perturbation ansatz: Instead of comparing the percolation two-point function to simple random walk, a comparison is made to *nonbacktracking random walk* yielding a nonbacktracking lace expansion (as explained in more detail below). Here is the result, which improves Thm. 5.1:

Theorem 10.1 (Infrared bound [107, 108]). *For percolation with $d \geq 11$, there exists a constant $A(d)$ (the "amplitude") such that*

$$\hat{\tau}_p(k) \leq \frac{A(d)}{1 - \widehat{D}(k)} \,, \tag{10.1.1}$$

uniformly for $p \leq p_c$.

The infrared bound of course also immediately implies that $\beta = \gamma = 1$ and $\delta = 2$, so that Thm. 10.1 shows the mean-field behavior of percolation for $d \geq 11$. Below, we also state some other consequences of Thm. 10.1.

The method in [107, 108] requires a detailed analysis of both the critical value and the amplitude $A(d)$. As a side result, the following bounds are also obtained:

Theorem 10.2 (Bounds on critical value and amplitude [107, 108]). *For percolation with $d \geq 11$, the upper bounds from Table 10.1 hold.*

We know that $(2d - 1)p_c(\mathbb{Z}^d) \geq 1$ (recall Exerc. 2.6). Therefore, the bound on $p_c(\mathbb{Z}^{11})$ in Thm. 10.2 is at most 1.32% off. The bounds on $p_c(\mathbb{Z}^d)$ are probably not optimal. In Table 10.2 are numerical estimates (nonrigorous, obtained via simulation) that have appeared in the literature.

We now explain in some more detail how Thms. 10.1–10.2 are proved. The proof for $d \geq 19$ by Hara and Slade [132] that we mentioned before relied on numerical computations like those used in their seminal paper [134] showing that five-dimensional self-avoiding walk (SAW) is diffusive. The result for SAW is *optimal*, in the sense that it is expected to be *false* in $d \leq 4$. See, e.g., the recent work by Bauerschmidt, Brydges, and Slade in [32] and the references therein for results in dimension $d = 4$.

For percolation, the upper critical dimension is believed to be $d_c = 6$, and therefore, Thm. 10.1 is expected to hold for all $d > 6$. However, the numerical control over the lace expansion, as performed in Chap. 6, was insufficient to prove Thm. 10.1 all the way down to $d \geq d_c + 1 = 7$. The Hara–Slade methods *did* prove mean-field behavior for $d \geq 19$, already a very impressive result. In private communication with Hara, we have learned that

Table 10.1 Numerical estimates of the critical percolation threshold $p_c(\mathbb{Z}^d)$ and the amplitude $A(d)$ for several d.

d	$(2d-1)p_c(\mathbb{Z}^d)$	$A(d)$
11	≤ 1.01315	≤ 1.02476
12	≤ 1.00861	≤ 0.995
13	≤ 1.006268	≤ 0.986
14	≤ 1.0048522	≤ 0.98243
15	≤ 1.00391	≤ 0.98088
20	≤ 1.00179	≤ 0.98115

Table 10.2 Numerical values of $p_c(\mathbb{Z}^d)$ taken from Grassberger [120, Table I]. Related numerical values can be found in [176, Table 3.6] and [6].

d	$p_c(\mathbb{Z}^d)$	$(2d-1)p_c(\mathbb{Z}^d)$
7	≈ 0.0786752	≈ 1.02278
8	≈ 0.06770839	≈ 1.01562585
9	≈ 0.05949601	≈ 1.01143217
10	≈ 0.05309258	≈ 1.00875902
11	≈ 0.04794969	≈ 1.00694349
12	≈ 0.04372386	≈ 1.00564878
13	≈ 0.04018762	≈ 1.0046905

he has been able to extend the methodology to $d \geq 15$, but that his method is restricted to $d \geq 15$, since it makes use of the finiteness of the heptagon (e.g., $\tau_{p_c}^{\star 7}(0)$). This analysis has not appeared in the literature.

The analysis of Fitzner and van der Hofstad in [107, 108] heavily relies on the methods of Hara and Slade in [132, 135], in particular to compute the random walk's Green's function numerically, as we explain in the sequel. It also contains novel ideas, particularly since it relies on a perturbation around nonbacktracking walk rather than simple random walk. We next define nonbacktracking random walk.

10.2 Nonbacktracking Walk

In this section, we introduce nonbacktracking random walk and relate its Green's function to that of simple random walk. We follow [107, Sect. 1.2.2] closely. We start by introducing some notation. An n-step path $\pi = (\pi_0, \ldots, \pi_n) \in (\mathbb{Z}^d)^n$ is a simple random walk (SRW) path when $|\pi_i - \pi_{i-1}| = 1$, $i = 1, \ldots, n$. Let $b_n^{\text{SRW}}(x)$ denote the number of n-step SRW paths starting in 0 and ending in $x \in \mathbb{Z}^d$, i.e., with $\pi_0 = 0$ and $\pi_n = x$. We exclusively use the Greek letters ι and κ for values in $\{-d, -d+1, \ldots, -1, 1, 2, \ldots, d\}$ and denote the unit vector in direction ι by $e_\iota \in \mathbb{Z}^d$, e.g., $(e_\iota)_i = \text{sign}(\iota)\delta_{|\iota|,i}$. Clearly, by summing over the direction $-e_\iota$ in the first step,

$$b_n^{\text{SRW}}(x) = \sum_{\iota \in \{\pm 1, \dots, \pm d\}} b_{n-1}^{\text{SRW}}(x + e_\iota) , \qquad (10.2.1)$$

from which we immediately obtain that

$$b_n^{\text{SRW}}(x) = (2d)^n D^{\star n}(x). \qquad (10.2.2)$$

If an n-step SRW path π satisfies $\pi_i \neq \pi_{i+2}$ for all $i = 0, 1, 2, \dots, n - 2$, then we call π a *nonbacktracking* path. In order to analyze nonbacktracking walk (NBW), we derive an equation similar to (10.2.1). The same equation does not hold for NBW as it neglects the condition that the walk does not revisit the origin after the second step. For this reason, we introduce the condition that a walk should not go in a certain direction ι with its first step.

Let $b_n(x)$ be the number of n-step NBW paths with $\pi_0 = 0, \pi_n = x$. Further, we define $b_n^\iota(x)$ as the number of n-step NBW paths π with $\pi_n = x$ and $\pi_1 \neq e_\iota$. Summing over the direction $-e_\iota$ of the first step we obtain, for $n \geq 1$,

$$b_n(x) = \sum_{\iota \in \{\pm 1, \dots, \pm d\}} b_{n-1}^\iota(x + e_\iota) . \qquad (10.2.3)$$

Further, we distinguish between the case that the walk visits $-e_\iota$ in the first step or not to obtain, for $n \geq 1$,

$$b_n(x) = b_n^{-\iota}(x) + b_{n-1}^\iota(x + e_\iota) . \qquad (10.2.4)$$

The NBW two-point functions B_z and B_z^ι are defined as the generating function of b_n and b_n^ι, respectively, i.e., for every $z \in \mathbb{C}$ for which the generating function converges,

$$B_z(x) = \sum_{n=0}^{\infty} b_n(x) z^n , \qquad B_z^\iota(x) = \sum_{n=0}^{\infty} b_n^\iota(x) z^n . \qquad (10.2.5)$$

We later come back to the question for which $z \in \mathbb{C}$ the definitions in (10.2.5) actually do make sense. Using (10.2.3) and (10.2.4) for the two-point functions gives

$$B_z(x) = \delta_{0,x} + z \sum_{\iota \in \{\pm 1, \dots, \pm d\}} B_z^\iota(x + e_\iota) ,$$
$$B_z(x) = B_z^{-\iota}(x) + z B_z^\iota(x + e_\iota) . \qquad (10.2.6)$$

Taking the Fourier transform, we obtain

$$\widehat{B}_z(k) = 1 + z \sum_{\iota \in \{\pm 1, \dots, \pm d\}} \widehat{B}_z^\iota(k) e^{-ik_\iota} ,$$
$$\widehat{B}_z(k) = \widehat{B}_z^{-\iota}(x) + z \widehat{B}_z^\iota(k) e^{-ik_\iota} . \qquad (10.2.7)$$

Here, for a clear distinction between scalar-, vector-, and matrix-valued quantities, we always write $\mathbb{C}^{2d}$-valued functions with a vector arrow (e.g., $\vec{v}$) and matrix-valued functions with bold capital letters (e.g., **M**). We use $\{-d, -d+1, \ldots, -1, 1, 2, \ldots, d\}$ for the index set of the elements of a vector or a matrix. Further, for $k \in (-\pi, \pi]^d$ and negative index $\iota \in \{-d, -d+1, \ldots, -1\}$, we write $k_\iota = -k_{|\iota|}$.

In the following lemma, we relate the NBW and SRW two-point or Green's functions to each other:

Lemma 10.3 (NBW Green's functions). *The NBW and SRW Green's functions are related by*

$$
\widehat{B}_z(k) = \frac{1 - z^2}{1 + (2d-1)z^2} \frac{1}{1 - 2dz/(1 + (2d-1)z^2)\widehat{D}(k)}
$$

$$
= \frac{1 - z^2}{1 + (2d-1)z^2} \widehat{C}_{2dz/(1+(2d-1)z^2)}(k) , \tag{10.2.8}
$$

so that $\widehat{B}_z(k)$ converges as long as $|z| < 1/(2d-1)$, and for $z = z_c = 1/(2d-1)$ for $k \neq 0$. Further,

$$
\widehat{B}_{1/(2d-1)}(k) = \frac{2d-2}{2d-1} \widehat{C}_1(k) = \frac{2d-2}{2d-1} \cdot \frac{1}{1 - \widehat{D}(k)} . \tag{10.2.9}
$$

The link between the NBW and the SRW Green's functions allows for a computation of NBW two-point function values in x- and in k-space. This is a crucial ingredient behind the proof in [107, 108]. These ideas are also applied to lattice trees and animals. A detailed analysis of the NBW, based on the ideas behind Lem. 10.3, can be found in [109].

Note that the representation in Lem. 10.3 shows that working with NBW has *two* benefits. Foremost, the critical point of the NBW is $1/(2d-1)$ rather than that for the generating function of the SRW sequence $b_n^{\mathrm{SRW}}(x)$, which is $1/(2d)$. Secondly, we see that, at the critical point, the NBW Green's function is a factor $(2d-2)/(2d-1)$ smaller than its SRW counterpart. When we wish to go to dimensions that are not too large, such factors can be quite useful, and we make as much use of them as possible. The extra factor $(2d-2)/(2d-1)$ also partly explains the surprising nonmonotonicity of the bounds on $A(d)$ in Table 10.1.

Lemma 10.3 allows to use the methodology of Hara and Slade in [132, 135], which is a crucial ingredient in the proof in [107, 108]. In particular, Hara and Slade develop a rigorous numerical scheme to bound simple random walk integrals such as

$$
I_{n,l}(x) = (D^{\star l} \star C_1^{\star n})(x) = \int_{(-\pi,\pi]^d} e^{-ik \cdot x} \frac{\widehat{D}(k)^l}{[1 - \widehat{D}(k)]^n} \frac{dk}{(2\pi)^d} . \tag{10.2.10}
$$

These integrals are used to bound similar integrals, but then for the percolation two-point function $\tau_p(x)$, using a bootstrap argument. We can see that $I_{n,l}(x)$ for $n = 3$ is closely related to a simple random walk triangle diagram.

Proof of Lem. 10.3. It is most convenient to prove the lemma using matrix notation. We denote the identity matrix by $\mathbf{I} \in \mathbb{C}^{2d \times 2d}$ and the all-one vector by $\vec{1}^T = (1, 1, \ldots, 1)^T \in \mathbb{C}^{2d}$. All our vectors are column vectors. Moreover, we define the matrices $\mathbf{J}, \widehat{\mathbf{D}}(k) \in \mathbb{C}^{2d \times 2d}$ by

$$(\mathbf{J})_{\iota,\kappa} = \delta_{\iota,-\kappa} \quad \text{and} \quad (\widehat{\mathbf{D}}(k))_{\iota,\kappa} = \delta_{\iota,\kappa} e^{ik_\iota} , \tag{10.2.11}$$

so that

$$\mathbf{J} = \begin{pmatrix} & & 1 \\ & \cdot\cdot & \\ 1 & & \end{pmatrix}, \quad \widehat{\mathbf{D}}(k) = \begin{pmatrix} e^{ik_d} & & \\ & \cdot\cdot & \\ & & e^{ik_{-d}} \end{pmatrix} \tag{10.2.12}$$

provided that the indices are decreasingly ordered. The following identities are valid for these matrices:

$$\widehat{\mathbf{D}}(k)\mathbf{J} = \mathbf{J}\widehat{\mathbf{D}}(-k), \quad \left(\widehat{\mathbf{D}}(k)\right)^{-1} = \widehat{\mathbf{D}}(-k), \quad \text{and} \quad \mathbf{J}\mathbf{J} = \mathbf{I} . \tag{10.2.13}$$

Exercise 10.1 (Matrix inverses for NBW). Prove (10.2.13).

We define the vector $\overrightarrow{B}_z(k)$ with entries $\left(\overrightarrow{B}_z(k)\right)_\iota = \overrightarrow{B}_z^\iota(k)$ and rewrite (10.2.7) as

$$\widehat{B}_z(k) = 1 + z\vec{1}^T\widehat{\mathbf{D}}(-k)\overrightarrow{B}_z(k) ,$$
$$\widehat{B}_z(k)\vec{1} = \mathbf{J}\overrightarrow{B}_z(k) + z\widehat{\mathbf{D}}(-k)\overrightarrow{B}_z(k) . \tag{10.2.14}$$

We use $\mathbf{J}\mathbf{J} = \mathbf{I}$ and $\widehat{\mathbf{D}}(k)\widehat{\mathbf{D}}(-k) = \mathbf{I}$ to rewrite the last equation as

$$\widehat{B}_z(k)\vec{1} = \mathbf{J}\widehat{\mathbf{D}}(k)\widehat{\mathbf{D}}(-k)\overrightarrow{B}_z(k) + z\mathbf{J}\mathbf{J}\widehat{\mathbf{D}}(-k)\overrightarrow{B}_z(k)$$
$$= \mathbf{J}(\widehat{\mathbf{D}}(k) + z\mathbf{J})\widehat{\mathbf{D}}(-k)\overrightarrow{B}_z(k) . \tag{10.2.15}$$

Using (10.2.13) again, this implies

$$\overrightarrow{B}_z(k) = \widehat{B}_z(k)\left[\mathbf{I} + z\widehat{\mathbf{D}}(k)\mathbf{J}\right]^{-1}\mathbf{J}\vec{1} . \tag{10.2.16}$$

We use $\mathbf{J}\vec{1} = \vec{1}$ and then combine (10.2.16) with the first equation in (10.2.14) to obtain

$$\widehat{B}_z(k) = \frac{1}{1 - z\vec{1}^T\left[\widehat{\mathbf{D}}(k) + z\mathbf{J}\right]^{-1}\vec{1}} . \tag{10.2.17}$$

We use again (10.2.13) to compute that

$$\left[\widehat{\mathbf{D}}(k) + z\mathbf{J}\right]^{-1} = \frac{1}{1 - z^2}\left(\widehat{\mathbf{D}}(-k) - z\mathbf{J}\right) , \tag{10.2.18}$$

so that, using $\vec{1}^T \widehat{\mathbf{D}}(-k)\vec{1} = 2d\,\widehat{D}(k)$, we conclude that

$$\widehat{B}_z(k) = \frac{1}{1 - 2dz\big(\widehat{D}(k) - z\big)/(1 - z^2)}$$

$$= \frac{1 - z^2}{1 + (2d - 1)z^2 - 2dz\,\widehat{D}(k)} . \qquad (10.2.19)$$

Lemma. 10.3 follows directly from (10.2.19). □

We next explain that the percolation two-point function satisfies an equation similar to (10.2.17).

10.3 The Nonbacktracking Lace Expansion (NoBLE)

In this section, we explain the shape of the Nonbacktracking Lace Expansion (NoBLE), which is a perturbative expansion of the two-point function. Next to the usual two-point function $\tau_p(x)$, we use a slight adaptation of it. For a direction $\iota \in \{\pm 1, \pm 2, \dots, \pm d\}$, we define

$$\tau_p^\iota(x) = \mathbb{P}_p(0 \leftrightarrow x \text{ occurs when all bonds containing } e_\iota \text{ are made vacant}). \qquad (10.3.1)$$

In the following, we drop the subscript p when possible, and write, e.g., $\tau(x) = \tau_p(x)$ and $\tau^\iota(x) = \tau_p^\iota(x)$.

The analysis in [107, 108] relies on two expansion identities relating $\tau_p(x)$ and $\tau_p^\iota(x)$, which are perturbations of (10.2.6). More precisely, the NoBLE is a perturbative expansion for $\tau_p(x)$ and $\tau_p^\iota(x)$ that reads as follows: For every $x \in \mathbb{Z}^d$, $\iota \in \{\pm 1, \pm 2, \dots, \pm d\}$, and $M \geq 1$, the following recursion relations hold for $p < p_c$:

$$\tau(x) = \delta_{0,x} + \mu_p \sum_{y \in \mathbb{Z}^d, \kappa} (\delta_{0,y} + \Psi_M^\kappa(y))\tau^\kappa(x - y + e_\kappa) + \varXi_M(x) , \qquad (10.3.2)$$

$$\tau(x) = \tau^\iota(x) + \mu_p \tau^{-\iota}(x - e_\iota)$$
$$+ \sum_{y \in \mathbb{Z}^d, \kappa} \Pi_M^{\iota,\kappa}(y)\tau^\kappa(x - y + e_\kappa) + \varXi_M^\iota(x) , \qquad (10.3.3)$$

where the sum over κ is over $\{\pm 1, \dots, \pm d\}$, and we define

$$\Pi_M^{\iota,\kappa}(y) = \sum_{N=0}^M (-1)^N \Pi^{(N),\iota,\kappa}(y) ,$$

$$\varXi_M(x) = R_M(x) + \sum_{N=0}^M (-1)^N \varXi^{(N)}(x) , \qquad (10.3.4)$$

$$\Psi_M^\kappa(x) = \sum_{N=0}^{M} (-1)^N \psi^{(N),\kappa}(y) \, ,$$

$$\Xi_M^\iota(x) = R_M^\iota(x) + \sum_{N=0}^{M} (-1)^N \Xi^{(N),\iota}(x) \, , \tag{10.3.5}$$

$$\mu_p = p\, \mathbb{P}(e_1 \notin \mathcal{C}(0) \mid \{0, e_1\} \text{ vacant}) \, . \tag{10.3.6}$$

Let us informally describe how these lace-expansion relations come about. We aim to perturb around the nonbacktracking walk, for which the Green's function satisfies (10.2.7). Here, (10.3.2) is a perturbation of the first identity in (10.2.7), while (10.3.3) is a perturbation of the second identity in (10.2.7). We again rely on descriptions using pivotal bonds.

Informal explanation of (10.3.2). The starting point for (10.3.2) is (6.2.18), combined with (6.2.8), which, for convenience, we restate as

$$\mathbb{P}(0 \leftrightarrow x) = \delta_{0,x} + \Pi^{(0)}(x) + \sum_{(u,v)} J(u, v)\, \mathbb{E}\left(\mathbb{1}_{\{0 \Longleftrightarrow u\}}\, \tau^{\widetilde{\mathcal{C}}^{(u,v)}(0)}(v, x)\right). \tag{10.3.7}$$

The main change with respect to the expansion in Chap. 6 is that for any A with $u \in A$ and $u \sim v$

$$\tau^A(v, x) = \tau^u(v, x) - \left(\tau^u(v, x) - \tau^A(v, x)\right)$$

$$= \mathbb{1}_{\{v \notin \widetilde{\mathcal{C}}^{(u,v)}(0)\}}\left[\tau^u(x - v) - \mathbb{P}\left(\left\{v \overset{A}{\leftrightarrow} x\right\} \text{ in } \mathbb{Z}^d \setminus \{u\}\right)\right] \tag{10.3.8}$$

where $\tau^u = \tau^{\{u\}}$, and this identity is used everywhere where we arrive at a restricted two-point function.

Exercise 10.2 (Percolation inclusion–exclusion revisited). Prove (10.3.8).

By definiton, $\widetilde{\mathcal{C}}^{(u,v)}(0)$ contains u but not v, and therefore, we can apply (10.3.8) for $A = \widetilde{\mathcal{C}}^{(u,v)}(0)$ and get

$$\tau(x) = \delta_{0,x} + \Pi^{(0)}(x) + \sum_{(u,v)} \left(\delta_{0,u} + \Pi^{(0)}(u)\right) J(u, v)\tau^u(v, x)$$

$$- \sum_{(u,v)} J(u, v)\, \mathbb{E}_0\left(\mathbb{1}_{\{0 \Longleftrightarrow u\}}\, \mathbb{P}_1\left(\left\{v \xrightarrow{\widetilde{\mathcal{C}}_0^{(u,v)}(0)} x\right\} \text{in } \mathbb{Z}^d \setminus \{u\}\right)\right). \tag{10.3.9}$$

Note that $\tau^u(v, x) = \tau^\iota(x - v)$ whenever $u - v = e_\iota$. Therefore, the second term in (10.3.9) is precisely of the form of the second term in (10.3.2).

Repeating the above ideas, in a similar way as in Chap. 6, but now relying on (10.3.8) instead of (6.2.19), leads to (10.3.2). This completes our informal explanation of (10.3.2). We omit further details.

Informal explanation of (10.3.3). Fix $x \in \mathbb{Z}^d$ and $\iota \in \{\pm 1, \pm 2, \ldots, \pm d\}$. We note that

$$\tau(x) = \tau^\iota(x) + \mathbb{P}\left(0 \overset{e_\iota}{\longleftrightarrow} x\right), \tag{10.3.10}$$

where we recall that $0 \overset{e_\iota}{\longleftrightarrow} x$ denotes that all occupied paths from 0 to x pass through e_ι. The second contribution can arise in two ways. Either the bond $(0, e_\iota)$ is occupied and pivotal for the connection from 0 to x, or not. This leads to

$$\tau(x) = \tau^\iota(x) + \mathbb{P}((0, e_\iota) \text{ occ. and piv. for } 0 \leftrightarrow x)$$
$$+ \mathbb{P}\left(0 \overset{e_\iota}{\longleftrightarrow} x, (0, e_\iota) \text{ not piv. for } 0 \leftrightarrow x\right). \tag{10.3.11}$$

In the second term, we have found our pivotal bond and can repeat as for (10.3.2), again each time using (10.3.8) instead of (6.2.19). In the third term in (10.3.11), we look for a cutting bond b that is (1) pivotal for $0 \leftrightarrow x$; and (2) is such that all other requirements, i.e., $0 \overset{e_\iota}{\longleftrightarrow} x$ and $(0, e_\iota)$ not being pivotal for $0 \leftrightarrow x$, occur before the cutting bond b. The main contribution to this event arises when b is incident to e_ι, but is different from $(0, e_\iota)$. It is in this term that the contribution $\mu_p = p \, \mathbb{P}(e_1 \notin \mathcal{C}(0) \mid \{0, e_1\} \text{ vacant})$ originates. We refrain from giving more details.

Lace-expansion diagrams consist of percolation paths or two-point functions that satisfy several self-intersection restrictions. For example, $\widehat{\Pi}_p^{(0)}(0) = \sum_x \mathbb{P}_p(0 \Leftrightarrow x)$ enforces a loop in the percolation configuration consisting of two disjoint paths connecting 0 and x. Similarly, when we bound $\widehat{\Pi}_p^{(0)}(0) \leq \sum_x \tau_p(x)^2$, we can display this as two lines connecting 0 and x, where now the lines correspond to two-point functions. We conclude that the lace-expansion coefficients enforce the existence of several loops in configurations, or in their two-point function upper bounds.

The effect of the change of the NoBLE compared to the classical lace expansion is that loops that are present in the lace-expansion coefficients are now suddenly all loops of size at least 4. Since loops of size at least 4 decay like $O(1/d^2)$ rather than $O(1/d)$ as loops of length at least two, this is a serious improvement. For example, the largest contribution to $\widehat{\Pi}_{p_c}(0)$, which is from $\widehat{\Pi}_{p_c}^{(1)}(0)$, is $O(1/d)$ for the classical lace expansion as derived in Chap. 6. In the NoBLE it is $O(1/d^2)$:

Exercise 10.3 (Main term lace-expansion coefficient). Verify that, for the classical lace expansion, $\widehat{\Pi}_{p_c}^{(1)}(0) = O(1/d)$.

In general, the philosophy in high dimensions is that when bounds become better in very high dimension, they probably also improve in slightly lower dimensions. Thus, the effect of the change of the NoBLE compared to the classical lace expansion can be expected to lead to (slightly) improved numerical bounds also when d is not too large. It is this effect that allows for a decrease in the numerical dimension above which the expansion works for the NoBLE compared to the classical lace expansion.

Using the NoBLE expansion of (10.3.2)–(10.3.3), we now discuss the NoBLE equation
that stands at the heart of the NoBLE analysis. The NoBLE equation rewrites $\hat{\tau}(k)$ in a form
that is a perturbation of (10.2.17). We take the Fourier transforms of (10.3.2) and (10.3.3) to
obtain

$$\hat{\tau}(k) = 1 + \widehat{\Xi}_M(k) + \mu_p \sum_\kappa \left(1 + \hat{\psi}_M^\kappa(k)\right)e^{-ik_\kappa}\hat{\tau}^\kappa(k) , \tag{10.3.12}$$

$$\hat{\tau}(k) = \hat{\tau}^\iota(k) + \mu_p e^{-ik_\iota}\hat{\tau}^{-\iota}(k) + \sum_\kappa \hat{\Pi}_M^{\iota,\kappa}(k)e^{-ik_\kappa}\hat{\tau}^\kappa(k) + \widehat{\Xi}_M^\iota(k) . \tag{10.3.13}$$

We write $\vec{\hat{\tau}}(k) \in \mathbb{R}^{2d}$ for the (column-)vector with entries

$$\left(\vec{\hat{\tau}}(k)\right)_\iota = \left(\hat{\tau}^\iota(k)\right) . \tag{10.3.14}$$

and note that, by $\widehat{\mathbf{D}}(k)\mathbf{J} = \mathbf{J}\widehat{\mathbf{D}}(-k)$ (recall Exerc. 10.1),

$$e^{-ik_\iota}\hat{\tau}^{-\iota}(k) = \left(\mathbf{J}\widehat{\mathbf{D}}(k)\vec{\hat{\tau}}(k)\right)_\iota = \left(\widehat{\mathbf{D}}(-k)\mathbf{J}\vec{\hat{\tau}}(k)\right)_\iota . \tag{10.3.15}$$

Defining the vectors $\vec{\hat{\psi}}(k)$, $\vec{\widehat{\Xi}}_M(k)$, and the matrix $\widehat{\boldsymbol{\Pi}}_M(k)$, with entries

$$\left(\vec{\hat{\psi}}(k)\right)_\kappa = \hat{\psi}^\kappa(k) , \quad \left(\vec{\widehat{\Xi}}_M(k)\right)_\iota = \widehat{\Xi}_M^\iota(k) ,$$
$$\left(\widehat{\boldsymbol{\Pi}}_M(k)\right)_{\iota,\kappa} = \hat{\Pi}^{\iota,\kappa}(k) , \tag{10.3.16}$$

we can rewrite (10.3.13) as

$$\hat{\tau}(k)\vec{1} = \vec{\hat{\tau}}(k) + \mu_p\widehat{\mathbf{D}}(k)\mathbf{J}\vec{\hat{\tau}}(k) + \widehat{\boldsymbol{\Pi}}_M(k)\widehat{\mathbf{D}}(-k)\vec{\hat{\tau}}(k) + \vec{\widehat{\Xi}}_M(k) , \tag{10.3.17}$$

so that

$$\vec{\hat{\tau}}(k) = \widehat{\mathbf{D}}(k)\left[\widehat{\mathbf{D}}(k) + \mu_p\mathbf{J} + \widehat{\boldsymbol{\Pi}}_M(k)\right]^{-1}\left(\hat{\tau}(k)\vec{1} - \vec{\widehat{\Xi}}_M(k)\right) . \tag{10.3.18}$$

In turn, by (10.3.12), the above gives rise to the relation

$$\hat{\tau}(k) = 1 + \widehat{\Xi}_M(k) + \mu_p\left(\vec{1}^T + \vec{\hat{\psi}}_M^T(k)\right)\widehat{\mathbf{D}}(-k)\vec{\hat{\tau}}(k)$$
$$= 1 + \hat{\Phi}_{M,p}(k) + \hat{\tau}(k)\left(\vec{1}^T + \vec{\hat{\psi}}_M^T(k)\right)$$
$$\times \left[\widehat{\mathbf{D}}(k) + \mu_p\mathbf{J} + \widehat{\boldsymbol{\Pi}}_M(k)\right]^{-1}\vec{1} , \tag{10.3.19}$$

with

$$\hat{\Phi}_{M,p}(k) = \widehat{\Xi}_M(k)$$
$$+ \mu_p\left(\vec{1}^T + \vec{\Psi}_M^T(k)\right) \times \left[\widehat{D}(k) + p\mathbf{J} + \widehat{\Pi}_M(k)\right]^{-1}\vec{\Xi}_M(k) . \qquad (10.3.20)$$

Exercise 10.4 (Proof of (10.3.19)). Verify that (10.3.19) holds with $\hat{\Phi}_{M,p}(k)$ as in (10.3.20).

Thus, we can solve (10.3.19) as

$$\hat{\tau}(k) = \frac{1 + \hat{\Phi}_M(k)}{1 - \mu_p\left(\vec{1}^T + \vec{\Psi}_M^T(k)\right)\left[\widehat{D}(k) + \mu_p\mathbf{J} + \widehat{\Pi}_M(k)\right]^{-1}\vec{1}} . \qquad (10.3.21)$$

Equation (10.3.21) is the NoBLE equation and is the workhorse behind the proof of Thms. 10.1 and 10.2. The goal of the NoBLE is to show that (10.3.21) is indeed a small perturbation of (10.2.17). This amounts to proving that $\hat{\Phi}_M(k)$, $\vec{\Psi}_M^T(k)$, and $\widehat{\Pi}_M(k)$ are small, which is only true in sufficiently high dimensions. As in Chap. 8, we again rely on a bootstrap argument, now with different bootstrap functions. In the next section, we give some more details.

10.4 The NoBLE Bootstrap Argument

As customary for lace-expansion analyses, the NoBLE analysis in [108] uses a bootstrap argument, which we explain next. We define the following set of bootstrap functions:

$$f_1(p) := \max\{(2d-1)p, c_\mu(2d-1)\mu_p\} , \qquad (10.4.1)$$

$$f_2(p) := \sup_{k \in (-\pi,\pi]^d} \frac{\hat{\tau}_p(k)}{\vec{B}_{1/(2d-1)}(k)}$$

$$= \frac{2d-1}{2d-2} \sup_{k \in (-\pi,\pi]^d} [1 - \widehat{D}(k)]\hat{\tau}_p(k) , \qquad (10.4.2)$$

$$f_3(p) := \max_{(n,l,S) \in \mathcal{S}} \frac{\sup_{x \in S} \sum_y \|y\|_2^2 \tau_p(y)(\tau_p^{\star n} \star D^{\star l})(x-y)}{c_{n,l,S}} , \qquad (10.4.3)$$

and

$$f(p) := \max\{c_1 f_1(p), c_2 f_2(p), c_3 f_3(p)\} , \qquad (10.4.4)$$

where $c_i > 0, c_\mu > 1$, and $c_{n,l,S} > 0$ are some well-chosen constants, and $\mathcal{S}$ is some finite set of indices of the form (n, l, S) for a finite number of values of $n, l \in \mathbb{N}$ and a finite number of different sets $S \subseteq \mathbb{Z}^d$. The constants $c_\bullet$ are introduced because the individual terms in $f_1(p)$ and $f_3(p)$ are incomparable in size. The $c_\bullet$ are chosen such that all terms contributing to $f_1(p)$ and $f_3(p)$ are of roughly the same size.

The choice of sets $S \subseteq \mathbb{Z}^d$ significantly improves the numerical accuracy of the method. For example, much better estimates are obtained in the case where $x = 0$, since this leads to closed diagrams, than for $x \neq 0$. For x being a neighbor of the origin, we can use symmetry to improve our bounds significantly. To obtain the infrared bound for percolation in $d \geq 11$, we use

$$\mathcal{S} = \{(0, 0, \mathcal{X}), (1, 0, \mathcal{X}), (1, 1, \mathcal{X}), (1, 2, \mathcal{X}), (1, 3, \mathcal{X}), (1, 4, \{0\})\}, \tag{10.4.5}$$

with $\mathcal{X} = \{x \in \mathbb{Z}^d : \|x\|_2 > 1\}$.

Let us relate the above NoBLE bootstrap functions to the classical ones in (8.2.7)–(8.2.8). The functions f_1 and f_2 are minor modifications of f_1 and f_2 in (8.2.7). The function f_3 is quite different. Indeed, instead of bounding objects in Fourier space, (10.4.3) bounds objects in terms of x-space. This is closer in spirit to the analysis of Hara and Slade for SAW [134]. The proof of Hara and Slade relies on the triangle and weighted bubbles for their bootstrap functions and turns out to be numerically much more efficient. This can be understood by (8.2.21) in Lem. 8.2, which contains quite some negative terms in the trigonometric approach. Unfortunately, we seem unable to make use of these negative terms, which is wasteful and makes the trigonometric approach numerically rather bad. This explains why x-space is used in (10.4.3) instead.

10.5 The Numerical Analysis

In this section, we explain how the numerical computations are performed using MATHE-MATICA notebooks that are available at the Web page of Fitzner [106]. The procedure starts by evaluating the notebook **SRW.nb**. This produces two files named **SRWCountData.nb** and **SRWIntegralsData.nb**, containing counts of SRWs of a given number of steps ending at various locations in $\mathbb{Z}^d$, and numerical values for SRW integrals, respectively. Running this program currently takes several hours.[1] Then, we evaluate the notebook **General.nb**. After this, we are ready to perform the NoBLE analysis for percolation by evaluating the notebook **Percolation.nb**. See Fig. 10.1 for the first output after evaluating it. Let us now explain this output in more detail.

---

[1] Alternatively, these two files can also be downloaded directly from [106], and put in the directory where MATHEMATICA looks for them. The command $InitialDirectory in MATHEMATICA will tell you what this directory is.

The dots in the bottom row of the table (appearing green in the MATHEMATICA note-book) mean that mean that the bootstrap has been successful for the parameters as chosen. This means that, using the NoBLE and relying on the bounds that $f_i(p) \leq \Gamma_i$ provide (the "assumed bounds"), we can improve the bound to $f_i(p) \leq \gamma_i$ with $\gamma_i < \Gamma_i$ for each $i = 1, 2, 3$ (the "concluded bounds"). The table in Fig. 10.1 gives the values of Γ_i on the first line of the table and the proved bound on $f_i(p)$ that the bootstrap functions imply on the second line of the table. These bounds are uniform in p. The bound on $f_i(p)$ is called

Result of the improvement of bounds in dimension 11

Bounds	$f_1(z_1)$	$f_2(z_1)$	$f_3(z_1)$	f_1	f_2	f_3
Assumed bound Γ_i	1.0131	1.076	1	1.01314412	1.076	1
Concluded bound	1.0113	1.0507	0.7475	1.01314411	1.07593619	0.999944
Comparison						

Detailed result of the improvement of bounds for f_3 in dimension 11

Bounds	F3−1,6,{0}	F3−0,0,x≠0	F3−1,0,x≠0	F3−1,1,x≠0	F3−1,2,x≠0	F3−1,3,x≠0
Assumed bound	0.008831	0.0677	0.1082	0.0537	0.041	0.02226
Concluded bound	0.0088305	0.0676421	0.108183	0.053684	0.0409087	0.0222542
Comparison						

Fig. 10.1 Output of the MATHEMATICA notebook Percolation in dimension $d = 11$

successful when the bound on $f_i(p)$ is smaller than Γ_i. As we see, the bounds on $f_i(p)$ in Fig. 10.1 are all successful, and then, the analysis of the improvement of the bootstrap argu-ment is completed. When evaluating the notebook, it is possible that some red dots appear, and this means that these improvements were not successful. In this case, either the NoBLE cannot be made to work in the dimension that we are investigating, or the constants involved (i.e., Γ_i for $i = 1, 2, 3$) need to be adapted so as to make the argument work.

Let us give some more details on the table in Fig. 10.1. The first 3 dots in the first table are the verifications that $f_i(1/(2d-1)) \leq \gamma_i$ for $i = 1, 2, 3$. The next three dots show that the improvement has been successful for all $p < p_c(\mathbb{Z}^{11})$. The values for $\Gamma_1, \Gamma_2, \Gamma_3$ are indicated in the first few lines. For example, $\Gamma_1 = 1.01314412$ means that $(2d-1)p \leq 1.01314412$. In the check, it turns out that γ_1 can be taken to be 1.01314411. Since this it true for all $p < p_c(\mathbb{Z}^{11})$, we obtain that $(2d-1)p_c(\mathbb{Z}^{11}) \leq 1.01314411$. This explains the value in the table in Thm. 10.2. Similarly, $\Gamma_2 = 1.076$. This implies that $A(11) \leq 1.076 \times 20/21 = 1.02476$. Anyone interested in obtaining improved constants can play with the notebook to optimize them.

The second table in Fig. 10.1 gives more details on the improvement of $f_3(p)$, which, as indicated in (10.4.3), consists of several contributions, over which the maximum is taken. The notebook Percolation.nb also includes a routine that optimizes the choices of Γ_1, Γ_2, and Γ_3. This makes it easier to find values for which the analysis works.

Exercise 10.5 (Bounds p_c and $A(d)$ in $d = 16$). Play with the MATHEMATICA notebooks to find good upper bounds on $p_c(\mathbb{Z}^d)$ and $A(d)$ for dimension $d = 16$.

The NoBLE analysis unfortunately does not work all the way down to dimension 6. This is primarily due to the fact that we still need that the lace expansion converges, which, numerically, is a stronger requirement than the triangle being finite. However, we have removed an important contribution to the triangle. Extending the analysis to all $d \geq d_c + 1 = 7$ is a major open problem for high-dimensional percolation:

Open Problem 10.1 (Percolation mean-field behavior all the way to dimension 6).
Extend the infrared bound in Thm. 10.1 to all dimensions $d > 6$.

Chapter 11
Further Critical Exponents

In this chapter, we discuss the existence of some further critical exponents. In Sect. 11.1, we discuss the correlation-length exponents ν, ν_2, as well as the gap exponent Δ. In Sect. 11.2, we discuss the two-point function exponent η in more detail, by discussing its sharp existence in Fourier space, as well as its existence in x-space. In Sect. 11.3, we discuss the arm exponents ρ_{in} and ρ_{ex}. We close this chapter in Sect. 11.4 by showing that $\eta = 0$ and $\rho_{ex} = \frac{1}{2}$ cannot simultaneously occur when $d < 6$, thus clarifying the role of the upper-critical dimension.

11.1 Correlation-Length Exponents ν and ν_2 and Gap Exponent Δ

We have proven in Thm. 4.1 that the triangle conditions implies that the critical exponents γ, β and δ take on their mean-field values. In fact, these are not the only critical exponents relying on the triangle condition; the following theorem gives two more examples:

Theorem 11.1 ($\nu_2 = \frac{1}{2}$ and $\Delta = 2$ [132, 220]). *For percolation in dimension $d \geq 11$ there exist constants $C_1, C_2, C_3, C_4 > 0$ such that for any $p < p_c$ and $k = 1, 2, 3, \ldots,$*

$$C_1|p_c - p|^{-1/2} \leq \xi_2(p) \leq C_2|p_c - p|^{-1/2}, \tag{11.1.1}$$

$$C_3|p_c - p|^{-2} \leq \frac{\mathbb{E}_p[|\mathscr{C}(0)|^{k+1}]}{\mathbb{E}_p[|\mathscr{C}(0)|^{k}]} \leq C_4|p_c - p|^{-2}. \tag{11.1.2}$$

Thus, $\nu_2 = \frac{1}{2}$ and $\Delta = 2$ in the bounded-ratio sense.

The bound (11.1.2) is a direct consequence of the triangle condition, as proven by Nguyen [220]. The former bound (11.1.1) comes as well from the triangle condition, as pointed out by Hara and Slade [132], see also Exerc. 11.1 below. Since the triangle condition holds for $d \geq 11$ by Thm. 10.1, and Thm. 11.1 follows.

Indeed, even the divergence of the correlation length is characterized in the subcritical regime:

© Springer International Publishing Switzerland 2017
M. Heydenreich and R. van der Hofstad, *Progress in High-Dimensional Percolation and Random Graphs*, CRM Short Courses, DOI 10.1007/978-3-319-62473-0_11

Theorem 11.2 ($v = \frac{1}{2}$ [129]). *For every $d \geq 19$, there exists constants $C_5, C_6 > 0$ such that for all $p \in [p_c/2, p_c)$,*

$$C_5|p_c - p|^{-1/2} \leq \xi(p) \leq C_6|p_c - p|^{-1/2} . \tag{11.1.3}$$

Consequently, the critical exponent v equals $\frac{1}{2}$ in the bounded-ratio sense.

This theorem is due to Hara [129], in which d_0 needs to be sufficiently large. In personal communication, Hara confirms that the methods in [130] imply that $d \geq 19$ is enough. Whether the results are valid for $d \geq 11$ using [107, 108] is unclear to us.

We next explain Thm. 11.2. To this end, denote by $m_p := \xi(p)^{-1}$ the *mass* associated with p. The proof proceeds by multiplying the identity (5.3.1) by e^{mx_1} yielding

$$\tau_p^{(m)}(x) = \delta_{0,x} + 2dp\big(D^{(m)} \star \tau_p^{(m)}\big)(x)$$
$$+ 2dp\big(\Pi_p^{(m)} \star D^{(m)} \star \tau_p^{(m)}\big)(x) + \Pi_p^{(m)}(x) , \tag{11.1.4}$$

where $f^{(m)}(x) := f(x)e^{mx_1}$. Furthermore, let $\chi^{(m)}(p) := \sum_{x \in \mathbb{Z}^d} \tau_p^{(m)}(x)$. Hara's proof is based on two main pillars: the first one is the observation that $\chi^{(m)}(p) \nearrow \infty$ as $m \nearrow m_p$, the second is that the couples Π and $\Pi^{(m)}$, τ and $\tau^{(m)}$, D and $D^{(m)}$ are all very close to each other (in Fourier space). With these observations at hand, Hara demonstrates that

$$m_p^2/(3d) \leq \chi(p) \leq m_p^2/d , \tag{11.1.5}$$

which, together with $\gamma = 1$ due to Cor. 5.2, implies the result. Mind that in the proof p ($< p_c$) is kept fixed, and a bootstrap argument in m is applied (with m playing the role of p in our proof in Chap. 8).

Exercise 11.1 (*You prove $v_2 = \frac{1}{2}$*). Prove $\xi_2(p)^2 \sim \chi(p)$ for $p \in (\bar{p}, p_c)$ in bounded-ratio sense for some $\bar{p} < p_c$ and conclude that (11.1.1) holds. To this end, express (1.1.12) for $p < p_c$ in Fourier terms as

$$\xi_2(p)^2 = -\sum_{j=1}^{d} \frac{\partial_{jj}\hat{\tau}_p(0)}{\hat{\tau}_p(0)} , \tag{11.1.6}$$

where ∂_j represents the partial derivative w.r.t. the jth component, and ∂_{jj} the double derivative. Argue that $\partial_j \hat{D}(0) = 0$ and $\partial_j \hat{\Pi}_p(0) = 0$, and show that

$$\xi_2(p) = -\sum_{j=1}^{d}\left(\frac{\partial_{jj}\hat{\Pi}_p(0)}{1 + \hat{\Pi}_p(0)} + 2dp\,\hat{\tau}_p(0)\,\frac{\partial_{jj}\hat{D}(0)(1 + \hat{\Pi}_p(0)) + \partial_{jj}\hat{\Pi}_p(0)}{1 + \hat{\Pi}_p(0)}\right) . \tag{11.1.7}$$

Finally, derive the conclusion using Prop. 8.3.

Mind that all the exponents derived so far characterize critical behavior either *at* the critical point (exponent δ) or in the *subcritical* regime (exponents γ, v, v_2, Δ). The only exception is the exponent β in (1.2.1), which, however, is linked to behavior *at* p_c via the extrapolation technique in the proof of Thm. 9.5.

The sparsity of results about critical behavior in the supercritical regime (i.e., when $p \searrow p_c$) is rooted in the lace-expansion technology. In order to see this, let us recall the main steps in the lace-expansion proof: First we derive the expansion, secondly we derive bounds on the lace-expansion coefficients $\Pi(x)$ valid throughout the entire subcritical regime, and finally, uniformity of these bounds allows us to deduce the same bounds at the critical point.

Generally speaking, the lace expansion is a fantastic tool to control critical behavior at, or slightly below, p_c, but it fails in the supercritical regime. One difficulty that arises in this context is that events of the form $\{x \leftrightarrow y, |\mathcal{C}(x)| < \infty\}$ are not increasing (rather, intersections of increasing and decreasing events), so that the lace expansion in the form of Chap. 6 cannot be carried out for such events. Indeed, it is an important open problem to control (almost) critical behavior in the *supercritical regime*:

> **Open Problem 11.1 ('Super' critical exponents).** Prove that the (super-)critical exponents γ', ν', and ν_2', as defined in (1.2.6)–(1.2.8), exist and that their values coincide with γ, ν and ν_2, respectively.

11.2   The Two-Point-Function Critical Exponent η

Here we informally explain the proof that $\eta = 0$ holds in the asymptotic sense, both in k-space, as proven by Hara and Slade [139] and in our paper with Hulshof [145], as well as in x-space by Hara in [130], improving the result to $d \geq 11$ using the NoBLE work of Chap. 10, without giving all details.

Let us start by sharpening the infrared bound in Thm. 5.1 to sharp asymptotics:

Theorem 11.3   ($\eta = 0$ asymptotically in k-space [139, 145]). *For percolation with d sufficiently large, there exists a constant $A = A(d)$ and $\varepsilon > 0$ such that*

$$\hat{\tau}_{p_c}(k) = \frac{A}{|k|^2}\left(1 + O(|k|^\varepsilon)\right). \tag{11.2.1}$$

Consequently, the critical exponent η exists in the asymptotic sense and takes on the mean-field value $\eta = 0$.

Theorem 11.3 is proved by Hara and Slade in [139, Thm. 1.1], where also the dependence on γ in $\hat{\tau}_{p,\gamma}(k)$ was included. The latter is a much more difficult problem than the asymptotics of $\hat{\tau}_{p_c}(k)$ alone. Further, it also follows from our results with Hulshof [145], where we prove that there exists $\varepsilon > 0$ such that

$$\sum_x |x|^{2+\varepsilon} |\Pi_{p_c}(x)| < \infty. \tag{11.2.2}$$

From this bound, (11.2.1) easily follows by Taylor expansion:

Exercise 11.2 (Sharp asymptotics of η through Taylor expansion). Prove that (11.2.2) implies (11.2.1), and compute the constant A in terms of $\Pi_{p_c}(x)$.

We continue by discussing η in x-space. The main results is the following:

Theorem 11.4 ($\eta = 0$ in x-space [108, 130, 131]). *For percolation with $d \geq 11$, there exists a constant $A_2 = A_2(d)$ such that*

$$\tau_{p_c}(x) = \frac{A_2}{|x|^{d-2}}\left(1 + O\left(|x|^{-2/d}\right)\right). \tag{11.2.3}$$

Consequently, the critical exponent η exists in the asymptotic sense and takes on the mean-field value $\eta = 0$.

Theorem 11.4 has proved to be a very important result. Indeed, the x-space asymptotics for $\tau_{p_c}(x)$ has been used as an assumption in various papers. We see examples in Sect. 11.3 below. Thus, it can be seen as one of the crucial results in high-dimensional percolation.

Both Thm. 11.4 as well as Thm. 11.3 are valid for the spread-out model as discussed in Sect. 5.2 for $d > 6$ and $L > L_0$, with L_0 sufficiently large (depending on d).

We informally discuss the proof by Hara in [130]. We note that, by (7.1.1),

$$\tau_p(x) = \big((1 + \Pi_p) \star V_p\big)(x). \tag{11.2.4}$$

The proof again relies on a bootstrap argument, now using

$$f(p) = \max_{x \in \mathbb{Z}^d}(|x| \vee 1)^{d-2}\tau_p(x). \tag{11.2.5}$$

The fact that $p \mapsto f(p)$ is continuous is left as an exercise:

Exercise 11.3 (Continuity of bootstrap function). Prove that the bootstrap function $f(p)$ in (11.2.5) is continuous for $p \in [0, p_c - \varepsilon]$ for every $\varepsilon > 0$. [*Hint:* Use the fact that $x \mapsto \tau_p(x)$ is exponentially small for $|x| \to \infty$.]

Assuming the bootstrap bound that $f(p) \leq \Gamma$, we obtain that

$$\tau_p(x) \leq \Gamma(|x| \vee 1)^{-(d-2)}, \tag{11.2.6}$$

which in turn can be used to prove that the lace expansion coefficients $\Pi_p(x)$ are even smaller. For example, by (11.2.6),

$$\Pi_p^{(0)}(x) \leq \Gamma^2(|x| \vee 1)^{-2(d-2)}. \tag{11.2.7}$$

In particular, for $d > 6$,

$$\sum_x |x|^2 \Pi_p^{(0)}(x) < \infty, \tag{11.2.8}$$

which is a good sign when trying to obtain Gaussian behavior. In the following exercise, you are asked to extend this to $N = 1$:

Exercise 11.4 (Bound on $\Pi_p^{(1)}(x)$). Extend the proof of (11.2.7) to $N = 1$ by proving that (11.2.6) implies that there exists a constant $C < \infty$ such that

$$\Pi_p^{(1)}(x) \leq C(|x| \vee 1)^{-2(d-2)} . \tag{11.2.9}$$

The key ingredient in Hara's proof is an analysis showing that if $U(x)$ is the Fourier inverse of

$$\widehat{U}(k) = \frac{1}{1 - \hat{J}(k)} , \tag{11.2.10}$$

where J satisfies the following restrictions:

(1) $|J(x)| = O\big((|x| \vee 1)^{-(d+2+\alpha)}\big)$ for some $\alpha > 0$;
(2) $\sum_x J(x) = 1$;
(3) $1 - \hat{J}(k) \geq K_0|k|^2$;

then there exists $B > 0$ such that

$$U(x) = B(|x| \vee 1)^{-(d-2)} + O\big((|x| \vee 1)^{-(d-2+(\alpha \wedge 2)/d)}\big) . \tag{11.2.11}$$

The asymptotics in (11.2.11) may not *appear* a surprising result, as they typically apply to random walk Green's function (see e.g., the extremely sharp asymptotics proved by Uchiyama in [256]). However, the point is that Hara does *not* require that $J(x) \geq 0$, so that $U(x)$ does *not* have the interpretation of a random walk Green's function. As a result, an analytical proof is necessary. This proof is based on the integral representation

$$U(x) = \int_0^\infty I_t(x) \, dt , \quad \text{where } I_t(x) = \int_{[-\pi,\pi]^d} e^{ik \cdot x} e^{-t[1-\hat{J}(k)]} \frac{dk}{(2\pi)^d} . \tag{11.2.12}$$

This is combined with a careful analysis of $I_t(x)$, showing that $I_t(x)$ is close to a Gaussian density when $t \geq T$ is large, while it is small when $t \leq T$ is small. The truncation value T is chosen as $T = \varepsilon|x|^2$. The analysis of $I_t(x)$ by Hara is similar in spirit to that performed for spread-out random walk by Hara, the second author and Slade in [131].

Exercise 11.5 (Representation Green's function). Verify that (11.2.12) holds.

Using the bootstrap argument, Hara proves that there exists $\Gamma < \infty$ such that (11.2.6) holds for $p = p_c$. In turn, we have that $\widehat{V}_{p_c}(k) = [1 - 2dp_c\widehat{D}(k)[1 + \widehat{\Pi}_{p_c}(k)]]^{-1}$, so that it is of the form (11.2.10) with $J(x) = 2dp_c(D \star [\delta_{0,x} + \Pi_{p_c}])(x)$. By the bootstrap argument, it follows that (see also (11.2.7)–(11.2.9))

$$|\Pi_{p_c}(x)| \leq C(|x| \vee 1)^{-2(d-2)} \tag{11.2.13}$$

for a constant $C > 0$, which implies that, for another constant $C' > 0$,

$$|J(x)| \leq C'(|x| \vee 1)^{-2(d-2)} . \tag{11.2.14}$$

This implies the main assumption above with $\alpha = 2(d-2) - (d+2) = d - 6 \geq 2$ for $d \geq 11$. Therefore, (11.2.11) implies that $V_{p_c}(x) = B(|x| \vee 1)^{-(d-2)} + O((|x| \vee 1)^{-(d-2+2/d)})$. In turn, it can be seen that this asymptotics for $V_{p_c}(x)$, together with (11.2.13), implies

$$\tau_p(x) = \left((1 + \Pi_p) \star V_p\right)(x)$$
$$= A\left(1 + \widehat{\Pi}_{p_c}(0)\right)(|x| \vee 1)^{-(d-2)} + O\left((|x| \vee 1)^{-(d-2+1/d)}\right). \tag{11.2.15}$$

Exercise 11.6 (Convolution bound). Prove that if

$$V(x) = B(|x| \vee 1)^{-(d-2)} + O((|x| \vee 1)^{-(d-2+\alpha)}) \text{ for some } \alpha > 0,$$

and $|g(x)| \leq K(|x| \vee 1)^{-(d+\alpha)}$, then

$$(g \star V)(x) = B\hat{g}(0)(|x| \vee 1)^{-(d-2)} + O\left((|x| \vee 1)^{-(d-2+\alpha)}\right). \tag{11.2.16}$$

The original proof by Hara [130] applies to $d \geq 19$. The extension to $d \geq 11$ follows from the NoBLE analysis by Fitzner and the second author, which is discussed in Chap. 10.

11.3   Arm Exponents ρ_{in} and ρ_{ex}

The term *arm exponent* refers to the critical exponent characterizing the decay rate of the probability that the origin is connected to the boundary of a ball of radius n. We speak of an *intrinsic* arm exponent when considering an n-ball in the intrinsic distance $d_{\mathcal{C}(0)}$, whereas for the *extrinsic* arm exponent we consider the ball in the ℓ^∞ (or some other extrinsic) distance on $\mathbb{Z}^d$.

The identification of arm exponents in high dimension has been pioneered by Kozma and Nachmias [201, 202]. Here we give the full analysis of the proof that $\rho_{in} = 1$ and sketch the main ideas in the proof that $\rho_{ex} = \frac{1}{2}$.

Theorem 11.5 ($\rho_{in} = 1$ and $\rho_{ex} = \frac{1}{2}$ [201, 202]). *For percolation with $d \geq 11$, there exist constants $0 < c_{in} < C_{in} < \infty$ and $0 < c_{ex} < C_{ex} < \infty$ such that*

$$\frac{c_{in}}{n} \leq \mathbb{P}_{p_c}(\exists x \in \mathbb{Z}^d : d_{\mathcal{C}(0)}(0, x) \geq n) \leq \frac{C_{in}}{n}, \tag{11.3.1}$$

and

$$\frac{c_{ex}}{n^2} \leq \mathbb{P}_{p_c}(0 \leftrightarrow \partial \Lambda_n) \leq \frac{C_{ex}}{n^2}. \tag{11.3.2}$$

Consequently, the critical exponents ρ_{in} and ρ_{ex} exist in the bounded-ratio sense and take on the mean-field values $\rho_{in} = 1$ and $\rho_{ex} = \frac{1}{2}$.

The proof of (11.3.1) is also valid for spread-out models. It further extends to long-range percolation, as pointed out in joint work with Hulshof [145]. The proof of (11.3.2) crucially relies on Thm. 11.4, and is limited to finite-range percolation. Very recently, Hulshof [177] proved a version for long-range percolation with appropriately adjusted exponent, see Sect. 15.4.

A remark is due on *multiarm* or k-arm exponents. The k-arm exponent characterizes polynomial decay of the k-arm event (=occurrence of k disjoint connections from the origin to $\partial \Lambda_n$). Indeed, with a certain regularization around the origin, Kozma and Nachmias [202, Thm. 3] prove that the k-arm exponent in the extrinsic metric equals $1/k$ times the 1-arm exponent ρ_{ex}. One might note that the upper bound follows readily from the BK inequality. The result might therefore be interpreted as saying that the BK inequality is (in a certain sense) *sharp* in high dimensions.

For comparison, we mention that in two dimension, $\rho_{\text{ex}} = 48/5$ (cf. the seminal work of Lawler, Schramm and Werner [205], see also Table 1.1 in Sect. 1.2), whereas ρ_{in} has not been identified. Multi-arm exponents in two dimensions have been shown to exist by Beffara and Nolin, but could not be computed explicitly [33].

Observe that Thm. 11.5 provides upper and lower bounds of the same order. One would expect that that upper and lower bound (on the leading order term) coincide, which however is not provided by the current proof techniques:

Open Problem 11.2 (Sharp arm exponents). Prove that the limits

$$a_{\text{in}} := \lim_{n \to \infty} n \, \mathbb{P}_{p_c}(\exists x \in \mathbb{Z}^d : d_{\mathcal{C}(0)}(0, x) \geq n)$$

and

$$a_{\text{ex}} := \lim_{n \to \infty} n^2 \, \mathbb{P}_{p_c}(0 \leftrightarrow \partial \Lambda_n)$$

exist.

It follows from Thm. 11.5 that $0 < a_{\text{in}}, a_{\text{ex}} < \infty$ provided that the limits exist.

11.3.1 Proof of the Intrinsic Arm Exponent

We give a full proof of the intrinsic arm exponent in (11.3.1) due to Kozma and Nachmias [201] with a considerable shortcut due to Sapozhnikov [235]. Mind that the proof crucially uses other critical exponents that we have derived earlier (in particular γ, ν, and δ). This gives rise to relations between the critical exponents that are known as *scaling and hyperscaling relations*. Scaling relations involve only the critical exponents, while hyperscaling relations also involve the dimension d. The scaling relations are believed to hold for every dimension, while the hyperscaling relations are expected to hold only below the upper critical dimension. We give an example of such scaling relations in Sect. 11.4 below.

To describe the proof of the intrinsic one-arm exponent, we define the *intrinsic ball* by

$$B(n) := \left\{ x \in \mathbb{Z}^d : d_{\mathcal{C}(0)}(0, x) \leq n \right\}. \tag{11.3.3}$$

The expected growth of $B(n)$ is expressed in the following lemma:

Lemma 11.6 (Growth of intrinsic balls). *For percolation in dimension $d \geq 11$, there exists constants $c, C > 0$ such that for all $n \in \mathbb{N}$,*

$$cn < \mathbb{E}_{p_c}|B(n)| < Cn . \tag{11.3.4}$$

Proof. We start with the upper bound and follow Sapozhnikov [235]. Let $p < p_c$. We consider the following coupling of percolation with parameters p and p_c: Starting with a critical percolation configuration (edges are occupied with probability p_c), make every occupied edge vacant with probability $1 - (p/p_c)$. This construction implies that for any $x \in \mathbb{Z}^d$, $p < p_c$, and $n \in \mathbb{N}$,

$$\mathbb{P}_p(d_{\mathcal{C}(0)}(0, x) \leq n) \geq \left(\frac{p}{p_c}\right)^n \mathbb{P}_{p_c}(d_{\mathcal{C}(0)}(0, x) \leq n) .$$

Summing over x and using the inequality $\mathbb{P}_p(d_{\mathcal{C}(0)}(0, x) \leq n) \leq \mathbb{P}_p(0 \leftrightarrow x)$, we obtain

$$\mathbb{E}_{p_c}|B(n)| \leq \left(\frac{p_c}{p}\right)^n \chi(p) \leq C\left(\frac{p_c}{p}\right)^n (p_c - p)^{-1} ,$$

where the last line comes from $\gamma = 1$ in Thm. 4.1, see also Cor. 5.2. The upper bound follows by taking $p = p_c(1 - 1/(2n))$.

For the lower bound, we follow [201] and first estimate $\mathbb{E}_{p_c}[d_{\mathcal{C}(0)}(0, x) \mid 0 \leftrightarrow x]$. Indeed, if $0 \leftrightarrow x$, then $d_{\mathcal{C}(0)}(0, x)$ is no more than the number of $y \in \mathbb{Z}^d$ such that the events $\{0 \leftrightarrow y\}$ and $\{y \leftrightarrow x\}$ occur disjointly. Consequently, by the BK inequality and Thm. 11.4,

$$\mathbb{E}_{p_c}[d_{\mathcal{C}(0)}(0, x)\mathbb{1}_{\{0\leftrightarrow x\}}] \leq C \sum_{y \in \mathbb{Z}^d} |y|^{2-d}|x - y|^{2-d} \leq C|x|^{4-d} .$$

Hence, Thm. 11.4 implies that $\mathbb{E}_{p_c}[d_{\mathcal{C}(0)}(0, x) \mid 0 \leftrightarrow x] \leq C|x|^2$. We thus may conclude that if $|x| \leq \sqrt{n/2C}$ then $\mathbb{P}_{p_c}(d_{\mathcal{C}(0)}(0, x) \leq n \mid 0 \leftrightarrow x) \geq \frac{1}{2}$. Consequently,

$$\mathbb{E}_{p_c}|B(n)| \geq \sum_{x:|x|\leq\sqrt{n/2C}} \mathbb{P}_{p_c}(d_{\mathcal{C}(0)}(0, x) \leq n \text{ and } 0 \leftrightarrow x)$$

$$\geq \frac{c}{2} \sum_{x:|x|\leq\sqrt{n/2C}} |x|^{2-d} \geq cn$$

for a small constant $c > 0$. $\qquad\square$

Exercise 11.7 (Proof of Lem. 11.6). Fill in the details in the proof of Lem. 11.6.

Proof of the lower bound in (11.3.1). The proof of the lower bound proceeds via the well-known second-moment method. The second-moment method uses that, for any nonnegative random variable Z,

$$\mathbb{P}(Z > 0) \geq (\mathbb{E}\, Z)^2 / \mathbb{E}\, Z^2 . \tag{11.3.5}$$

To this end, let $\lambda = 2C/c$, where C and c are the constants appearing in Lem. 11.6, so that Lem. 11.6 yields

$$\mathbb{E}_{p_c} |B(\lambda n) \setminus B(n)| \geq c\lambda n - Cn = Cn .$$

We now estimate the second moment of $B(\lambda n)$. Indeed,

$$\{d_{\mathcal{C}(0)}(0, x) \leq \lambda n\} \cap \{d_{\mathcal{C}(0)}(0, y) \leq \lambda n\}$$
$$\subseteq \bigcup_z \{d_{\mathcal{C}(0)}(0, z) \leq \lambda n\} \circ \{d_{\mathcal{C}(0)}(z, x) \leq \lambda n\} \circ \{d_{\mathcal{C}(0)}(z, y) \leq \lambda n\} .$$

Consequently, the BK inequality (1.3.4) yields

$$\mathbb{E}_{p_c} |B(\lambda n)|^2 = \sum_{x,y} \mathbb{P}_{p_c}(d_{\mathcal{C}(0)}(0, x) \leq \lambda n, d_{\mathcal{C}(0)}(0, y) \leq \lambda n)$$
$$\leq \sum_{x,y,z} \mathbb{P}_{p_c}(d_{\mathcal{C}(0)}(0, z) \leq \lambda n)\, \mathbb{P}_{p_c}(d_{\mathcal{C}(0)}(z, x) \leq \lambda n)$$
$$\times \mathbb{P}_{p_c}(d_{\mathcal{C}(0)}(z, y) \leq \lambda n)$$
$$= \left[\sum_{z \in \mathbb{Z}^d} \mathbb{P}_{p_c}(d_{\mathcal{C}(0)}(0, z) \leq \lambda n) \right]^3 \leq C'n^3 , \tag{11.3.6}$$

for some constant $C' > 0$. Consequently, the bound in (11.3.5) yields

$$\mathbb{P}_{p_c}(\exists x \in \mathbb{Z}^d : d_{\mathcal{C}(0)}(0, x) \geq n) \geq \mathbb{P}_{p_c}(|B(\lambda n) \setminus B(n)| > 0) \geq \frac{c^2 n^2}{C' n^3} \geq \frac{c^2}{C'} \frac{1}{n} ,$$

which proves the statement with $c_{\mathrm{in}} = c^2/C'$. $\square$

Exercise 11.8 Verify the steps in (11.3.6).

We proceed with the lower bound in (11.3.1):
Proof of the upper bound in (11.3.1). The upper bound uses a clever induction argument. For subgraphs $\mathcal{G} \subseteq \mathcal{E}(\mathbb{Z}^d)$ of the infinite lattice $\mathbb{Z}^d$, we define

$$H(n; \mathcal{G}) := \{\partial B(n; \mathcal{G}) \neq \varnothing\}$$

for the "one-arm event" on the graph $\mathcal{G}$. We further define

$$\Gamma(n) = \sup_{\mathcal{G} \subseteq \mathcal{E}(\mathbb{Z}^d)} \mathbb{P}_{p_c}(H(n; \mathcal{G})) .$$

It turns out that working with $\Gamma(n)$ rather than $\mathbb{P}_{p_c}(H(n; \mathbb{Z}^d))$ enables us to apply a regeneration argument, which would not work for $\mathbb{P}_{p_c}(H(n; \mathbb{Z}^d))$ since it is not *monotone*.

To this end, choose $A \geq 1$ large enough so that

$$3^3 A^{2/3} + C_\delta A^{2/3} \leq A, \tag{11.3.7}$$

where C_δ is from Thm. 9.2. We claim that, for any integer $k \geq 0$,

$$\Gamma(3^k) \leq \frac{A}{3^k}. \tag{11.3.8}$$

This readily implies the upper bound in (11.3.1). Indeed, for any n we choose k such that $3^{k-1} \leq n < 3^k$ and then

$$\mathbb{P}_{p_c}(H(n; \mathbb{Z}^d)) \leq \Gamma(n) \leq \Gamma(3^{k-1}) \leq \frac{A}{3^{k-1}} \leq \frac{3A}{n}.$$

The proof of (11.3.8) is via induction in k. The claim is trivial for $k = 0$ since $A \geq 1$. For the inductive step we assume (11.3.8) for $k - 1$ and prove it for k. Let $|\mathcal{C}_\mathcal{G}(0)|$ denote the size of the cluster of 0 in the graph $\mathcal{G}$. Depending on the size of $|\mathcal{C}_\mathcal{G}(0)|$ for arbitrary $\mathcal{G} \subseteq \mathcal{E}(\mathbb{Z}^d)$, we estimate

$$\mathbb{P}_{p_c}(H(3^k; \mathcal{G})) \leq \mathbb{P}_{p_c}(H(3^k; \mathcal{G}), |\mathcal{C}_\mathcal{G}(0)| \leq A^{-4/3} 9^k)$$
$$+ \mathbb{P}_{p_c}(|\mathcal{C}_\mathcal{G}(0)| > A^{-4/3} 9^k). \tag{11.3.9}$$

For the second summand, we use that $\delta = 2$ in Thm. 9.2 to obtain

$$\mathbb{P}_{p_c}(|\mathcal{C}_\mathcal{G}(0)| > A^{-4/3} 9^k) \leq \mathbb{P}_{p_c}(|\mathcal{C}_{\mathcal{E}(\mathbb{Z}^d)}(0)| > A^{-4/3} 9^k)$$
$$\leq C_\delta A^{2/3} 3^{-k}. \tag{11.3.10}$$

For the former, on the other hand, we claim that

$$\mathbb{P}_{p_c}(H(3^k; \mathcal{G}), |\mathcal{C}_\mathcal{G}(0)| \leq A^{-4/3} 9^k) \leq A^{-4/3} 3^{k+1} (\Gamma(3^{k-1}))^2. \tag{11.3.11}$$

Indeed, if $|\mathcal{C}_\mathcal{G}(0)| \leq A^{-4/3} 9^k$, then there exists $j \in [\frac{1}{3} 3^k, \frac{2}{3} 3^k]$ such that $|\partial B(j; \mathcal{G})| \leq A^{-4/3} 3^{k+1}$. Denote the first such level by j. Then, on the right hand side, we get a factor $\Gamma(j)$ (which is bounded by $\Gamma(3^{k-1})$) from the probability of a connection from the origin to level j, and $A^{-4/3} 3^{k+1}$ times the probability to go from level j to level 3^k (each of these probabilities is again bounded above by $\Gamma(3^{k-1})$), which shows (11.3.11). There are a few technical points involved, which we have ignored in our proof.[1]

---

[1]*Hint:* One needs to condition on the precise form of $B(j; \mathcal{G})$, and then exploit that Γ gives a *uniform* bound on all subgraphs.

We combine (11.3.9), (11.3.10), (11.3.11) with the induction hypothesis, and finally (11.3.7), to obtain

$$\Gamma(3^k) \le A^{-4/3}3^{k+1}\left(\frac{A}{3^{k-1}}\right)^2 + \frac{C_\delta A^{2/3}}{3^k} = \frac{3^3 A^{2/3} + C_\delta A^{2/3}}{3^k} \le \frac{A}{3^k},$$

thus proving (11.3.8). ☐

11.3.2 The Extrinsic Exponent

Similar to the intrinsic exponent, the lower bound in (11.3.2) follows straightforwardly using the second moment method.

Exercise 11.9 (Lower bound in (11.3.2)**).** Prove the lower bound in (11.3.2) by applying the second moment method (11.3.5) to the random variable

$$Z = |\{x \in \Lambda_{2n} \setminus \Lambda_n : 0 \leftrightarrow x\}|,$$

and using Thm. 11.4.

For the upper bound of (11.3.2), an induction scheme similar to the proof of the intrinsic exponent has been derived by Kozma and Nachmias [202], which we explain briefly now.

It is instructive to revisit the proof of (11.3.8) first. For the induction step, we distinguished between two cases, depending on whether the size of the cluster is "large" (which we bound by Thm. 9.2), or it is "small" (but then the ball has sparse intermediate shells, which make its probability small). A careful distinction between "large" and "small" provides the finishing touch.

Similar ideas are at the basis for the upper bound of (11.3.2), but details are severely more complicated. Here is a brief outline of the argument. Suppose $\{0 \leftrightarrow \partial\Lambda_{3n}\}$. Then there are three possibilities:

(1) The cluster of the origin is "not-too-small", say $|\mathcal{C}(0)| \ge n^4/100$. By Thm. 9.2, the probability of this is at most c/n^2.
(2) There exists some $j \in [n, 2n]$ such that

$$|\{x \in \partial\Lambda_j : 0 \leftrightarrow x \text{ through } \Lambda_j\}| \le n^2$$

("a thin intermediate shell"). The probability of this is $\mathbb{P}_{p_c}(0 \leftrightarrow \Lambda_j^c) \le \mathbb{P}_{p_c}(0 \leftrightarrow \partial\Lambda_n)$ for the connection to $\partial\Lambda_j$ multiplied by n^2 times the probability that $x \in \partial\Lambda_j$ is connected to $\partial\Lambda_{3n}$. Together, this gives an upper bound $n^2 \, \mathbb{P}_{p_c}(0 \leftrightarrow \partial\Lambda_n)^2$.
(3) None of the two cases above, that is, a small cluster without thin intermediate shells. The absence of thin intermediate shells suggests that $|\mathcal{C}(0)|$ is at least n^4. In particular, we expect that the probability $\{|\mathcal{C}(0)| \ge n^4/100\}$ is small, say at most $1/20$. Therefore, we expect an upper bound $\mathbb{P}_{p_c}(0 \leftrightarrow \partial\Lambda_n)$ for this case.

Summarizing the three options, one (heuristically) obtains the inequality

$$\mathbb{P}_{p_c}(0 \leftrightarrow \partial \Lambda_{3n}) \leq \frac{c}{n^2} + n^2 \, \mathbb{P}_{p_c}(0 \leftrightarrow \partial \Lambda_n)^2 + \frac{1}{20} \, \mathbb{P}_{p_c}(0 \leftrightarrow \partial \Lambda_n) \,,$$

from which it is possible to derive that $\mathbb{P}_{p_c}(0 \leftrightarrow \partial \Lambda_n) < C/n^2$, as desired. Making the reasoning in case (3) rigorous is the key part of [202] and uses so-called *local regularization* arguments.

11.4   On the Percolation Upper Critical Dimension

In this section, we investigate for which dimensions mean-field critical exponents can occur. In general statistical physics models, there exists a so-called *upper critical dimension*, denoted by d_c, such that mean-field critical exponents hold for $d > d_c$. Often, at the critical dimension, the mean-field critical exponents are expected to have logarithmic corrections. There is very little work that rigorously proves such results. A remarkable exception is self-avoiding walks, for which we refer to the recent work by Bauerschmidt, Brydges and Slade [31] and the references therein.

Here we show that $\rho_{ex} = \frac{1}{2}$ and $\eta = 0$ in x-space imply $d \geq 6$. This immediately implies that $d_c \geq 6$. Since, at least for spread-out models as in Sect. 5.2, we know that $\rho_{ex} = \frac{1}{2}$ and $\eta = 0$ in x-space hold (recall Thms. 11.4 and 11.5), this provides very strong evidence for the statement that $d_c = 6$. The main result in this section is the following corollary:

Corollary 11.7 (The upper critical dimension satisfies $d_c \geq 6$). *The mean-field critical exponents $\rho_{ex} = \frac{1}{2}$ and $\eta = 0$ in x-space imply that $d \geq 6$. As a result, the percolation upper critical dimension d_c satisfies $d_c \geq 6$.*

The role of the critical dimension was already investigated early on by Chayes and Chayes in [74], but their work involves many more critical exponents, some of which we do not yet know the existence of, even in sufficiently high dimensions. Tasaki [253] proves a hyperscaling inequality stating that $d\nu \geq 2\Delta - \gamma$. Inserting the mean-field values $\nu = \frac{1}{2}, \Delta = 2$ and $\gamma = 1$ shows that $d \geq 6$ is required for mean-field behavior. The current proof is much simpler, so we stick to this.

Proof. Take n large, and let $e = (1, 0, \ldots, 0)$ denote the first basis vector in $\mathbb{Z}^d$. When $0 \leftrightarrow 2ne$, we must have that 0 is connected to $\partial \Lambda_n$ and, at the same time, $2ne$ is connected to $2ne + \partial \Lambda_n$. Thus,

$$\begin{aligned} \tau_{p_c}(2ne) &= \mathbb{P}_{p_c}(0 \leftrightarrow 2ne) \\ &\leq \mathbb{P}_{p_c}(\{0 \leftrightarrow \partial \Lambda_n\} \cap \{2ne \leftrightarrow (2ne + \partial \Lambda_n)\}) \,. \end{aligned} \tag{11.4.1}$$

Since the events $\{0 \leftrightarrow \partial \Lambda_n\}$ and $\{2ne \leftrightarrow (2ne + \partial \Lambda_n)\}$ rely on the occupation statuses of disjoint sets of bonds, these events are *independent*. Moreover, due to translation invariance, the probabilities of these events are equal, so that we arrive at

$$\tau_{p_c}(2ne) \leq \mathbb{P}_{p_c}(0 \leftrightarrow \partial \Lambda_n)^2 . \tag{11.4.2}$$

Since $\eta = 0$ in x-space, the left-hand side is at least $cn^{-(2-d)}$ for some $0 < c < \infty$. Further, since $\rho_{\mathrm{ex}} = \frac{1}{2}$, the right-hand side is at most C/n^4. We conclude that, for every $n \geq 1$,

$$cn^{-(2-d)} \leq C/n^4 , \tag{11.4.3}$$

which can only be true when $d \geq 6$. $\qquad\qquad\qquad\qquad\qquad\qquad\qquad\square$

Chapter 12
Kesten's Incipient Infinite Cluster

In this chapter, we introduce the high-dimensional *incipient infinite cluster*, henceforth abbreviated IIC, which is an infinite cluster at the critical threshold p_c as constructed by Kesten [195]. We first motivate the IIC in Sect. 12.1, and then discuss its construction and properties in high dimensions in Sect. 12.2.

12.1 Motivation for the Incipient Infinite Cluster

One of the most classical questions about percolation concerns the formation of infinite clusters *at* the critical point (recall Open Problem 1.1). When $\theta(p_c) = 0$, as proven in high dimensions and believed quite generally, this leaves us with a most remarkable situation: At the critical point p_c there are clusters at all length scales, which are, however, all finite. As we then make a density $\varepsilon > 0$ of closed edges open, the large clusters connect up to form a (unique) infinite cluster, no matter how small ε is. At criticality, the critical cluster is therefore *at the verge of appearing*. This observation motivated the introduction of an *incipient infinite cluster* (IIC) as a critical cluster that is *conditioned* to be infinite. Since at the critical value there does not exist an infinite component when $\theta(p_c) = 0$, the IIC can only be defined through an appropriate limiting scheme. The underlying idea is to construct a cluster that has all the remarkable features of a critical cluster, but on the other hand provides us with the advantage of being infinite. This last feature is particularly handy when studying random walks on critical structures.

Somewhat simplified, the *incipient infinite cluster* (IIC) is defined as the cluster of the origin under the critical measure $\mathbb{P}_{p_c}$ conditioned on $\{|\mathcal{C}(0)| = \infty\}$. Since this would condition on an event of probability 0, a rigorous construction of the IIC requires a limiting argument. The first mathematical construction has been carried out by Kesten [195] in $d = 2$. Kesten considers two limiting schemes:

© Springer International Publishing Switzerland 2017
M. Heydenreich and R. van der Hofstad, *Progress in High-Dimensional Percolation and Random Graphs*, CRM Short Courses, DOI 10.1007/978-3-319-62473-0_12

- under $\mathbb{P}_{p_c}$, condition on the event $\{\mathcal{C}(0) \cap \partial\Lambda_n \neq \varnothing\}$, and then let $n \to \infty$;
- under $\mathbb{P}_p$ with $p > p_c$, condition on the event $\{|\mathcal{C}(0)| = \infty\}$ and let $p \searrow p_c$.

Kesten proved that both limits exist in $d = 2$, and give rise to the *same* limiting measure. Járai [189, 190] proved that several other limiting schemes give rise to the same limit, illustrating the robustness of the IIC construction.

Naturally, this is not the only way of viewing large critical clusters. For example, one could also be interested in the *scaling limit* of critical clusters, and we return to that perspective from two different angles in Chaps. 13 and 15. Kesten's IIC is particularly relevant when studying random walks on critical clusters, as the IIC is an infinite graph with fractal properties. Random walks on such structures could behave rather differently compared to random walks on the full lattice, as we discuss in more detail in Sect. 14.3.

12.2   The Incipient Infinite Cluster in High Dimensions

We now turn towards the high-dimensional regime, where three different IIC constructions are known to exist. To this end, we recall the notion of a *cylinder event*, which is an event that only depends on a finite number of bonds. The main result concerning the existence and uniqueness of the IIC in high dimensions is the following:

Theorem 12.1 (IIC construction [145, 161]). *For $d \geq 11$ and any* cylinder event E, *the limits*

$$\mathbb{P}_{\text{IIC}}(E) := \lim_{|x|\to\infty} \mathbb{P}_{p_c}(E \mid 0 \leftrightarrow x) \tag{12.2.1}$$

and

$$\mathbb{Q}_{\text{IIC}}(E) := \lim_{p \nearrow p_c} \frac{\sum_{x\in\mathbb{Z}^d} \mathbb{P}_p(E \cap \{0 \leftrightarrow x\})}{\sum_{x\in\mathbb{Z}^d} \mathbb{P}_p(0 \leftrightarrow x)} \tag{12.2.2}$$

exist and $\mathbb{P}_{\text{IIC}}(E) = \mathbb{Q}_{\text{IIC}}(E)$. *If, furthermore, the extrinsic arm exponent exists in the asymptotic sense of Open Problem 11.2, then*

$$\mathbb{R}_{\text{IIC}}(E) := \lim_{n\to\infty} \mathbb{P}_{p_c}\left(E \mid 0 \leftrightarrow \partial\Lambda_n\right) \tag{12.2.3}$$

exists and $\mathbb{R}_{\text{IIC}}(E) = \mathbb{P}_{\text{IIC}}(E) = \mathbb{Q}_{\text{IIC}}(E)$.

A few remarks are in place concerning Thm. 12.1:
- Note that the cylinder events form an algebra that is stable under intersections, and that the consistency hypothesis of Kolmogorov's extension theorem is satisfied by the definition of $\mathbb{P}_{\text{IIC}}$ in Thm. 12.1 (see, e.g., [242]). Therefore, Kolmogorov's extension theorem implies that (12.2.1) determines a *unique* measure on the σ-fields of events, which we denote the *incipient infinite cluster measure*.
- The limit in (12.2.1) does not depend on the way in which $|x|$ diverges to infinity. This is related to the asymptotic rotational symmetry of $\tau_{p_c}(x)$ in Thm. 11.4.

• It is straightforward to see that indeed $\mathbb{P}_{\mathrm{IIC}}(|\mathcal{C}(0)| = \infty) = 1$, as desired. Since $\theta(p_c) = 0$, the IIC is also *one-ended*, in the sense that the removal of any finite region of the IIC leaves one infinite part. It can be seen that the infinite path is *essentially unique* in the sense that any pair of infinite self-avoiding paths in the IIC share infinitely many bonds.

• The existence of the extrinsic arm exponent in the bounded-ratio form in Thm. 11.5 is sufficient to deduce that the limit in (12.2.3) exists *along subsequences*; the strong form of Open Problem 11.2 is needed in a regularity argument to show that the limit is unique.

Mind that the measure $\mathbb{P}_{\mathrm{IIC}}$ has lost the translation invariance of the percolation measures $\mathbb{P}_p$. Indeed, the point 0 plays a special role, since we have enforced that the cluster $\mathcal{C}(0)$ is infinite. The cluster $\mathcal{C}(0)$ under the IIC measure $\mathbb{P}_{\mathrm{IIC}}$ is clearly larger than the law of $\mathcal{C}(0)$ under the critical percolation measure $\mathbb{P}_{p_c}$. One consequence is that the two-point function changes under the measure $\mathbb{P}_{\mathrm{IIC}}$:

Theorem 12.2 (IIC properties [161]). *Under the assumptions of Thm. 12.1, there exist constants $C_{\mathrm{IIC}}, c_{\mathrm{IIC}} > 0$ such that*

$$c_{\mathrm{IIC}}|y|^{4-d} \leq \mathbb{P}_{\mathrm{IIC}}(0 \leftrightarrow y) \leq C_{\mathrm{IIC}}|y|^{4-d} \quad as\ |y| \to \infty\,. \qquad (12.2.4)$$

Comparison with Thm. 11.4 shows that the conditioning indeed changes the two-point function. To illustrate this difference, we now prove the upper bound in Thm. 12.2.

Proof of the Upper Bound in Thm. 12.2. Denote

$$\mathbb{Q}_p(E) = \frac{\sum_{x \in \mathbb{Z}^d} \mathbb{P}_p(E \cap \{0 \leftrightarrow x\})}{\sum_{x \in \mathbb{Z}^d} \mathbb{P}_p(0 \leftrightarrow x)}, \qquad (12.2.5)$$

so that, by (12.2.2) and Thm. 12.1,

$$\mathbb{P}_{\mathrm{IIC}}(0 \leftrightarrow y) = \lim_{p \nearrow p_c} \mathbb{Q}_p(0 \leftrightarrow y). \qquad (12.2.6)$$

The careful reader might note that Thm 12.1 requires a cylinder event, and thus cannot be applied to $\{0 \leftrightarrow y\}$. In such cases, we approximate $\{0 \leftrightarrow y\}$ as the increasing limit of $\{0 \leftrightarrow y$ in $\Lambda_n\}$, apply the construction for these events, and subsequently take the limit $n \to \infty$. We note that

$$\mathbb{Q}_p(0 \leftrightarrow y) = \frac{\sum_{x \in \mathbb{Z}^d} \mathbb{P}_p(\{0 \leftrightarrow y\} \cap \{0 \leftrightarrow x\})}{\sum_{x \in \mathbb{Z}^d} \mathbb{P}_p(0 \leftrightarrow x)}\,. \qquad (12.2.7)$$

By (4.2.24),

$$\mathbb{P}_p(\{0 \leftrightarrow y\} \cap \{0 \leftrightarrow x\}) \leq \sum_{z \in \mathbb{Z}^d} \mathbb{P}_p(0 \leftrightarrow z)\,\mathbb{P}_p(z \leftrightarrow y)\,\mathbb{P}_p(z \leftrightarrow x)\,. \qquad (12.2.8)$$

Therefore,

$$\mathbb{Q}_p(0 \leftrightarrow y) \leq \frac{\sum_{x,z \in \mathbb{Z}^d} \mathbb{P}_p(0 \leftrightarrow z) \, \mathbb{P}_p(z \leftrightarrow y) \, \mathbb{P}_p(z \leftrightarrow x)}{\sum_{x \in \mathbb{Z}^d} \mathbb{P}_p(0 \leftrightarrow x)}$$

$$= \sum_{z \in \mathbb{Z}^d} \mathbb{P}_p(0 \leftrightarrow z) \, \mathbb{P}_p(z \leftrightarrow y) = (\tau_p \star \tau_p)(y) \, . \tag{12.2.9}$$

Letting $p \nearrow p_c$ proves that $\mathbb{P}_{\mathrm{IIC}}(0 \leftrightarrow y) \leq (\tau_{p_c} \star \tau_{p_c})(y)$, which is upper bounded by $c|y|^{-(d-4)}$ by Thm. 11.4, as required.   □

We see that the limiting scheme in (12.2.2) is particularly convenient to prove properties of the limiting IIC measure $\mathbb{P}_{\mathrm{IIC}}$. It is possible to adapt the proof for the limiting scheme in (12.2.1), but this is a little more involved and left as an exercise:

Exercise 12.1 (Upper bound IIC two-point function). Prove that $\mathbb{P}_{\mathrm{IIC}}(0 \leftrightarrow y) \leq (\tau_{p_c} \star \tau_{p_c})(y)$ by using the limiting scheme in (12.2.1) instead.

In order to understand the high-dimensional IIC, it is worthwhile to investigate its mean-field model. Just as critical branching random walk (BRW) is the mean-field model for percolation, critical BRW conditioned on nonextinction is the mean-field model for the high-dimensional IIC:

The IIC for BRW. We consider critical BRW with $\mathsf{Bin}(2d, p)$-offspring distribution where $p = p_c = 1/(2d)$, and we condition the total progeny to be infinite. Since critical BRW dies out a.s., also here we need to take an appropriate limit. The nice thing is that we can perform this limit explicitly. Let $\mathbb{P}_{p_c}$ denote the BRW measure at $p = p_c$, and $\mathbb{E}_{p_c}$ its corresponding expectation. Let N_m denote the number of particles in generation m (with $N_0 = 1$) and $\mathscr{F}_m$ be the σ-algebra generated by all events that are determined by the BRW up to time m.

A natural candidate for the BRW IIC arises when we condition on survival until time n and then take the limit $n \to \infty$. Thus, we define

$$\mathbb{P}_{\mathrm{IIC}}(E) = \lim_{n \to \infty} \mathbb{P}_{p_c}(E \mid N_n \geq 1) \, , \tag{12.2.10}$$

assuming that this limit exists.

Lemma 12.3. *The limit in (12.2.10) exists for any cylinder event E, and if E is measurable w.r.t. $\mathscr{F}_m$, then*

$$\mathbb{P}_{\mathrm{IIC}}(E) = \mathbb{E}_{p_c}[\mathbb{1}_E N_m] \, . \tag{12.2.11}$$

Proof. Clearly, (12.2.11) defines a probability measure, since $(N_m)_{m \geq 0}$ is a nonnegative martingale with $\mathbb{E}_{p_c}[N_m] = 1$. This also implies that $\mathbb{P}_{\mathrm{IIC}}$ in (12.2.11) is consistent, since $\mathbb{E}_{p_c}[N_n \mid N_m] = N_m$ for $n \geq m$.

Since $\theta_n = \mathbb{P}_{p_c}(N_n \geq 1)$ satisfies that $n\theta_n \to C_{\mathrm{in}}$ (recall Thm. 2.1), we obtain

$$\mathbb{P}_{\mathrm{IIC}}(E) = (1/C_{\mathrm{in}}) \lim_{n \to \infty} n \, \mathbb{P}_{p_c}(E \cap \{N_n \geq 1\}) \, . \tag{12.2.12}$$

Since E is $\mathscr{F}_m$-measurable and the N_m particles present at time m are all independent,

$$\mathbb{P}_{p_c}(E \cap \{N_n \geq 1\}) = \mathbb{E}_{p_c}\left[\mathbb{1}_E\left(1 - (1 - \theta_{n-m})^{N_m}\right)\right]. \qquad (12.2.13)$$

When $n \to \infty$, therefore,

$$n\,\mathbb{P}_{p_c}(E \cap \{N_n \geq 1\}) = n\mathbb{E}_{p_c}\left[\mathbb{1}_E\left(1 - (1 - \theta_{n-m})^{N_m}\right)\right]$$
$$\to C_{\text{in}}\mathbb{E}_{p_c}[\mathbb{1}_E\,N_m]\,, \qquad (12.2.14)$$

which, combined with (12.2.12), proves (12.2.11). □

Exercise 12.2 (Equivalent limit scheme for BRW IIC). Define

$$\mathbb{Q}_{\text{IIC}}(E) := \lim_{n\to\infty} \mathbb{E}_{p_c}[\mathbb{1}_E\,N_n]\,. \qquad (12.2.15)$$

Prove that the limit in (12.2.15) exists, and that $\mathbb{Q}_{\text{IIC}} = \mathbb{P}_{\text{IIC}}$.

The Existence of the Percolation IIC. Here we sketch the proof of Thm. 12.1. Let E be an event that is determined by the bonds in Λ_m. Since E is determined by the occupation status of a finite number of bonds, such an m must exist. Then, we recall (12.2.5) to see that we need to investigate $\sum_{x\in\mathbb{Z}^d} \mathbb{P}_p(E \cap \{0 \leftrightarrow x\})$. For $x \in \mathbb{Z}^d$, we split depending on whether there is a pivotal bond for $\Lambda_m \leftrightarrow x$ or not. When there is a pivotal bond, we use adaptations of the Hara–Slade inclusion–exclusion proof to show that there exists $\Pi_p(x; E)$ such that

$$\mathbb{P}_p(E \cap \{0 \leftrightarrow x\}) = \Pi_p(x; E) + 2dp \sum_{u,v} \Pi_p(u; E)D(v - u)\tau_p(x - v)\,. \qquad (12.2.16)$$

Now we can sum out over all x, and use that $\sum_x |\Pi_p(x; E)| < \infty$ to arrive at

$$\mathbb{Q}_p(E) = \frac{1}{\chi(p)} \sum_{x\in\mathbb{Z}^d} \mathbb{P}_p(E \cap \{0 \leftrightarrow x\})$$
$$= \frac{\widehat{\Pi}_p(0; E)}{\chi(p)} + 2dp\widehat{\Pi}_p(0; E)\,, \qquad (12.2.17)$$

where we note that the factor $\chi(p) = \sum_{x\in\mathbb{Z}^d} \tau_p(x - v)$ cancels in the last term. Letting $p \nearrow p_c$, we see that the first term vanishes, so that

$$\mathbb{Q}_{\text{IIC}}(E) = 2dp_c\widehat{\Pi}_{p_c}(0; E)\,. \qquad (12.2.18)$$

The other limiting schemes in (12.2.1) and (12.2.3) can be seen to converge to the same limit, which implies that $\mathbb{P}_{\text{IIC}} = \mathbb{Q}_{\text{IIC}} = \mathbb{R}_{\text{IIC}}$. □

Chapter 13
Finite-Size Scaling and Random Graphs

So far, we have considered percolation on the *infinite lattice* $\mathbb{Z}^d$. When, instead, considering percolation on a *bounded* domain, the clusters are restricted by the boundary of the domain under consideration. This leads to so-called *finite-size effects*. For example, in the supercritical regime, there is a large connected component, but this component cannot be infinite. Additionally, in any finite domain, the probability of a certain event and expectations of random variables are continuous functions of the percolation threshold p (even polynomials). Thus, discontinuities as for $p \mapsto \mathbb{E}_p[|\mathcal{C}(0)|]$ or in the derivative of $p \mapsto \theta(p)$, such as present in high-dimensional percolation, cannot occur. Instead, one has to deal with *asymptotic* phase transitions, for example, in the proportion of vertices in the largest connected component.

In finite domains, one can expect that the boundary conditions play hardly any role when the value of p is far away from the critical value. When p is close to the critical value, clusters become fractal and self-similar, and clusters feel the boundary of the domain. In this chapter, we investigate these finite-size effects.

Additionally, when dealing with, say, a finite domain that is a subset of $\mathbb{Z}^d$ of the volume of order n^d, we can investigate how the size of the clusters depends on p more closely, and take $p = p_n$ such that $p_n \to p_c$ at a certain rate. Now, when this convergence is sufficiently slow, then it is as if the value of p is sub- or supercritical, while if it is very quick, then it is *as if* the value is p_c. The values of p for which the behavior of percolation quantities, such as the expected cluster size or the largest connected component, is as in the critical case are sometimes called the *scaling window* or *critical window*. We aim to derive the asymptotic properties of the critical window in high-dimensional percolation on tori. In statistical mechanics, the setting of percolation on a torus is also termed percolation on a cube with periodic boundary conditions.

The main advantage of working with the torus is that it is a *transitive* graph. As such, the boundary of the graph is "equally far away for every vertex." This is different for *zero* boundary conditions, where only connections inside the cube are allowed. There, one may expect large clusters to be more likely to sit close to the center of the cube. See Sect. 13.6 where the role of boundary conditions is investigated in more detail.

© Springer International Publishing Switzerland 2017
M. Heydenreich and R. van der Hofstad, *Progress in High-Dimensional Percolation and Random Graphs*, CRM Short Courses, DOI 10.1007/978-3-319-62473-0_13

In this chapter, we focus on percolation on the high-dimensional torus, with the following main aim:

> Show that the mean-field model for percolation on the high-dimensional torus is the Erdős–Rényi random graph, in the sense that the phase transition for percolation on the high-dimensional torus mimics that of percolation on the complete graph.

Investigating mean-field critical behavior on various graphs is a highly active field of research. In general, for geometric graphs, in many settings the subcritical behavior and (with considerably more effort) the critical behavior are fairly well understood. Interestingly, the supercritical regime is not. In particular, the "barely supercritical" regime (i.e., close to the critical value) is a challenge, and it is approached by investigating specific examples, such as the complete graph, products of it, the hypercube, and various random graphs.

This chapter is organized as follows. In Sect. 13.1, we draw inspiration from the Erdős–Rényi random graph, which is percolation on the complete graph. We rigorously prove a number of statements for the Erdős–Rényi random graph, whose proof can be modified to the setting of high-dimensional tori. In Sect. 13.2, we then proceed to critical percolation on high-dimensional tori. In Sect. 13.3, we extend this to more general tori including the hypercube. In Sect. 13.4, we focus exclusively on the hypercube, where also the supercritical regime is now well understood. In Sect. 13.5, we discuss scaling limits of critical percolation on random graphs. We close this chapter in Sect. 13.6 by discussing the role of boundary conditions beyond the periodic boundary conditions that give rise to high-dimensional tori.

13.1   Inspiration: The Erdős–Rényi Random Graph

A leitmotif in the text so far is that for percolation in high dimensions, we can think of far away clusters as being close to independent, and thus, geometry plays a less important role. This is why percolation on $\mathbb{Z}^d$ shares many features with critical branching random walk on $\mathbb{Z}^d$. The only *finite* graph that is transitive and has no geometry is the complete graph K_n. For K_n, the vertex set is $[n] = \{1, \ldots, n\}$ and the edge set $\mathcal{E}$ consists of all pairs $\{i, j\}$ with $i, j \in [n]$ such that $i \neq j$. For simplicity of notation, we often write $ij = \{i, j\}$ in this chapter, so that $ij = ji$.

Percolation on the complete graph is obtained by letting every edge in $\mathcal{E}$ be occupied with probability p and vacant with probability $1 - p$, independently across the edges. This model is also called the *Erdős–Rényi random graph*, named after Paul Erdős and Alfréd Rényi who were the first to deduce scaling properties of this model for p close to criticality [98]. We denote the Erdős–Rényi random graph with n vertices and edge probability p by $\mathrm{ER}_n(p)$. In fact, this model, which is sometimes called the *binomial* model, was introduced by Gilbert [117], while Erdős and Rényi investigated the closely related setting in which a *fixed* number of uniformly chosen edges is added.

Write $\mathcal{C}_{(j)}$ for the jth largest component, and $|\mathcal{C}_{(j)}|$ for the number of vertices in $\mathcal{C}_{(j)}$, so that $|\mathcal{C}_{(1)}| \geq |\mathcal{C}_{(2)}| \geq \cdots$. Additionally, we often write $\mathcal{C}_{(1)} = \mathcal{C}_{max}$. An inspiring discovery of Erdős and Rényi [98] is that this model exhibits a phase transition when p is scaled like $p = \lambda/n$. When $\lambda < 1$, we have $|\mathcal{C}_{(1)}| = \Theta_{\mathbb{P}}(\log n)$ whp while $|\mathcal{C}_{(1)}| = \Theta_{\mathbb{P}}(n)$ whp when $\lambda > 1$. This can be understood by noting that, for $p = \lambda/n$, the average degree is close to λ. Thus, for $\lambda > 1$, locally and in expectation, boundaries of clusters in the graph distance are growing exponentially in the graph distance, while for $\lambda < 1$, their expectations decay exponentially. This leads us to the above prediction. More precisely, in $ER_n(p)$, the scaling of the largest connected components can be subdivided into the following three cases:

The Subcritical Phase. Let $\varepsilon = \varepsilon_n = o(1)$ be a nonnegative sequence with $\varepsilon \gg n^{-1/3}$ and put $p = (1 - \varepsilon)/n$. Then, for any fixed integer $j \geq 1$,

$$\frac{|\mathcal{C}_{(j)}|}{2\varepsilon^{-2} \log(\varepsilon^3 n)} \xrightarrow{\mathbb{P}} 1. \tag{13.1.1}$$

The scaling in (13.1.1) can be interpreted as that the largest connected components in $ER_n(\lambda/n)$ are obtained from a maximum of n i.i.d. random variables $(X_i)_{i \in [n]}$, each having an exponential tail $\mathbb{P}(X_i \geq \ell) \sim e^{-\ell \varepsilon^2/2}$. This is almost true, since the cluster sizes $(|\mathcal{C}(i)|)_{i \in [n]}$ are close to being independent and do satisfy that $\mathbb{P}(|\mathcal{C}(i)| \geq \ell) \sim e^{-\ell \varepsilon^2/2}$, but their dependence is crucial to obtain the factor $\log(\varepsilon^3 n)$, which the above crude argument does not yield. When inserting more precise estimates in $\mathbb{P}(|\mathcal{C}(i)| \geq \ell) \sim e^{-\ell \varepsilon^2/2}$, one can get the factor $2\varepsilon^{-2}$ correctly, but not the factor $\log(\varepsilon^3 n)$. To see this additional factor, instead, one needs to take into account that if $|\mathcal{C}(i)| = m$, then there are $m - 1$ other vertices apart from i that have the same cluster size. We see such effects in more detail below.

The Supercritical Phase. Let $\varepsilon = \varepsilon_n = o(1)$ be a nonnegative sequence with $\varepsilon \gg n^{-1/3}$ and put $p = (1 + \varepsilon)/n$. Then,

$$\frac{|\mathcal{C}_{(1)}|}{2\varepsilon n} \xrightarrow{\mathbb{P}} 1, \tag{13.1.2}$$

and, for any fixed integer $j \geq 2$,

$$\frac{|\mathcal{C}_{(j)}|}{2\varepsilon^{-2} \log(\varepsilon^3 n)} \xrightarrow{\mathbb{P}} 1. \tag{13.1.3}$$

Equations (13.1.2)–(13.1.3) show that the largest connected component has a size that is *concentrated*, and all other connected components are much smaller. It is not hard to extend (13.1.2)–(13.1.3) to the (easier) setting where $p = \lambda/n$ with λ fixed and $\lambda > 1$, where the limit in (13.1.2) needs to be replaced by the survival probability $\zeta = \zeta(\lambda)$ of a Poisson branching process with mean offspring λ. When $\lambda \searrow 1$, then $\zeta(\lambda) \sim 2(\lambda - 1)$, which explains the factor 2ε in (13.1.2). The proof of (13.1.2)–(13.1.3) relies on *branching process approximations* that we explain in some detail below.

In both the sub- and the supercritical phase, the limit of $|\mathcal{C}_{(1)}| = |\mathcal{C}_{max}|$, properly normalized, is *deterministic*. At the critical point, one can expect that such a scaling limit is *random*. This is reflected in the following asymptotics:

The Critical Window When $p = (1 + \theta n^{-1/3})/n$ for some $\theta \in \mathbb{R}$, for any fixed integer $j \geq 1$,

$$\left(n^{-2/3}|\mathcal{C}_{(1)}|, \ldots, n^{-2/3}|\mathcal{C}_{(j)}|\right) \overset{d}{\to} (\gamma_1, \ldots, \gamma_j), \tag{13.1.4}$$

where $(\gamma_i)_{i=1}^{j}$ are nondegenerate random variables supported on $(0, \infty)$, and $\overset{d}{\to}$ denotes convergence in distribution. Thus, we see that large critical clusters obey nontrivial scaling, in that they are of order $n^{2/3}$. The power $2/3$ was first identified by Erdős and Rényi in their seminal work [98]. It is sometimes called the *double jump*, since the maximal cluster size jumps first from $2\varepsilon^{-2} \log(\varepsilon^3 n)$ to $n^{2/3}$ and then from $n^{2/3}$ to $2\varepsilon n$. This "jump," however, turns out to be quite smooth indeed, the phase transition only becoming sharp when $p = \lambda/n$ with λ fixed, and $n \to \infty$. We discuss the critical window in much more detail in Sect. 13.5 below.

In this text, we focus on the critical behavior in percolation, and we thus focus on the critical window as described in (13.1.4). In this section, we do not give a full proof of (13.1.4), but instead show that both $n^{-2/3}|\mathcal{C}_{\max}|$ and $n^{2/3}/|\mathcal{C}_{\max}|$ are *tight* sequences of random variables. This is because the proof of these statements is the most robust and can be adapted to high-dimensional tori, as explained in more detail in this section, as well as to *inhomogeneous random graphs* in Sect. 13.5. In that section, we also investigate the scaling limits of the largest connected components in the random graph setting.

We summarize that the prominent qualitative features of the phase transition on the Erdős–Rényi random graph are:

• The emergence of the giant component occurs just above the scaling window. That is, only in the supercritical phase $|\mathcal{C}_{(2)}| \ll |\mathcal{C}_{\max}|$, and $|\mathcal{C}_{\max}|/n$ increases suddenly but smoothly above the critical value (in mathematical physics terminology, the phase transition is of *second order*).

• Concentration of the size of the largest connected components outside the scaling window and nonconcentration inside the window.

• Duality: $|\mathcal{C}_{(2)}|$ in the supercritical phase has the same asymptotics as $|\mathcal{C}_{\max}|$ in the corresponding subcritical phase.

The aim of this chapter is to state and partly derive similar results for percolation on high-dimensional tori.

In the remainder of this section, we make the discussion of the behavior of the largest connected component of $\mathrm{ER}_n(\lambda/n)$, for p close to the critical value $1/n$, precise. The main result is the following, proving that indeed $n^{-2/3}|\mathcal{C}_{\max}|$ and $n^{2/3}/|\mathcal{C}_{\max}|$ form tight sequences of random variables:

Theorem 13.1 (Largest critical cluster). *Take $p = (1 + \theta n^{-1/3})/n$, where $\theta \in \mathbb{R}$. There exists a constant $b = b(\theta) > 0$ such that, for all $A > 1$,*

$$\mathbb{P}_p\left(A^{-1}n^{2/3} \leq |\mathcal{C}_{\max}| \leq An^{2/3}\right) \geq 1 - \frac{b}{A}. \tag{13.1.5}$$

In the next section, we describe the strategy of proof of Thm. 13.1, giving quite some details.

13.1.1 Strategy of the Proof of Thm. 13.1

We follow the second author [154]. A key ingredient to the proofs is the ingenious choice of a certain family of random variables $Z_{\geq k}$. We denote the number of vertices in connected components of size at least k by

$$Z_{\geq k} = \sum_{v \in [n]} \mathbb{1}_{\{|\mathscr{C}(v)| \geq k\}} . \tag{13.1.6}$$

Since

$$\{|\mathscr{C}_{\max}| \geq k\} = \{Z_{\geq k} \geq k\}, \tag{13.1.7}$$

we can prove bounds on $|\mathscr{C}_{\max}|$ by investigating $Z_{\geq k}$ for appropriately chosen values of k. This strategy has been successfully applied in several related settings, and we see some more examples below.

Exercise 13.1 (Relation $|\mathscr{C}_{\max}|$ and $Z_{\geq k}$). Prove (13.1.7).

The nice thing about $Z_{\geq k}$ is that it is a *sum* of indicators. While these indicators are *dependent*, they turn out to be sufficiently weakly dependent to make their analysis possible using first and second moment methods. This is in sharp contrast to $|\mathscr{C}_{\max}| = \max_{v \in [n]} |\mathscr{C}(v)|$, which is the maximum of dependent random variables and as such much more difficult to deal with. We note that $\mathbb{E}[Z_{\geq k}] = n\,\mathbb{P}(|\mathscr{C}(1)| \geq k)$, so that the study of $Z_{\geq k}$ quickly leads us to study tail probabilities of cluster sizes.

We see that we need to understand the cluster tails to study the first moment of $Z_{\geq k}$. We next investigate the variance of $Z_{\geq k}$, which allows us to apply a *second moment method* on $Z_{\geq k}$. We state this result more generally, since we later wish to apply it to other settings. To state the result, we say that a random graph is an *inhomomegeneous random graph* with edge probabilities $\mathbf{p} = (p_{ij})_{1 \leq i < j \leq n}$ when the edge ij is occupied with probability p_{ij} and the occupation statuses of different edges are independent. Then, the main variance estimate on $Z_{\geq k}$ is as follows:

Proposition 13.2 (A variance estimate for $Z_{\geq k}$). *For an inhomogeneous random graph with edge probabilities $\mathbf{p} = (p_{ij})_{1 \leq i < j \leq n}$, every n and $k \geq 1$,*

$$\mathrm{Var}(Z_{\geq k}) \leq \sum_{i \in [n]} \mathbb{E}\big[|\mathscr{C}(i)| \mathbb{1}_{\{|\mathscr{C}(i)| \geq k\}}\big] .$$

Proof We use the fact that

$$\mathrm{Var}(Z_{\geq k}) = \sum_{i,j \in [n]} \big[\mathbb{P}(|\mathscr{C}(i)| \geq k, |\mathscr{C}(j)| \geq k) - \mathbb{P}(|\mathscr{C}(i)| \geq k)\,\mathbb{P}(|\mathscr{C}(j)| \geq k)\big] . \tag{13.1.8}$$

We split the probability $\mathbb{P}(|\mathcal{C}(i)| \geq k, |\mathcal{C}(j)| \geq k)$, depending on whether $i \leftrightarrow j$ or not, i.e.,

$$
\mathbb{P}(|\mathcal{C}(i)| \geq k, |\mathcal{C}(j)| \geq k) = \mathbb{P}(|\mathcal{C}(i)| \geq k, |\mathcal{C}(j)| \geq k, i \leftrightarrow j)
$$
$$
+ \mathbb{P}(|\mathcal{C}(i)| \geq k, |\mathcal{C}(j)| \geq k, i \nleftrightarrow j) . \quad (13.1.9)
$$

We condition on the edges and vertices in $\mathcal{C}(i)$ to obtain

$$
\mathbb{P}(|\mathcal{C}(i)| \geq k, |\mathcal{C}(j)| \geq k, i \nleftrightarrow j)
$$
$$
= \sum_{S:|S| \geq k} \mathbb{P}(\mathcal{C}(i) = S) \mathbb{P}(|\mathcal{C}(j)| \geq k, i \nleftrightarrow j \mid \mathcal{C}(i) = S) , \quad (13.1.10)
$$

where the sum is over all collections of vertices $S \subseteq [n]$ satisfying $i \in S$ and $|S| \geq k$. Now, in order for $|\mathcal{C}(j)| \geq k$ and $i \nleftrightarrow j$ to occur, $|\mathcal{C}(j)| \geq k$ must happen without using any of the vertices in S, which clearly has a smaller probability than $\mathbb{P}(|\mathcal{C}(j)| \geq k)$. Therefore,

$$
\mathbb{P}(|\mathcal{C}(j)| \geq k, i \nleftrightarrow j \mid \mathcal{C}(i) = S) \leq \mathbb{P}(|\mathcal{C}(j)| \geq k) . \quad (13.1.11)
$$

Applying this yields

$$
\text{Var}(Z_{\geq k}) \leq \sum_{i,j \in [n]} \mathbb{P}(|\mathcal{C}(i)| \geq k, |\mathcal{C}(j)| \geq k, i \leftrightarrow j) , \quad (13.1.12)
$$

and we arrive at the fact that

$$
\text{Var}(Z_{\geq k}) \leq \sum_{i,j \in [n]} \mathbb{P}(|\mathcal{C}(i)| \geq k, |\mathcal{C}(j)| \geq k, i \leftrightarrow j)
$$
$$
= \sum_{i \in [n]} \sum_{j \in [n]} \mathbb{E}\left[\mathbb{1}_{\{|\mathcal{C}(i)| \geq k\}} \mathbb{1}_{\{j \in \mathcal{C}(i)\}} \right]
$$
$$
= \sum_{i \in [n]} \mathbb{E}\left[\mathbb{1}_{\{|\mathcal{C}(i)| \geq k\}} \sum_{j \in [n]} \mathbb{1}_{\{j \in \mathcal{C}(i)\}} \right]
$$
$$
= \sum_{i \in [n]} \mathbb{E}\left[|\mathcal{C}(i)| \mathbb{1}_{\{|\mathcal{C}(i)| \geq k\}} \right] . \quad (13.1.13)
$$

$\square$

13.1.2 Critical Scaling of Cluster Sizes in the Erdős–Rényi Random Graph

In this section, we investigate cluster tails and expected cluster sizes of the Erdős–Rényi random graph within the scaling window. First, we show that the cluster tail is, within the critical window, of the order $1/\sqrt{k}$ just as for critical high-dimensional percolation:

Proposition 13.3 (Critical cluster tails). *Take $p = (1 + \theta n^{-1/3})/n$, where $\theta \in \mathbb{R}$, and let $r > 0$. For $k \leq rn^{2/3}$, there exist constants $0 < c_1 < c_2 < \infty$ with $c_1 = c_1(r, \theta)$ satisfying $\min_{r \leq 1} c_1(r, \theta) > 0$, and c_2 independent of r and θ, such that, for n sufficiently large,*

$$\frac{c_1}{\sqrt{k}} \leq \mathbb{P}_p(|\mathcal{C}(1)| \geq k) \leq c_2\left((\theta \vee 0)n^{-1/3} + \frac{1}{\sqrt{k}}\right). \tag{13.1.14}$$

Proposition 13.3 implies that the tail of the critical cluster size distribution obeys similar asymptotics as the tail of the total progeny of a critical branching process (recall the "$\delta = 2$" result in Thm. 2.1). The lower bound on the tail in (13.3.23) can only be valid for values of k that are not too large. Indeed, when $k > n$, then $\mathbb{P}_\lambda(|\mathcal{C}(v)| \geq k) = 0$. Therefore, there must be a cutoff above which the asymptotics fails to hold. As it turns out, this cutoff is given by $rn^{2/3}$. The upper bound in (13.3.23) holds for a wider range of k, and in fact, the proof yields that the upper bound in (13.3.23) is valid for *all k*.

Sketch proof of Prop. 13.3. Let T_m denote the total progeny of a branching process with a $\mathsf{Bin}(m, p)$ offspring distribution. Then, we claim that

$$\mathbb{P}_p(T_{n-k} \geq k) \leq \mathbb{P}_p(|\mathcal{C}(1)| \geq k) \leq \mathbb{P}_p(T_n \geq k), \tag{13.1.15}$$

The proof of (13.1.15) is not very hard, and we leave it to the reader. A full proof can be found in [155, Proof of Thms. 4.2 and 4.3].

Exercise 13.2 (Stochastic domination of $|\mathcal{C}(1)|$ in terms of T_n). Prove the upper bound in (13.1.15) by proving that $|\mathcal{C}(1)|$ is stochastically dominated by T_n.

Exercise 13.3 (Stochastic relation of $|\mathcal{C}(1)|$ in terms of T_{n-k}). Prove the lower bound in (13.1.15) by using that, when exploring a cluster up to its kth element, the number of potential neighbors on the complete graph is always at least $n - k$.

Special attention needs to be paid to the case where $T_n = \infty$, which occurs with positive probability when $p > 1/n$. We split

$$\mathbb{P}_p(T_n \geq k) = \mathbb{P}_p(T_n = \infty) + \sum_{l=k}^{\infty} \mathbb{P}_p(T_n = l). \tag{13.1.16}$$

By the Random Walk Hitting Time Theorem (recall (2.1.24)),

$$\mathbb{P}_p(T_m = l) = \frac{1}{l}\, \mathbb{P}_p(X_1 + \cdots + X_l = l - 1)\,, \qquad (13.1.17)$$

where $X_i \sim \mathsf{Bin}(m, p)$. Then, it is not hard to use Stirling's formula to prove that

$$\mathbb{P}_p(T_n \geq k) \leq c_2\left((\theta \vee 0)n^{-1/3} + \frac{1}{\sqrt{k}}\right). \qquad (13.1.18)$$

Exercise 13.4 (Tails of T_n). Complete the proof of (13.1.18).

We continue by investigating the expected cluster size at the bottom part of the scaling window:

Lemma 13.4 (Bound on critical expected cluster size). *Take $p = (1 + \theta n^{-1/3})/n$ with $\theta < 0$. Then, for all $n \geq 1$,*

$$\mathbb{E}_p[|\mathcal{C}(1)|] \leq n^{1/3}/|\theta|. \qquad (13.1.19)$$

Proof The upper bound in (13.1.15) gives that $|\mathcal{C}(1)|$ is stochastically dominated by T_n, where T_n is the total progeny of a branching process with a $\mathsf{Bin}(n, \lambda/n)$ offspring distribution, and where $p = (1 + \theta n^{-1/3})/n$. As a result, for $\theta < 0$, since $\mathbb{E}[T] = 1/(1 - \mu)$ when T is the total progeny of a branching process with mean $\mu < 1$ offspring, we obtain

$$\mathbb{E}_p[|\mathcal{C}(1)|] \leq \mathbb{E}[T_n] = 1/(1 - \lambda) = n^{1/3}/|\theta|. \qquad (13.1.20)$$

This proves the claim. $\qquad \qquad \qquad \qquad \qquad \qquad \qquad \qquad \qquad \qquad \qquad \square$

Lemma 13.4 is intuitively consistent with Thm. 13.1. Indeed, in the critical regime, one can expect the largest cluster to contribute substantially to the expected cluster size. This suggests that

$$\mathbb{E}_p[|\mathcal{C}(1)|] \approx \mathbb{E}_p[|\mathcal{C}(1)|\mathbb{1}_{\{1 \in \mathcal{C}_{\max}\}}]$$
$$= \mathbb{E}_p[|\mathcal{C}_{\max}|\mathbb{1}_{\{1 \in \mathcal{C}_{\max}\}}] = \frac{1}{n}\, \mathbb{E}_p[|\mathcal{C}_{\max}|^2]\,, \qquad (13.1.21)$$

where $\approx$ denotes asymptotic equality with an uncontrolled error. When $|\mathcal{C}_{\max}| = \Theta_{\mathbb{P}}(n^{2/3})$, intuitively,

$$\mathbb{E}_p[|\mathcal{C}(1)|] \approx \frac{1}{n}\, \mathbb{E}_p[|\mathcal{C}_{\max}|^2] \approx n^{1/3}\,. \qquad (13.1.22)$$

The above heuristic is confirmed by Lem. 13.4, at least when $\theta < 0$. With a little more effort, we can show that Lem. 13.4 remains to hold for *all* $\theta \in \mathbb{R}$. We refrain from proving this here and return to this question in the next section.

Exercise 13.5 (Critical expected cluster size). Prove that Prop. 13.3 also implies that

$$\mathbb{E}_p[|\mathcal{C}(1)|] \geq cn^{1/3} \tag{13.1.23}$$

for some $c > 0$. Therefore, for $p = (1 + \theta n^{-1/3})/n$ with $\theta < 0$, the bound in Lem. 13.4 is asymptotically sharp.

Proof of Thm. 13.1. We start by proving the upper bound on $|\mathcal{C}_{\max}|$. We use the first moment method (or Markov inequality) to bound

$$\mathbb{P}_p(|\mathcal{C}_{\max}| \geq k) = \mathbb{P}_p(Z_{\geq k} \geq k)$$
$$\leq \frac{1}{k}\mathbb{E}_p[Z_{\geq k}] = \frac{n}{k}\mathbb{P}_p(|\mathcal{C}(1)| \geq k) . \tag{13.1.24}$$

Taking $k = An^{2/3}$ and using the upper bound in Prop. 13.3 then leads to

$$\mathbb{P}_p(|\mathcal{C}_{\max}| \geq An^{2/3}) \leq \frac{c_2 n}{An^{2/3}}\left((\theta \vee 0)n^{-1/3} + \frac{1}{n^{1/3}\sqrt{A}}\right)$$
$$\leq \frac{c_2((\theta \vee 0) + 1/\sqrt{A})}{A} , \tag{13.1.25}$$

as required.

To prove the matching lower bound on $|\mathcal{C}_{\max}|$, we rely on a second moment method for $Z_{\geq k}$. We use that it suffices to study $p = (1 + \theta n^{-1/3})/n$ with $\theta < 0$, since increasing θ makes the event $|\mathcal{C}_{\max}| < k_n$ less likely. We use the fact that $|\mathcal{C}_{\max}| < k$ precisely when $Z_{\geq k} = 0$, to obtain that, with $k = k_n = A^{-1}n^{2/3}$,

$$\mathbb{P}_p\left(|\mathcal{C}_{\max}| < k_n\right) = \mathbb{P}_p\left(Z_{\geq k_n} = 0\right) \leq \frac{\mathrm{Var}_p(Z_{\geq k_n})}{\mathbb{E}_p[Z_{\geq k_n}]^2} . \tag{13.1.26}$$

By the lower bound in Prop. 13.3,

$$\mathbb{E}_p[Z_{\geq k_n}] = n\,\mathbb{P}_p(|\mathcal{C}(1)| \geq k_n) \geq \frac{nc_1 A^{1/2}}{n^{1/3}} = c_1 A^{1/2}n^{2/3} . \tag{13.1.27}$$

Also, by Prop. 13.2 and Lem. 13.4, with $|\theta| \geq 1$,

$$\mathrm{Var}_p(Z_{\geq k_n}) \leq n\,\mathbb{E}_p[|\mathcal{C}(1)|] \leq n^{4/3} . \tag{13.1.28}$$

Substituting (13.1.26)–(13.1.28), for n sufficiently large, leads to

$$\mathbb{P}_p\left(|\mathcal{C}_{\max}| < k_n\right) \leq \frac{n^{4/3}}{c_1^2 An^{4/3}} = \frac{1}{c_1^2 A} , \tag{13.1.29}$$

as required. $\square$

In the next section, we continue to investigate the critical behavior of percolation on high-dimensional tori. In the sequel, we are extending Thm. 13.1 to several high-dimensional tori.

13.2   Critical High-Dimensional Tori

In this section, we return to the study of percolation on high-dimensional tori. To this end, we call two vertices $x, y \in \mathbb{Z}^d$ n-connected (and write $x \overset{n}{\sim} y$) whenever $x - y \in n\mathbb{Z}^d$ and define $\mathbb{T}_{n,d} = \mathbb{Z}^d / \overset{n}{\sim}$. It is common to think of $\mathbb{T}_{n,d}$ as the cube $\{0, \dots, n-1\}^d$ with periodic boundary conditions. Mind that $\mathbb{T}_{n,d}$ is a *transitive* graph, that is, every vertex plays the same role.

We investigate the size of the maximal cluster on the torus, $|\mathscr{C}_{\max}| = \max_{x \in \mathbb{T}_{n,d}} |\mathscr{C}_{\mathbb{T}}(x)|$, at, or close to, the critical percolation threshold on the infinite lattice $p_c = p_c(\mathbb{Z}^d)$. Here, to avoid confusion, we write $\mathscr{C}_{\mathbb{T}}(x)$ for the cluster of x on the torus $\mathbb{T}_{n,d}$. Later, we also encounter its $\mathbb{Z}^d$ analog $\mathscr{C}_{\mathbb{Z}}(x)$. Additionally, we write $\tau_{\mathbb{T},p}(x)$ for the probability that $x \in \mathscr{C}_{\mathbb{T}}(0)$ and $V = n^d$ for the number of vertices or volume of the torus $\mathbb{T}_{n,d}$. The main result in this section is the following theorem:

Theorem 13.5 (Random graph asymptotics of the largest cluster size [143, 144]). *There is $d_0 > 6$ such that for percolation on the torus $\mathbb{T}_{n,d}$ with $d \geq d_0$ there exists a constant $b > 0$, such that for all $A \geq 1$ and all $n \geq 1$,*

$$\mathbb{P}_{p_c(\mathbb{Z}^d)}\left(A^{-1}V^{2/3} \leq |\mathscr{C}_{\max}| \leq AV^{2/3}\right) \geq 1 - \frac{b}{A}. \tag{13.2.1}$$

Theorem 13.5 is identical to Thm. 13.1 for the Erdős–Rényi random graph, and therefore, the asymptotics in Thm. 13.5 is sometimes called *random graph asymptotics*. Theorem 13.5 suggests that the Erdős–Rényi random graph is the mean-field model for critical percolation on high-dimensional tori.

The upper bound in (13.2.1) in Thm. 13.5 is proved in [143, Thm. 1.1]. That theorem also contains a lower bound that involves an extra logarithmic factor. This extra factor was removed in [144].

We next extend the above result to the other large clusters $\mathscr{C}_{(2)}, \mathscr{C}_{(3)}, \dots$ Our next result implies that the scaling of these clusters is similar to that of $|\mathscr{C}_{\max}|$:

Theorem 13.6 (Random graph asymptotics of the ordered cluster sizes [144]). *There is $d_0 > 6$ such that for percolation on the torus $\mathbb{T}_{n,d}$ with $d \geq d_0$ and every $m = 1, 2, \dots$ there exist constants $b_1, \dots, b_m > 0$, such that for all $A \geq 1$, $n \geq 1$, and all $j = 1, \dots, m$,*

$$\mathbb{P}_{p_c(\mathbb{Z}^d)}\left(A^{-1}V^{2/3} \leq |\mathscr{C}_{(j)}| \leq AV^{2/3}\right) \geq 1 - \frac{b_j}{A}. \tag{13.2.2}$$

Consequently, the expected cluster sizes satisfy $\mathbb{E}_{p_c(\mathbb{Z}^d)}[|\mathcal{C}_{(j)}|] \geq b'_j V^{2/3}$ *for certain constants* $b'_j > 0$. *Moreover, there are positive constants* c_1 *and* c_2 *such that*

$$\mathbb{P}_{p_c(\mathbb{Z}^d)}\left(|\mathcal{C}_{\max}| > AV^{2/3}\right) \leq \frac{c_1}{A^{3/2}} e^{-c_2 A} . \qquad (13.2.3)$$

Consequently, also $\mathbb{E}_{p_c(\mathbb{Z}^d)}[|\mathcal{C}_{\max}|] \leq bV^{2/3}$ *for some* $b > 0$.

We see below that Thms. 13.5 and 13.6 extend to various values of p sufficiently close to p_c. These values form the *critical* or *scaling window*. We defer this discussion to the next section.

The proofs of Thms. 13.5 and 13.6 are somewhat indirect. Indeed, we aim to apply the ideas for the critical Erdős–Rényi random graph to the setting of percolation on a high-dimensional torus. For the upper bound, this turns out to be relatively straightforward, as we indicate now. The lower bound is much more involved and is discussed in more detail in the next section.

Proposition 13.7 (Stochastic domination of clusters on torus by those on lattice, [40]).
Consider percolation on $\mathbb{T}_{n,d}$ *with* $n \geq 3$ *and any dimension* $d \geq 1$. *Then, the size of the cluster of the origin on the torus denoted by* $|\mathcal{C}_\mathbb{T}(0)|$ *is stochastically dominated by the size of the cluster of the origin on* $\mathbb{Z}^d$ *denoted by* $|\mathcal{C}_\mathbb{Z}(0)|$.

The proof is carried out via a coupling of the two clusters, where we use a technique known as *cluster exploration*. Here, we follow the original proof by Benjamini and Schramm [40, Thm. 1], which generalizes to any quotient graph. In [143, Prop. 2.1], we do the coupling in a different way, which not only gives the upper bound formulated in the proposition, but also a highly useful lower bound. This lower bound is the key ingredient to the lower bound of (13.2.1).

Proof We prove Prop. 13.7 in the generalized setup of quotient graphs. To this end, let Γ be a group of automorphisms of a graph $\mathcal{G}$. The quotient graph $\mathcal{G}' = \mathcal{G}/\Gamma$ is the graph whose vertices $\mathcal{V}(\mathcal{G}/\Gamma)$ are the equivalence classes $\mathcal{V}(\mathcal{G})/\Gamma = \{\Gamma v : v \in \mathcal{V}(\mathcal{G})\}$, and an edge $\{\Gamma u, \Gamma v\}$ appears in $\mathcal{G}/\Gamma$ if there are representatives $\{u_0 \in \Gamma u, v_0 \in \Gamma v\}$ that are neighbors in $\mathcal{G}$, i.e., $\{u_0, v_0\} \in \mathcal{E}(\mathcal{G})$. The setting on the torus $\mathbb{T}_{n,d}$ arises when we take $\mathcal{G} = \mathbb{Z}^d$ and Γ is the set of translations over $n\mathbb{Z}^d$.

We construct a coupling between percolation on $\mathcal{G}'$ and on $\mathcal{G}$. We start by describing an inductive procedure for constructing the percolation cluster of $v' \in \mathcal{V}(\mathcal{G}')$ known as *cluster exploration*. For a set of edges $B' \subset \mathcal{E}(\mathcal{G}')$, we write $\partial B'$ for the set of those edges that have precisely one endpoint in $\mathcal{V}(B')$, where $\mathcal{V}(B')$ is the union of all endpoints of edges in B'. Let C'_1 denote all the occupied edges incident to v' and W'_1 the vacant edges incident to v'. For $t \geq 2$, we proceed as follows. If $\partial C'_{t-1}$ is contained in W'_{t-1}, set $C'_t = C'_{t-1}, W'_t = W'_{t-1}$. Otherwise, choose an edge $b'_t \in \mathcal{E}(\mathcal{G}')$ that is not in $C'_{t-1} \cup W'_{t-1}$, but is in $\partial C'_{t-1}$. If b'_t is occupied, then let $C'_t = C'_{t-1} \cup \{b'_t\}$ and $W'_t = W'_{t-1}$; if vacant, let $C'_t = C'_{t-1}$ and $W'_t = W'_{t-1} \cup \{b'_t\}$. Then, $C' = \bigcup_t C'_t$ is the edge set of the percolation cluster of v', and its vertex set is denoted by $\mathcal{C}(v')$ in $\mathcal{G}'$.

We now describe the coupling with the percolation process in $\mathcal{G}$. Let f be a quotient map from $\mathcal{G}$ to $\mathcal{G}'$ with $f(v) = v'$. Again let C_1 denote all the occupied edges incident to v and W_1 the vacant edges incident to v. Assume that $t \geq 2$ and C_{t-1}, W_{t-1} were defined and satisfy $f(C_{t-1}) = C'_{t-1}$, $f(W_{t-1}) = W'_{t-1}$. If the construction of $\mathcal{C}(v)$ in $\mathcal{G}'$ is stopped at stage t, that is, if $C'_t = C'_{t-1}$ and $W'_t = W'_{t-1}$, then let $C_t = C_{t-1}$, $W_t = W_{t-1}$. Otherwise, let b_t be some edge in $f^{-1}(b'_t) \cap \partial C_{t-1}$. Let b_t be occupied if and only if b'_t is occupied and define C_t and W_t accordingly. Then, $\bigcup_t C_t$ is a connected set of occupied edges contained in the edge set of the percolation cluster $\mathcal{C}(v)$ of v. Hence, $f(\mathcal{C}(v)) \supseteq \mathcal{C}(v')$, and the result follows.   □

Partial proof of Thm. 13.5. We prove the upper bound on $|\mathcal{C}_{\max}|$ in Thm. 13.5. By Prop. 13.7 and at $p = p_c(\mathbb{Z}^d)$, $|\mathcal{C}_{\mathbb{T}}(0)|$ is stochastically dominated by $|\mathcal{C}_{\mathbb{Z}}(0)|$. Therefore, by Thm. 9.2,

$$\mathbb{P}_{p_c(\mathbb{Z}^d)}(|\mathcal{C}_{\mathbb{T}}(0)| > k) \leq \mathbb{P}_{p_c(\mathbb{Z}^d)}(|\mathcal{C}_{\mathbb{Z}}(0)| > k) \leq C_8/\sqrt{k} \ . \tag{13.2.4}$$

Repeating the argument for the Erdős–Rényi random graph as in (13.1.24)–(13.1.25), there exists $b > 0$ such that for every $A \geq 1$,

$$\mathbb{P}_{p_c(\mathbb{Z}^d)}(|\mathcal{C}_{\max}| > AV^{2/3}) \leq \frac{b}{A} \ . \tag{13.2.5}$$

                                                                                                                                                      □

While Prop. 13.7 is a useful tool to prove upper bounds on percolation clusters on the torus, it does not help so much in proving the corresponding lower bounds. For this, it turns out to be helpful to extend the results in Chaps. 4–9 to percolation on general high-dimensional tori. This is the content of the next section.

13.3   General High-Dimensional Tori

We now consider percolation on various transitive graphs with vertex set $\mathcal{V} = \{0, \ldots, n-1\}^d$ in the asymptotic regime where the volume (= number of vertices) $V = n^d$ diverges to infinity. These graphs differ by their edge set:

(i)   the classical *torus* $\mathbb{T}_{n,d}$ of width n and fixed dimension d that was introduced at the beginning of Sect. 13.2;

(ii) the *hypercube* $\{0, 1\}^d$, which is the torus as in (i) but with $n = 2$ fixed;

(iii) the *complete graph* K_n, whose percolation yields the Erdős–Rényi random graph as discussed in Sect. 13.1;

(iv) the *Hamming graph* K_n^d, which is the Cartesian product of d complete graphs.

We write $\mathbb{T}_{n,d}$ for all of these cases.

In these settings, it is not at all obvious precisely *what* the appropriate critical value $p_c(\mathbb{T}_{n,d})$ is. Of course, on the torus $\mathbb{T}_{n,d}$ with d fixed and $n \to \infty$, one would expect that the critical value $p_c(\mathbb{T}_{n,d})$ can be taken as $p_c(\mathbb{Z}^d)$, which is the critical value of the set in which the torus can naturally be embedded. In other settings, however, such an obvious choice is not available. Since the graphs that we deal with are finite, one cannot expect the critical value to be *unique*, since any value p that is sufficiently close to $p_c(\mathbb{T}_{n,d})$ has similar scalings of the cluster sizes. This leads us to the notion of the *scaling window*, which informally consists of those values of p for which the scaling behavior of cluster sizes agrees with that at $p = p_c(\mathbb{T}_{n,d})$.

The BCHSS Definition of the Critical Value for High-Dimensional Tori. The aim in this section is to *define* the critical value $p_c(\mathbb{T}_{n,d})$ in the high-dimensional setting and then to show that indeed $p_c(\mathbb{T}_{n,d})$ is a sensible choice. This we achieve by describing results about the largest connected components below the scaling window, within the scaling window close to $p_c(\mathbb{T}_{n,d})$ and above the scaling window. We restrict ourselves to the high-dimensional or mean-field setting, where we expect the scaling behavior to be similar to that on the complete graph K_n. We follow Borgs et al. in [59]. Recall that $\mathcal{C}_{\mathbb{T}}(x)$ denotes the cluster of $x \in \mathbb{T}_{n,d}$. In view of the correspondence with Erdős–Rényi random graphs, Lem. 13.4 and Exerc. 13.5 suggest that a critical value can be obtained by equating the expected cluster size to the cube root of the volume of the graph. This is our point of departure for percolation on high-dimensional tori:

Fix $\lambda > 0$ independent of n and d. Define the critical value $p_c(\mathbb{T}_{n,d}) = p_c(\mathbb{T}_{n,d}; \lambda)$ to be the unique solution of

$$\chi_{\mathbb{T}}(p) := \mathbb{E}_p|\mathcal{C}_{\mathbb{T}}(0)| = \lambda V^{1/3} . \tag{13.3.1}$$

The appearance of $\lambda > 0$ in (13.3.1) may be somewhat surprising. It is there for technical reasons, and we often take λ to be sufficiently small. In fact, we show that *any* value of λ gives qualitatively similar results. Indeed, taking another value of λ simply shifts $p_c(\mathbb{T}_{n,d}; \lambda)$ within the scaling window. So, in hindsight, we could have taken $\lambda = 1$ in (13.3.1). However, this only comes out as a consequence of our results, so we stick to the above definition.

For the complete graph K_n, $\mathbb{E}_{1/n}[|\mathcal{C}_{\mathbb{T}}(0)|] = \Theta(n^{1/3}) = \Theta(V^{1/3})$ (cf. Lem. 13.4), so we can see (13.3.1) as the natural generalization of the critical value $p_c = 1/n$ for the complete graph. Also here we do not know what the value of λ is for which $\mathbb{E}_{1/n}|\mathcal{C}_{\mathbb{T}}(0)| = \lambda n^{1/3}$, which can be seen as another reason to define the critical value in (13.3.1) more generally.

Let us emphasize here that it is highly unclear that the definition of $p_c(\mathbb{T}_{n,d})$ in (13.3.1) is a natural choice. This can only be established by showing that the behavior of the largest components is quite different below, within and above the critical window defined by (13.3.1) for various λ. We return to alternative definitions of p_c on finite graphs at the end of this section.

Finite-Graph Triangle Condition. Of course, (13.3.1) aims to describe the critical behavior in the *high-dimensional setting*. On the full lattice, high-dimensional can be interpreted in terms of the triangle diagram being finite at the critical point. We define the finite-graph triangle diagram by

$$\Delta_{\mathbb{T},p}(x, y) = \sum_{u,v} \tau_{\mathbb{T},p}(v - x)\tau_{\mathbb{T},p}(u - v)\tau_{\mathbb{T},p}(y - u) . \tag{13.3.2}$$

Clearly, $\Delta_{\mathbb{T},p}(x, y) \le V^2$ for every x, y, so that the finiteness is guaranteed. However, we need that $\Delta_{\mathbb{T},p}(x, y)$ remains finite *uniformly* in the volume (like the statement that $\Delta(p_c) \le 1 + O(1/d)$ for high-dimensional percolation on the infinite lattice). More precisely, the finite-graph triangle condition is the following:

Fix $\lambda > 0$, and take $p_c = p_c(\mathbb{T}_{n,d}; \lambda)$ as in (13.3.1). The *finite-graph triangle condition* holds when there exists a sufficiently small constant a_0 such that

$$\Delta_{\mathbb{T},p}(x, y) = \delta_{x,y} + a_0 .\qquad(13.3.3)$$

We shall see that $a_0 = C\lambda^3 + O(1/m)$, where λ is defined in (13.3.1) and m denotes the degree of the graph in the settings that we investigate. Note that this explains why we wish to take $\lambda > 0$ small, rather than just $\lambda = 1$. It is instructive to see what happens on the complete graph $K_n = \mathbb{T}$, in which, for consistency, we now write the vertex set as $\{0, \ldots, n - 1\}$. By symmetry,

$$\tau_{\mathbb{T},p}(x) = \delta_{0,x} + (1 - \delta_{0,x})\frac{\chi_{\mathbb{T}}(p) - 1}{n - 1} ,\qquad(13.3.4)$$

so that

$$\Delta_{\mathbb{T},p_c}(x, y) \le \delta_{x,y} + (n - 2)(n - 3)\frac{(\chi_{\mathbb{T}}(p) - 1)^3}{(n - 1)^3}$$

$$+ 3(n - 2)\frac{(\chi_{\mathbb{T}}(p) - 1)^2}{(n - 1)^2} + 3\frac{\chi_{\mathbb{T}}(p) - 1}{n - 1}$$

$$\le \delta_{x,y} + 9\frac{\chi_{\mathbb{T}}(p)^3}{n} = \delta_{x,y} + 9\lambda^3 .\qquad(13.3.5)$$

As a result, it is natural to expect $a_0 \ge C\lambda^3$. For general tori, we expect extra contributions, and we see that also a contribution of the form $1/m$ is present in a_0, where m is the degree of the graph under consideration.

Subcritical and Critical Results on High-Dimensional Tori. Of course, the definition of $p_c(\mathbb{T}_{n,d}; \lambda)$ in (13.3.1) necessitates a proof that it actually *is* a correct critical value, at least when the finite-graph triangle condition holds. In the sequel, we give partial results in this direction. We start by investigating the subcritical phase, following Borgs et al. [59]:

Theorem 13.8 (Subcritical phase [59]). *There is a (small) constant $b_0 > 0$ such that the following statements hold for all positive λ and all p of the form $p = p_c(\mathbb{T}_{n,d}; \lambda)(1 - \varepsilon)$ with $\varepsilon \ge 0$. If the triangle condition holds for some $a_0 \le b_0$, if $V \ge \lambda^{-3}b_0^3$, then*

$$\mathbb{P}_p\left(|\mathscr{C}_{\max}| \le 2\chi_{\mathbb{T}}^2(p) \log(V/\chi_{\mathbb{T}}^3(p))\right) \ge 1 - \frac{\sqrt{e}}{\left[2\log(V/\chi_{\mathbb{T}}^3(p))\right]^{3/2}} ,\qquad(13.3.6)$$

and, for $A \ge 1$,

$$\mathbb{P}_p\left(|\mathscr{C}_{\max}| \ge \frac{\chi_{\mathbb{T}}^2(p)}{3600A}\right) \ge \left(1 + \frac{36\chi_{\mathbb{T}}^3(p)}{AV}\right)^{-1} .\qquad(13.3.7)$$

The next theorem gives results inside the scaling window:

Theorem 13.9 (Scaling window [59, 144]). *Let $\lambda > 0$ and $C < \infty$. Then, there are finite positive constants b_1, b_2, b_3 such that the following statements hold provided the triangle condition (13.3.3) is valid for some constant $a_0 \leq b_0$ and $V \geq \lambda^{-3} b_0^3$, with b_0 as in Thm. 13.8. Let $p = p_c(\mathbb{T}_{n,d}; \lambda)(1 + \varepsilon)$ with $|\varepsilon| \leq C V^{-1/3}$.*

(i) If $A \geq 1$, then

$$\mathbb{P}_p\left(A^{-1} V^{2/3} \leq |\mathscr{C}_{\max}| \leq A V^{2/3}\right) \geq 1 - \frac{b_1}{A} . \tag{13.3.8}$$

(ii)

$$b_2 V^{1/3} \leq \chi_{\mathbb{T}}(p) \leq b_3 V^{1/3} . \tag{13.3.9}$$

Theorem 13.8 is an adaptation of [59, Thm. 1.2]. Theorem 13.9(i)–(ii) are adaptations of [59, Thm. 1.3]. We do not give the entire proofs of Thms. 13.8 and 13.9, but rather explain the intuition behind them. We leave many partial proofs as exercises and refer to the original papers for other parts.

Role of λ in the Definition of $p_c = p_c(\mathbb{T}_{n,d}; \lambda)$ in (13.3.1). Theorem 13.9(ii) states that

$$\chi_{\mathbb{T}}(p) = \Theta(V^{1/3}) \tag{13.3.10}$$

for any p of the form $p = p_c(1 + \Lambda V^{-1/3})$. This suggests that indeed the precise value of λ is not so important. A similar statement follows from the following exercise:

Exercise 13.6 (Asymptotics expected cluster size below $p_c(\mathbb{T}_{n,d}; \lambda)$). Prove that if the triangle condition (13.3.3) holds for some $a_0 < 1$, then, for $p = p_c(\mathbb{T}_{n,d}; \lambda)(1 - \varepsilon)$ with $\varepsilon \geq 0$,

$$\frac{1}{\chi_{\mathbb{T}}(p_c)^{-1} + \varepsilon} \leq \chi_{\mathbb{T}}(p) \leq \frac{1}{\chi_{\mathbb{T}}(p_c)^{-1} + [1 - a_0]\varepsilon} . \tag{13.3.11}$$

Hint: Adapt the proof in Sect. 4.2.

Upper Bound on $|\mathscr{C}_{\max}|$ below the Scaling Window: Tree-Graph Inequalities. Here, we explain how the upper bound in (13.3.6) can be proved using the so-called *tree-graph inequalities* derived by Aizenman–Newman in [13]. We start by proving that $\mathbb{P}_p(|\mathscr{C}_{\mathbb{T}}(0)| \geq k)$ decays exponentially:

Lemma 13.10 (Exponential cluster size decay [13, Prop. 5.1]). *For every $p \in [0, 1]$ and every $k \geq \chi_{\mathbb{T}}(p)^2$,*

$$\mathbb{P}_p(|\mathscr{C}_{\mathbb{T}}(0)| \geq k) \leq \sqrt{\frac{e}{k}} e^{-k/(2\chi_{\mathbb{T}}(p)^2)} . \tag{13.3.12}$$

Proof We study the moments of $|\mathscr{C}_{\mathbb{T}}(0)|$ so as to bound the moment-generating function. For this, we use that

$$\mathbb{E}_p[|\mathscr{C}_{\mathbb{T}}(0)|^k] = \sum_{x_1, \ldots, x_k} \mathbb{P}_p(x_1, \ldots, x_k \in \mathscr{C}_{\mathbb{T}}(0)) . \tag{13.3.13}$$

Let us investigate this sum for small values of k. When $k = 2$, we can write

$$\{0 \leftrightarrow x_1, 0 \leftrightarrow x_2\} = \bigcup_z \{0 \leftrightarrow z\} \circ \{z \leftrightarrow x_1\} \circ \{0 \leftrightarrow x_2\}. \tag{13.3.14}$$

Using the BK inequality (1.3.4) and the union bound, we thus arrive to[1]

$$\mathbb{E}_p[|\mathscr{C}_{\mathbb{T}}(0)|^2] \leq \sum_{x_1, x_2} \sum_z \mathbb{P}_p(0 \leftrightarrow z)\, \mathbb{P}(z \leftrightarrow x_1)\, \mathbb{P}_p(z \leftrightarrow x_2)$$

$$= \left(\mathbb{E}_p[|\mathscr{C}_{\mathbb{T}}(0)|] \right)^3 = \chi_{\mathbb{T}}(p)^3 . \tag{13.3.15}$$

With a little more work, this can be extended to

$$\mathbb{E}_p[|\mathscr{C}_{\mathbb{T}}(0)|^k] = \sum_{x_1, \ldots, x_k} \mathbb{P}_p(0 \leftrightarrow x_1, \ldots, x_k)$$

$$\leq (2k-3)!! \left(\mathbb{E}_p[|\mathscr{C}_{\mathbb{T}}(0)|] \right)^{2k-1}$$

$$= (2k-3)!! \chi_{\mathbb{T}}(p)^{2k-1} , \tag{13.3.16}$$

where $(2k-3)!! = (2k-3)(2k-5)\cdots 3 \cdot 1 = 2^{-(k-1)}(2(k-1))!/(k-1)!$ counts the number of ways that the paths from 0 to $x_1, \ldots, x_k$ can occur, which is the same as the number of trees with $k + 1$ ordered leaves. We return to this counting problem in Sect. 15.1.

Exercise 13.7 (The Aizenman–Newman tree-graph inequality [13]). Prove (13.3.16).

We use the identity

$$\frac{1}{\sqrt{1-t}} = \sum_{k=0}^{\infty} t^k \binom{2k}{k} \left(\frac{1}{4}\right)^k \tag{13.3.17}$$

for $|t| < 1$. We use the tree-graph inequalities in (13.3.16) to thus obtain, for every $t \geq 0$,

$$\mathbb{E}_p\left[|\mathscr{C}_{\mathbb{T}}(0)|e^{t|\mathscr{C}_{\mathbb{T}}(0)|}\right] = \sum_{k=0}^{\infty} \frac{t^k}{k!} \mathbb{E}_p\left[|\mathscr{C}_{\mathbb{T}}(0)|^{k+1}\right]$$

$$\leq \sum_{k=0}^{\infty} \frac{t^k}{k!} \chi_{\mathbb{T}}(p)^{2k+1} \frac{(2k)!}{2^k k!} = \frac{\chi_{\mathbb{T}}(p)}{\sqrt{1 - 2t\chi_{\mathbb{T}}(p)^2}} . \tag{13.3.18}$$

---

[1]Interestingly, Aizenman and Newman achieved this *without* the BK inequality, which was only proved later.

Therefore,

$$
\mathbb{P}_p(|\mathcal{C}_{\mathbb{T}}(0)| \geq k) = \mathbb{P}_p(|\mathcal{C}_{\mathbb{T}}(0)|e^{t|\mathcal{C}_{\mathbb{T}}(0)|} \geq ke^{tk})
$$
$$
\leq \frac{1}{ke^{tk}} \mathbb{E}_p[|\mathcal{C}_{\mathbb{T}}(0)|e^{t|\mathcal{C}_{\mathbb{T}}(0)|}]
$$
$$
= \frac{\chi_{\mathbb{T}}(p)}{ke^{tk}\sqrt{1 - 2t\chi_{\mathbb{T}}(p)^2}}. \tag{13.3.19}
$$

We conclude that

$$
\mathbb{P}_p(|\mathcal{C}_{\mathbb{T}}(0)| \geq k) \leq \inf_{t \geq 0} \frac{\chi_{\mathbb{T}}(p)}{ke^{tk}\sqrt{1 - 2t\chi_{\mathbb{T}}(p)^2}}. \tag{13.3.20}
$$

The minimizer is $t = 1/[2\chi_{\mathbb{T}}(p)^2] - 1/[2k]$, which is indeed nonnegative when $k \geq \chi_{\mathbb{T}}(p)^2$. Substitution leads to (13.3.12) and thus completes the proof of Lem. 13.3.12. $\qquad\square$

Proof of the upper bound in (13.3.6). Take $k = 2\chi_{\mathbb{T}}^2(p)\log(V/\chi_{\mathbb{T}}^3(p))$. We use that $|\mathcal{C}_{\max}| \geq k$ precisely when $Z_{\geq k} \geq k$ (recall (13.1.7)). This leads to

$$
\mathbb{P}_{\mathbb{T},p}(|\mathcal{C}_{\max}| \geq k) = \mathbb{P}_{\mathbb{T},p}(Z_{\geq k} \geq k)
$$
$$
\leq \frac{1}{k}\mathbb{E}_{\mathbb{T},p}[Z_{\geq k}] = \frac{V}{k}\mathbb{P}_{\mathbb{T},p}(|\mathcal{C}_{\mathbb{T}}(0)| \geq k). \tag{13.3.21}
$$

By Lem. 13.10 and using that $k = 2\chi_{\mathbb{T}}^2(p)\log(V/\chi_{\mathbb{T}}^3(p))$, this can be more bounded by

$$
\mathbb{P}_{\mathbb{T},p}(|\mathcal{C}_{\max}| \geq k) \leq V(e/k^3)^{1/2}e^{-k/(2\chi_{\mathbb{T}}(p)^2)}
$$
$$
= V(e/k^3)^{1/2}e^{-\log(V/\chi_{\mathbb{T}}^3(p))}
$$
$$
= \frac{\sqrt{e}}{2^{3/2}(\log(V/\chi_{\mathbb{T}}^3(p)))^{3/2}}, \tag{13.3.22}
$$

as required. $\qquad\square$

Bounds on $|\mathcal{C}_{\max}|$ Inside Scaling Window: First and Second Moment Methods. The behavior inside the critical window follows to a large extent from the statement that there exists constants c_1, c_2, c_3 such that, for $p = p_c(\mathbb{T}_{n,d}; \lambda)(1 + \varepsilon)$ with $|\varepsilon| \leq \Lambda V^{-1/3}$,

$$
\frac{c_2}{\sqrt{k}} \leq \mathbb{P}_{\mathbb{T},p}(|\mathcal{C}_{\mathbb{T}}(0)| \geq k) \leq c_3\left(\frac{1}{\sqrt{k}} + (\varepsilon \vee 0)\right), \tag{13.3.23}
$$

where the lower bound holds for $k \leq c_1 V^{2/3}$. The lower bound is [59, Thm. 1.3(i)], and the upper bound is [59, (6.4) in Lem. 6.1 and (6.2)]. The proof of (13.3.23) follows the proofs of $\delta = 2$ and $\beta = 1$ in Chap. 9 and is omitted here.

Tightness proof for $|\mathcal{C}_{\max}|V^{-2/3}$ *in Thm.* 13.9(i). As for the Erdős–Rényi random graph, we again use (13.3.21). Now take $k = AV^{2/3}$ and apply the upper bound in (13.3.23) to obtain

$$\mathbb{P}_p(|\mathcal{C}_{\max}| \geq AV^{2/3}) \leq c_3 \frac{V}{k}\left(\frac{1}{\sqrt{k}} + (\varepsilon \vee 0)\right) \leq \frac{c_3}{A}\left(\frac{1}{\sqrt{A}} + \Lambda\right). \tag{13.3.24}$$

This proves the upper bound on $|\mathcal{C}_{\max}|$ inside the scaling window. For the lower bound, we first use that monotonicity in p implies that it suffices to prove the claim for $p = p_c(1 - \Lambda V^{-1/3})$. For this, we use the second moment method, as well as the bound $\mathrm{Var}_p(Z_{\geq k}) \leq V\chi_{\mathbb{T}}(p)$ in Prop. 13.2, to arrive at

$$\mathbb{P}_p(|\mathcal{C}_{\max}| < k) = \mathbb{P}_p(Z_{\geq k} = 0) \leq \frac{\mathrm{Var}_p(Z_{\geq k})}{\mathbb{E}_p[Z_{\geq k}]^2} \leq \frac{V\chi_{\mathbb{T}}(p)}{\mathbb{E}_p[Z_{\geq k}]^2}. \tag{13.3.25}$$

Now, $\chi_{\mathbb{T}}(p) \leq \chi_{\mathbb{T}}(p_c) = \lambda V^{1/3}$, and, by the lower bound in (13.3.23),

$$\mathbb{E}_p[Z_{\geq k}] = V\,\mathbb{P}_p(|\mathcal{C}_{\mathbb{T}}(0)| \geq k) \geq \frac{c_2 V}{\sqrt{k}}, \tag{13.3.26}$$

when we take $k = V^{2/3}/A$ for $A \geq 1$ sufficiently large. Thus,

$$\mathbb{P}_p(|\mathcal{C}_{\max}| < V^{2/3}/A) \leq \frac{V\chi_{\mathbb{T}}(p)}{(c_2 V/\sqrt{k})^2} = \frac{\lambda V^{4/3}k}{(c_2 V)^2} = \frac{\lambda}{c_2^2 A}. \tag{13.3.27}$$

This completes the proof of (13.3.8). $\qquad\qquad\qquad\qquad\qquad\qquad\qquad\qquad\qquad\qquad\quad$ $\square$

Exercise 13.8 (Exponential tails of $V^{-2/3}|\mathcal{C}_{\max}|$). Let $p = p_c(\mathbb{T}_{n,d}; \lambda)(1 + \varepsilon)$ with $|\varepsilon| \leq \Lambda V^{-1/3}$. Use (13.3.9) and Lem. 13.10 to prove that there exists $a = a(\Lambda), b = b(\Lambda) > 0$ such that

$$\mathbb{P}_p(V^{-2/3}|\mathcal{C}_{\max}| > x) \leq \frac{b}{x^{3/2}}e^{-ax}. \tag{13.3.28}$$

Proof of the Triangle Condition in (13.3.3). The proof of the finite-graph triangle condition in (13.3.3) is similar to that of the original triangle condition in (4.1.1), as performed in Chaps. 5–8. In fact, there we have adapted the argument by Borgs et al. in [60] that proved the finite-graph triangle condition to the infinite lattice. We omit additional details.

Back to High-Dimensional Tori of Fixed Dimension. In (13.3.1), we have given an alternative definition of a critical value for high-dimensional tori of fixed dimension and degree. Of course, one would believe that the infinite-lattice critical value $p_c = p_c(\mathbb{Z}^d)$ lies inside the critical window. In fact, the stochastic domination in Prop. 13.7 implies that $p_c(\mathbb{Z}^d)(1 - \Lambda V^{-1/3}) \leq p_c(\mathbb{T}_{n,d}; \lambda)$, as formalized in the following exercise:

Exercise 13.9   (Upper bound on $p_c(\mathbb{Z}^d)$ in terms of $p_c(\mathbb{T}_{n,d}; \lambda)$). Use Prop. 13.7 to prove that $p_c(\mathbb{Z}^d)(1 - \Lambda V^{-1/3}) \leq p_c(\mathbb{T}_{n,d}; \lambda)$ when $\Lambda > 0$ is sufficiently large.

The lower bound $p_c(\mathbb{Z}^d)(1 + \Lambda V^{-1/3}) \geq p_c(\mathbb{T}_{n,d}; \lambda)$ is more involved, and it is proved by us in [143, Thm. 1.1] by a careful coupling argument in order to lower bound $\chi_{\mathbb{T}}(p)$ in terms of $\chi_{\mathbb{Z}}(p)$. We omit the details.

Alternative Definitions of p_c on Finite Graphs. We mentioned earlier that it is a priori unclear what a valid and meaningful notion of a critical percolation threshold on finite transitive graphs is. One condition that every definition of a critical value (rather, a critical window) should satisfy is a separation into a sub- and a supercritical regime that shows clearly different scaling of cluster sizes.

The definition in (13.3.1) may seem surprising at first sight, but is motivated through the corresponding behavior of Erdős–Rényi random graphs. The validity of that definition for "high-dimensional" tori is verified through the results in Thms. 13.8 and 13.9, yet we are missing a result that above the critical value there is a giant component. A more fundamental problem with this definition is that it anticipated *mean-field* behavior, and is expected to be meaningless, for example, for lower-dimensional tori. It is a challenge to find a more general definition of p_c which is suitable for a large class of transitive graphs. One possible generalization is to replace (13.3.1) by $\mathbb{E}_{p_c}|\mathcal{C}_{\mathbb{T}}(0)| = \lambda V^{1/(\delta+1)}$, which should be the correct value with δ being the cluster tail exponent. Plugging in $\delta = 2$ from Thm. 9.2 then recovers the high-dimensional definition in (13.3.1). Here, we extend this discussion by proposing related definitions from the literature.

These related definitions rely on the fact that the phase transition of Erdős–Rényi random graphs exhibits a number of other intriguing properties, and these might also be used for formalizing the notion of criticality. Nachmias and Peres [216, Rem. 7.2] suggest to use the maximizer of the logarithmic derivative of the susceptibility

$$\frac{d}{dp} \log \chi_{\mathbb{T}}(p) = \frac{\frac{d}{dp} \mathbb{E}_p|\mathcal{C}_{\mathbb{T}}(v)|}{\mathbb{E}_p|\mathcal{C}_{\mathbb{T}}(v)|} . \tag{13.3.29}$$

Motivation for this choice comes from Russo's formula (1.3.9), as (13.3.29) expresses the expected number of edges that can affect the size of $\mathcal{C}_{\mathbb{T}}(v)$. Janson and Warnke [187] confirm that this definition is meaningful and valid for the Erdős–Rényi random graph, and even identify the width of the critical window correctly. It is an open problem to investigate this definition for percolation on other transitive graphs.

Other possible choices for defining the critical value p_c are the maximizer of the *second largest component*

$$p_c^{(2)} = \arg\max \mathbb{E}_p|\mathcal{C}_{(2)}| \tag{13.3.30}$$

or the maximizer of the (properly rescaled) cluster size variance

$$p_c^{(3)} = \arg\max \frac{\mathrm{Var}_p|\mathcal{C}_{\max}|}{(\mathbb{E}_p|\mathcal{C}_{\max}|)^2} . \tag{13.3.31}$$

The second choice is motivated by the fact that it is expected that the second largest component is maximized within the scaling window. The last choice is motivated by the intuition that only at the critical value the rescaled maximal cluster size $|\mathcal{C}_{max}|/\mathbb{E}_p|\mathcal{C}_{max}|$ has a nondegenerate scaling limit.[2] We consider an additional investigation into such notions an interesting open problem:

> **Open Problem 13.1 (Notions of criticality for high-dimensional tori).** *For percolation on the torus $\mathbb{T}_{n,d}$ with $d > 6$, show that*
>
> $$p_c^{(1)} = arg\,max\frac{\mathrm{d}}{\mathrm{d}p}\log\chi_{\mathbb{T}}(p)$$
>
> *satisfies $|p_c^{(1)} - p_c(\mathbb{T}_{n,d})| \leq O(V^{-1/3})$, and therefore (in view of Thm. 13.9) $p_c^{(1)}$ characterizes critical behavior correctly. Do the same for $p_c^{(2)}$ and $p_c^{(3)}$.*

The Lack of Results in the Supercritical Case We close this section by describing the supercritical phase, where the results are *not* complete:

Theorem 13.11 (Supercritical phase [59]). *Let $\lambda > 0$. The following statements hold provided the triangle condition (13.3.3) holds for some constant $a_0 \leq b_0$ and $\lambda V^{1/3} \geq b_0^{-1}$, with b_0 as in Thm. 13.8. Let $p = p_c(\mathbb{T}_{n,d};\lambda)(1 + \varepsilon)$ with $\varepsilon \geq 0$:*

(i)

$$\mathbb{E}_p\big(|\mathcal{C}_{max}|\big) \leq 21\varepsilon V + 7V^{2/3} \tag{13.3.32}$$

and, for all $A > 0$,

$$\mathbb{P}_p\big(|\mathcal{C}_{max}| \geq A(V^{2/3} + \varepsilon V)\big) \leq \frac{21}{A}\,. \tag{13.3.33}$$

(ii) *If $0 \leq \varepsilon \leq 1$, then*

$$\chi_{\mathbb{T}}(p) \leq 81(V^{1/3} + 81\varepsilon^2 V)\,. \tag{13.3.34}$$

All the estimates in Thm. 13.11 prove *upper* bounds on $|\mathcal{C}_{max}|$ above the scaling window. The missing ingredient in Thm. 13.11 is a *lower* bound on $|\mathcal{C}_{max}|$ above the scaling window. Without such a lower bound, we actually do not know rigorously that $p_c(\mathbb{T}_{n,d};\lambda)$ or $p_c(\mathbb{Z}^d)$ really is an appropriate critical value, as the results so far do not exclude the possibility that also above the scaling window centered at $p_c(\mathbb{T}_{n,d};\lambda)$ the *two* largest clusters are of about equal size, which would contradict the uniqueness of the giant component. Also, it might be that for some other $p_c'(\mathbb{T}_{n,d})$, while $\chi_{\mathbb{T}}\big(p_c'(\mathbb{T}_{n,d})\big) \gg V^{2/3}$, still the largest connected component is not concentrated. We believe that such a situation cannot occur:

---

[2] We learned of the possible choices $p_c^{(2)}$ and $p_c^{(3)}$ through private communication with Asaf Nachmias.

> **Open Problem 13.2 (Concentration and uniqueness giant component above the scaling window).** *Fix* $\mathbb{T}_{n,d}$ *with* $d > 6$ *and let* $n \to \infty$. *Prove that, for* p *such that* $p = p_c(\mathbb{T}_{n,d}; \lambda)(1 + \varepsilon)$ *with* $\varepsilon = o(1)$ *but* $\varepsilon \gg V^{-1/3}$, *there exists* $c > 0$ *such that*
>
> $$\frac{|\mathcal{C}_{\max}|}{\varepsilon V} \overset{\mathbb{P}}{\longrightarrow} c, \qquad \frac{|\mathcal{C}_{(2)}|}{\varepsilon V} \overset{\mathbb{P}}{\longrightarrow} 0. \tag{13.3.35}$$
>
> *Verify that* c *is indeed the right-derivative of* $\theta(p)$ *at* $p = p_c(\mathbb{Z}^d)$.

In order to resolve Open Problem 13.2, an improved understanding of the *supercritical* phase of percolation is necessary. Recall also Open Problem 11.1. Currently, we approach the critical phase always from the subcritical side, as our understanding is the best there. Unfortunately, our understanding of the supercritical phase is lacking, which so far prevents us from resolving Open Problem 13.2. In the next section, we discuss percolation on the hypercube $\{0, 1\}^d$, where we *can* prove the concentration and uniqueness of the giant component in (13.3.35).

13.4 Hypercube Percolation

The study of percolation on the hypercube $\mathbb{T}_{2,d} = \{0, 1\}^d$ was initiated by Erdős and Spencer in [100], where they showed that the connectivity transition of the random graph obtained by randomly and independently removing edges with probability p is close to $p = \frac{1}{2}$. They also raised the question what the percolation phase transition is, i.e., for which values of p is there a unique giant component, and for which values are all connected components small. This issue was substantially clarified by Ajtai, Komlós, and Szemerédi [15], who proved that for $p = \lambda/d$ with $\lambda < 1$ all connected components are much smaller than the volume $V = 2^d$ of the graph, while for $\lambda > 1$, there exists a giant component containing a positive proportion of the vertices of the graph.

Of course, this leaves open what the critical behavior is, or even what the precise critical value is (except for the rough asymptotics $p_c = (1/d)(1 + o(1))$ as $d \to \infty$). Bollobás, Kohayakawa, and Łuczak [57] picked up this topic by studying values of p that are allowed to be quite close to $1/d$. However, also in this work, the ratio between the scaling of $|\mathcal{C}_{\max}|$ in the sub- and the supercritical regimes was close to 2^d (in fact, 2^d modulo a power of d), which indicates that these values are not close enough to the true critical value to observe critical behavior.

The definition (13.3.1) was proposed by Borgs et al. in [61] (this paper applies the general results in [59, 60] to the special case of the hypercube). The results proved there *do* prove that close to what was believed to be the scaling window, $|\mathcal{C}_{\max}|$ has intricate scaling behavior of the form $V^{2/3} = 2^{2d/3}$. Unfortunately, as already remarked in the previous section (and in Open Problem 13.2), the control over the size of the largest connected component $|\mathcal{C}_{\max}|$ right above the scaling window is too weak to conclude that $p_c(\{0, 1\}^d; \lambda)$ in (13.3.1) really is the right critical value. Borgs et al. [61] prove that for $p = p_c(1 + \varepsilon)$ with $\varepsilon > e^{-ad^{1/3}}$ for some $a > 0$, there is a unique giant component of size $\Theta(\varepsilon V)$. Even though $e^{-ad^{1/3}}$ is *much* smaller than any inverse power of d, it is much larger than the size of the predicted scaling window, which is $V^{-1/3} = 2^{-d/3}$.

For the hypercube, as well as several other graphs that have sufficient symmetries, this problem was taken up afresh by the second author and Nachmias in [164, 165]. There, symmetry was defined in terms of random walks having sufficiently small *mixing time*. Interestingly, it is not the mixing time of *simple random walk* that is relevant, but rather that of *nonbacktracking random walk* (NBW), which already appeared in Chap. 10 to study percolation above, but somewhat close to, the upper critical dimension. Let us start by stating the main result of [165] on the hypercube:

Theorem 13.12 (The hypercube supercritical phase [165]). *Put $p = p_c(1 + \varepsilon_d)$ where $\varepsilon_d = o(1)$ is a positive sequence with $\varepsilon_d \gg 2^{-d/3}$. Then, as $d \to \infty$,*

$$\frac{|\mathcal{C}_{\max}|}{2\varepsilon_d 2^d} \overset{\mathbb{P}}{\longrightarrow} 1, \quad \mathbb{E}_p|\mathcal{C}(0)| = (4 + o(1))\varepsilon_d^2 2^d, \quad \frac{|\mathcal{C}_{(2)}|}{\varepsilon_d 2^d} \overset{\mathbb{P}}{\longrightarrow} 0. \tag{13.4.1}$$

Combined with the results in Thms. 13.8 and 13.9, Thm. 13.12 shows that the critical value $p_c(\{0, 1\}^d; \lambda)$ defined in (13.3.1) really *is* the appropriate critical value, and that any alternative definition lies within the scaling window of that in (13.3.1).

The proof of Thm. 13.12 follows from a more general result applying to finite transitive graphs. Let us start by introducing some necessary notation.

Let $\mathcal{G}$ be a finite transitive graph on V vertices and with degree m. Consider the nonbacktracking random walk (NBW) on it (recall that this is just a simple random walk not allowed to traverse back on the edge it just came from, see Chap. 10). As we explain later, the use of NBW is vital for the argument that we present here. For any two vertices x, y, we write $\mathbf{p}_n(x, y)$ for the probability that the NBW started at x visits y at time n. More precisely, recalling from Sect. 10.2 that $b_n(x)$ denotes the number of n-step NBWs starting at the origin and ending at x,

$$\mathbf{p}_n(x, y) = \frac{b_n(x - y)}{m(m - 1)^{n-1}}. \tag{13.4.2}$$

We write $T_{\mathrm{mix}}(\xi)$ for the ξ-*uniform mixing time* of the walk, that is,

$$T_{\mathrm{mix}}(\xi) = \min\left\{n : \max_{x,y} \tfrac{1}{2}[\mathbf{p}_n(x, y) + \mathbf{p}_{n+1}(x, y)] \leq (1 + \xi)V^{-1}\right\}. \tag{13.4.3}$$

Then, the generalization of Thm. 13.12 is as follows:

Theorem 13.13 (Supercritical phase general high-dimensional graphs [165]). *Let $\mathcal{G}$ be a transitive graph on V vertices with degree m and define p_c as in (13.3.1) with $\lambda = 1/10$. Assume that there exists a sequence $\xi = \xi_m \to 0$ as $m \to \infty$ such that the following conditions hold:*

(1) $m \to \infty$ *as* $V \to \infty$,
(2) $[p_c(m - 1)]^{T_{\mathrm{mix}}} = 1 + o(1)$ *with* $T_{\mathrm{mix}} = T_{\mathrm{mix}}(\xi_m)$,

(3) *For any vertices x, y,*

$$\sum_{u,v} \sum_{t_1,t_2,t_3=0}^{T_{mix}} \mathbb{1}_{\{t_1+t_2+t_3 \geq 3\}} \mathbf{p}_{t_1}(x,u)\mathbf{p}_{t_2}(u,v)\mathbf{p}_{t_3}(v,y) = o(1/\log V). \quad (13.4.4)$$

Then,

(a) *the finite triangle condition (13.3.3) holds (and hence the assertions of Thms. 13.8–13.9 follow),*

(b) *for any sequence $\varepsilon = \varepsilon_m$ satisfying $\varepsilon_m \gg V^{-1/3}$ and $\varepsilon_m = o(T_{mix}^{-1})$,*

$$\frac{|\mathcal{C}_{max}|}{2\varepsilon_m V} \xrightarrow{\mathbb{P}} 1, \quad \mathbb{E}_p|\mathcal{C}(0)| = (4+o(1))\varepsilon_m^2 V, \quad \frac{|\mathcal{C}_{(2)}|}{\varepsilon_m V} \xrightarrow{\mathbb{P}} 0. \quad (13.4.5)$$

Theorem 13.13 gives precise conditions for random graph asymptotics, i.e., critical behavior like that on the Erdős–Rényi random graph, to be valid on transitive graphs. It not only proves that the finite-graph triangle condition holds (so that the results from Borgs et al. [59] apply), but also shows that the *supercritical* regime is like that on the Erdős–Rényi random graph. While we believe this to be true much more generally for high-dimensional graphs, we are not able to show this. The fact that p_c is sufficiently close to $1/(m-1)$, as formalized in Assumption (2) and the fact that NBW mixes quite fast and has short loops, as formalized in Assumptions (2) and (3), provide just enough "symmetry" to push the argument through. Remarkably, as we show in more detail below, the application of Thm. 13.13 to the hypercube does not need the lace expansion!

We continue to explain the basic philosophy of the proof of Thm. 13.13, which consists of five key steps. This overview is similar in spirit as the one provided in [164], some parts of which are copied verbatim.

Step 1: The Number of Vertices in Large Clusters Is what it Should Be. We start by investigating the number of vertices in large clusters:

Theorem 13.14 (Bounds on the cluster tail). *Let $\mathcal{G}$ be a finite transitive graph of degree m on V vertices such that the finite triangle condition (13.3.3) holds and put $p = p_c(1 + \varepsilon_m)$ where $\varepsilon_m = o(1)$ and $\varepsilon_m \gg V^{-1/3}$. Then, for the sequence $k_0 = \varepsilon_m^{-2}(\varepsilon_m^3 V)^{1/4}$,*

$$\mathbb{P}_p(|\mathcal{C}(0)| \geq k_0) = 2\varepsilon_m(1 + o(1)). \quad (13.4.6)$$

Further,

$$\frac{Z_{\geq k_0}}{2\varepsilon_m V} \xrightarrow{\mathbb{P}} 1, \quad (13.4.7)$$

where we recall that $Z_{\geq k}$ is the number of vertices having cluster size at least k.

This theorem resembles the statement that a branching process with Poisson offspring distribution of mean $1 + \varepsilon$ has survival probability of $2\varepsilon(1 + O(\varepsilon))$. The choice of k_0 is such that $k_0 \gg 1/\varepsilon_m^2$. This choice is inspired by the fact that $\mathbb{P}(T \geq k_0) = 2\varepsilon(1 + o(1))$

precisely when $k_0 \gg 1/\varepsilon^2$, where T is the total progeny (=total population size) of a Poisson branching process with mean $1 + \varepsilon$ offspring. Thus, Thm. 13.13 can be interpreted by saying that the majority of vertices in clusters of size at least k_0 are in the *same* giant component.

Upper and lower bounds of order ε for the cluster tail were proved already by Borgs et al. in [60] using Barsky and Aizenman's differential inequalities [28] as discussed in Chaps. 3 and 9, and were sharpened in [165, App. A] to obtain the right constant 2. We omit additional details.

Step 2: Uniform Connection Probabilities, Nonbacktracking Random Walk and the Triangle. One of the most useful estimates on percolation connection probabilities relies on *symmetry*. To give this symmetry a quantitative shape, a simple key connection between percolation and the mixing time of the nonbacktracking walk is revealed. In the analysis of the Erdős–Rényi random graph, symmetry plays a special role. One instance of this symmetry is that the function $\mathbb{P}_p(v \leftrightarrow x)$ is constant whenever $x \neq v \in [n]$ and its value is precisely $(V - 1)^{-1}(\mathbb{E}_p|\mathcal{C}(v)| - 1)$, while it equals 1 when $x = v$. Such a statement clearly does not hold on the hypercube at p_c: The probability that two neighbors are connected is at least $p_c \geq 1/d$, while the probability that 0 is connected to one of the vertices in the barycenter of the cube is at most $\sqrt{d}\,2^{-d}\,\mathbb{E}_p|\mathcal{C}(0)|$ by symmetry.

A key observation in the proof of Thm. 13.13 in [165] is that one can recover this symmetry as long as we require the connecting paths to be longer than the mixing time of the NBW, as shown in [165, Lem. 3.12]. In its statement, we write $x \xleftrightarrow{[a,b]} y$ for the event that the graph distance $d_{\mathcal{E}(x)}(x, y) \in [a, b]$. We also write $x \xleftrightarrow{b} y$ for $x \xleftrightarrow{[0,b]} y$, i.e., the event that there exists a path of at most b occupied bonds connecting x and y. Additionally, we write $B_x(r)$ for the balls of intrinsic radius r around x.

Lemma 13.15 (Uniform connection estimates). *Perform edge percolation on a graph $\mathcal{G}$ satisfying the assumptions of Thm. 13.13. Then, for every $r \geq m_0$ and any vertex $x \in \mathcal{G}$,*

$$\mathbb{P}_{p_c}\left(0 \xleftrightarrow{[T_{\text{mix}},r]} x\right) \leq \left(1 + o(1)\right)\frac{1}{V}\,\mathbb{E}_{p_c}|B(r)|\,, \tag{13.4.8}$$

where T_{mix} is the uniform mixing time as defined above Thm. 13.13. In particular,

$$\mathbb{P}_{p_c}\left(0 \xleftrightarrow{[T_{\text{mix}},\infty)} x\right) \leq \left(1 + o(1)\right)\frac{1}{V}\,\mathbb{E}_{p_c}|\mathcal{C}(0)|\,. \tag{13.4.9}$$

The proof of the above lemma is short and elementary, see [165]. There, it is also shown how to obtain similar estimates for $p = p_c(1 + \varepsilon)$ (with an error depending on ε). The uniformity of this lemma allows us to decouple the sum in the triangle diagram and yields a simple proof of the strong triangle condition, as we now show:

Proof of the Triangle Condition in Part (a) of Thm. 13.13 Let $p \leq p_c$. If one of the connections in the sum $\triangle_p(x, y)$ is of length in $[T_{\text{mix}}, \infty)$, say between x and u, then we may estimate

$$\sum_{u,v} \mathbb{P}_p\left(x \xleftrightarrow{[T_{\mathrm{mix}},\infty)} u\right) \mathbb{P}_p(u \leftrightarrow v) \mathbb{P}_p(v \leftrightarrow y)$$

$$\leq (1 + o(1)) \frac{1}{V} \mathbb{E}_p|\mathcal{C}(0)| \sum_{u,v} \mathbb{P}_p(u \leftrightarrow v) \mathbb{P}_p(v \leftrightarrow y)$$

$$= (1 + o(1)) \frac{1}{V} (\mathbb{E}_p|\mathcal{C}(0)|)^3 \,, \tag{13.4.10}$$

where we have used Lem. 13.15 for the first inequality. Thus, we are only left to deal with short connections:

$$\triangle_p(x, y) \leq \sum_{u,v} \mathbb{P}_p\left(x \xleftrightarrow{T_{\mathrm{mix}}} u\right) \mathbb{P}_p\left(u \xleftrightarrow{T_{\mathrm{mix}}} v\right) \mathbb{P}_p\left(v \xleftrightarrow{T_{\mathrm{mix}}} y\right)$$
$$+ O(\chi(p)^3/V) \,. \tag{13.4.11}$$

We write, using the notation $x \xleftrightarrow{=t} y$ to denote that x is at graph distance *precisely* t from y,

$$\mathbb{P}_p\left(x \xleftrightarrow{T_{\mathrm{mix}}} u\right) = \sum_{t_1=0}^{T_{\mathrm{mix}}} \mathbb{P}_p\left(x \xleftrightarrow{=t_1} u\right) , \tag{13.4.12}$$

and do the same for all three terms so that

$$\triangle_p(x, y) \leq \sum_{u,v} \sum_{t_1,t_2,t_3=0}^{T_{\mathrm{mix}}} \mathbb{P}_p\left(x \xleftrightarrow{=t_1} u\right) \mathbb{P}_p\left(u \xleftrightarrow{=t_2} v\right) \mathbb{P}_p\left(v \xleftrightarrow{=t_3} y\right)$$
$$+ O(\chi(p)^3/V) \,. \tag{13.4.13}$$

We bound

$$\mathbb{P}_p\left(x \xleftrightarrow{=t_1} u\right) \leq m(m-1)^{t_1-1} \mathbf{p}_{t_1}(x, u) p^{t_1} \,, \tag{13.4.14}$$

simply because $m(m-1)^{t_1-1} \mathbf{p}_{t_1}(x, u) = b_{t_1}(u - x)$ is an upper bound on the number of simple paths of length t_1 starting at x and ending at u. Hence,

$$\triangle_p(x, y) \leq \frac{m^3}{(m-1)^3} \sum_{u,v} \sum_{t_1,t_2,t_3=0}^{T_{\mathrm{mix}}} [p(m-1)]^{t_1+t_2+t_3} \mathbf{p}_{t_1}(x, u) \mathbf{p}_{t_2}(u, v) \mathbf{p}_{t_3}(v, y)$$
$$+ O(\chi(p)^3/V) \,. \tag{13.4.15}$$

Since $p \leq p_c$ and $t_i \leq T_{\mathrm{mix}}$, assumption (2) of Thm. 13.13 gives that $[p(m-1)]^{t_1+t_2+t_3} = 1 + o(1)$, and it is a simple consequence of assumption (3) of Thm. 13.13 that

$$\sum_{u,v} \sum_{t_1,t_2,t_3=0}^{T_{\mathrm{mix}}} [p(m-1)]^{t_1+t_2+t_3} \mathbf{p}_{t_1}(x, u) \mathbf{p}_{t_2}(u, v) \mathbf{p}_{t_3}(v, y) \leq \delta_{x,y} + o(1) \,, \tag{13.4.16}$$

where $o(1)$ vanishes as $m \to \infty$, concluding the proof. $\qquad \square$

Step 3: Most Large Cluster Share Large Boundary. Since this is the most technical part of the proof, at the expense of being precise, we have chosen to simplify notation and suppress several parameters. We ignore several dependencies between parameters and refer to [165] for details. However, we do emphasize the role of two important parameters. We choose r and r_0 so that $r \gg \varepsilon_m^{-1}$ but just barely, and $r_0 \gg r$ in a way that becomes clear later on. For vertices x, y, define the random variable

$$S_{r+r_0}(x, y) = \left| \left\{ (u, u') \in \mathcal{E} : \left\{ x \xleftrightarrow{r+r_0} u \right\} \circ \left\{ y \xleftrightarrow{r+r_0} u' \right\}, \right. \right.$$

$$\left. \left. |B_u(r + r_0)| \cdot |B_{u'}(r + r_0)| \le \varepsilon^{-2} (e_p |B_0(r_0)|)^2 \right\} \right|, \tag{13.4.17}$$

where $B_u(r)$ denotes the intrinsic ball of radius r around u, so that $B_0(r) = B(r)$ (recall (11.3.3)). The important part of the definition of $S_{r+r_0}(x, y)$ is the first requirement $\{ x \xleftrightarrow{r+r_0} u \} \circ \{ y \xleftrightarrow{r+r_0} u' \}$ (the second requirement is more technical). The edges contributing to $S_{r+r_0}(x, y)$, if made occupied, they enforce a connection between x and y, thus merging $\mathcal{C}(x)$ and $\mathcal{C}(y)$. These edges are used in the next step to *sprinkle* them.

Informally, a pair of vertices (x, y) is *good* when their clusters are large and $S_{r+r_0}(x, y)$ is large, so that their clusters have many edges between them. This is made quantitative in the following definition. In its statement, we use $\partial B_x(r) = B_x(r) \setminus B_x(r - 1)$ to denote the vertices at graph distance *equal* to r:

Definition 13.16 ((r, r_0)-good pairs). We say that x, y are (r, r_0)-good if all of the following occur:

(1) $\partial B_x(r) \ne \varnothing$, $\partial B_y(r) \ne \varnothing$ and $B_x(r) \cap B_y(r) = \varnothing$,
(2) $|\mathcal{C}(x)| \ge (\varepsilon_m^3 V)^{1/4} \varepsilon_m^{-2}$ and $|\mathcal{C}(y)| \ge (\varepsilon_m^3 V)^{1/4} \varepsilon_m^{-2}$,
(3) $S_{2r+r_0}(x, y) \ge V^{-1} m \varepsilon_m^{-2} (\mathbb{E}_p |B(r_0)|)^2$.

Write P_{r,r_0} for the number of (r, r_0)-good pairs.

In the following theorem, the asymptotics of the number of good pairs is investigated:

Theorem 13.17 (Most large clusters share many boundary edges). *Let $\mathcal{G}$ be a graph on V vertices and degree m satisfying the assumptions of Thm. 13.13. Assume that ε_m satisfies $\varepsilon_m \gg V^{-1/3}$ and $\varepsilon_m = o(T_{\mathrm{mix}}^{-1})$. Then,*

$$\frac{P_{r,r_0}}{(2\varepsilon_m V)^2} \xrightarrow{\mathbb{P}} 1 .$$

Theorem 13.17 is the crucial ingredient to the proof of Thm. 13.13. It shows that most large clusters have many edges between them, which is similar in spirit to the statement that two (distinct) clusters of sizes s_1 and s_2 have $s_1 s_2$ edges between them on the complete graph. While this is not *deterministically* true in our general setting, it turns out to be true for *most* pairs of clusters.

In light of Thm. 13.14, we expect that the number of pairs of vertices (x, y) with $|\mathcal{C}(x)| \ge (\varepsilon_m^3 V)^{1/4} \varepsilon_m^{-2}$ and $|\mathcal{C}(y)| \ge (\varepsilon_m^3 V)^{1/4} \varepsilon_m^{-2}$ is close to $(2\varepsilon_m V)^2$. Thm. 13.17 shows that almost all of these pairs have clusters that share many edges between them. Theorem 13.17 allows us to prove Thm. 13.13, as is described in more detail in the next step.

The difficulty in the proof of Thm. 13.17 is the requirement (3) in Def. 13.16. Indeed, conditioned on survival (i.e., on $\partial B_x(r) \neq \emptyset$, $\partial B_y(r) \neq \emptyset$ and that the balls are disjoint), the random variable $S_{r+r_0}(x, y)$ is not concentrated, and hence, it is hard to prove that it is large. In fact, even the variable $|B_0(r_0)|$ is not concentrated. This is not a surprising fact: The number of descendants at generation n of a branching process with mean $\mu > 1$ divided by μ^n converges as $n \to \infty$ to a nontrivial random variable. Nonconcentration occurs because the first generations of the process have a strong and lasting effect on the future of the population. In [165], this nonconcentration is counteracted by conditioning on the whole structure of $B_x(r)$ and $B_y(r)$. Since r is bigger than the correlation length ($r \gg \varepsilon_m^{-1}$), under this conditioning the variable $S_{r+r_0}(x, y)$ *is* concentrated (as one would expect from the branching process analogy).

Step 4: Sprinkling and Improved Sprinkling. The sprinkling technique was invented by Ajtai, Komlós, and Szemerédi [15] to show that $|\mathcal{C}_{max}| = \Theta_{\mathbb{P}}(2^d)$ when $p = (1 + \varepsilon)/d$ for fixed $\varepsilon > 0$ and can be described as follows. Fix some small $\theta > 0$ and write $p_1 = (1 + (1 - \theta)\varepsilon)/d$ and $p_2 \geq \theta\varepsilon/d$ such that $(1 - p_1)(1 - p_2) = 1 - p$. Let $\mathcal{G}_p$ be the subgraph of $\mathcal{G}$ obtained by removing each edge independently with probability $1 - p$. It is clear that $\mathcal{G}_p$ is distributed as the union of the edges in two independent copies of $\mathcal{G}_{p_1}$ and $\mathcal{G}_{p_2}$. The sprinkling method consists of two steps. The first step is performed in $\mathcal{G}_{p_1}$ and uses a branching process comparison argument together with an Azuma–Hoeffding concentration inequality to obtain that whp at least $c_2 2^d$ vertices are contained in connected components of size at least $2^{c_1 d}$ for some small but fixed constants $c_1, c_2 > 0$. In the second step, we add the edges of $\mathcal{G}_{p_2}$ (these are the "sprinkled" edges) and show that they connect many of the clusters of size at least $2^{c_1 d}$ into a giant cluster of size $\Theta_{\mathbb{P}}(2^d)$.

Let us give some details on how the last step is done. A key tool here is the *isoperimetric inequality* for the hypercube stating that two disjoint subsets of the hypercube of size at least $c_2 2^d/3$ have at least $2^d/d^{100}$ disjoint paths of length $C(c_2)\sqrt{d}$ connecting them, for some constant $C(c_2) > 0$. (The d^{100} in the denominator is not sharp, but this is immaterial as long as it is a polynomial in d.) This fact is used in the following way. Write V' for the set of vertices that are contained in a component of size at least $2^{c_1 d}$ in $\mathcal{G}_{p_1}$, so that $|V'| \geq c_2 d^d$. We say that *sprinkling fails* when $|\mathcal{C}_{max}| \leq c_2 2^d/3$ in the union $\mathcal{G}_{p_1} \cup \mathcal{G}_{p_2}$. If sprinkling fails, then we can partition V' as the disjoint union of A and B such that both A and B have cardinality at least $c_2 2^d/3$ and *any* path of length at most $C(c_2)\sqrt{d}$ between them has an edge that is p_2-closed. The number of such partitions is at most $2^{2^d/2^{c_1 d}}$. The probability that a path of length k has a p_2-closed edge is $1 - p_2^k$. Applying the isoperimetric inequality and using that the paths guaranteed to exist by it are disjoint so that the edges in them are independent, the probability that sprinkling fails is at most

$$2^{2^d/2^{c_1 d}} \cdot \left(1 - \left(\frac{\theta\varepsilon}{d}\right)^{C(c_2)\sqrt{d}}\right)^{2^d/d^{100}} = e^{-2^{(1+o(1))d}}, \qquad (13.4.18)$$

which tends to 0 (even rather quickly).

The sprinkling argument above is not optimal due to the use of the isoperimetric inequality. It is wasteful because it assumes that large percolation clusters can be "worst-case" sets, that is, sets that saturate the isoperimetric inequality (e.g., two balls of radius $d/2 - \sqrt{d}$ around two vertices at graph distance d). However, it is in fact very improbable for percolation clusters to be similar to this kind of worst-case sets. In [165], this is replaced by Thm. 13.17 showing that percolation clusters are "close" to uniform random sets of similar size, so that two large clusters share many closed edges with the property that if we open even *one* of them, then the two clusters connect. Additionally, the above argument is quite specific to the hypercube, while Thm. 13.13 aims for more general universal result in Thm. 13.17 allows for this, as it allows us to improve the above crude sprinkling argument.

Let us now describe the heuristics of the improved sprinkling argument. While previously there were paths of length $\sqrt{d}$ connecting the two clusters, by Thm. 13.17, instead, there are paths of length precisely 1. The final line of the proof, replacing (13.4.18), then becomes

$$2^{2\varepsilon V/(k_m \varepsilon^{-2})} \cdot \left(1 - \frac{\theta \varepsilon}{m}\right)^{m\varepsilon^2 V} \leq e^{-\theta \varepsilon^3 V(1+o(1))} \, , \tag{13.4.19}$$

where k_m is some sequence with $k_m \to \infty$ very slowly as $V \to \infty$. The right-hand side of (13.4.19) tends to 0 since $\varepsilon^3 V \to \infty$. Compared with the logic leading to (13.4.18), this line is rather suggestive: whp $2\varepsilon V$ vertices are in components of size at least $k_m \varepsilon^{-2}$, explaining the $2^{2\varepsilon V/(k_d \varepsilon^{-2})}$ term in (13.4.19).

The most difficult part in [165] is justifying the second term showing that for any partition of these vertices into two sets of the size of order εV, the number of closed edges between them is at least $\varepsilon^2 m V$. This is implied by Thm. 13.17. Perhaps not surprisingly, $\varepsilon^2 m V$ is the expected number of edges that two random uniform sets of size εV have between them.

Step 5: Unlacing Hypercube Percolation. In this step, we return to the special setting of the hypercube. In order to apply Thm. 13.17, we need to prove several estimates. For the hypercube, $m = d$, which indeed tends to infinity when $V = 2^d \to \infty$. Thus, condition (1) in Thm. 13.13 holds. Conditions (2) and (3) both rely on two key ingredients: properties of NBWs on the hypercube (appearing in terms of T_{mix} in condition (2) and in condition (3)) and an estimate on $(d-1)p_c$. We explain the necessary ingredients in this order below. As we shall see, these estimates also allow us to avoid the use of the lace expansion altogether. We start by investigating T_{mix} following the work of Fitzner and the second author [109]:

Theorem 13.18 (Mixing time NBW on hypercube [109]). *For NBW on the hypercube* $\{0, 1\}^d$ *and for every* $\varepsilon > 0$, *there exists* d_0 *such that for all* $d \geq d_0$,

$$T_{\mathrm{mix}} \leq (1 + \varepsilon)\frac{d}{2} \log d \, . \tag{13.4.20}$$

As we explain in more detail below, $(d/2) \log d$ also arises as the mixing time of simple random walk on the hypercube $\{0, 1\}^d$ (see, e.g., the work on mixing times by Levin, Peres, and Wilmer [207], or Aldous [16, Lem. 2.5(a)] where it is shown that the mixing times of discrete-time random walk are similar to that in continuous time). Thus, the nonbacktracking condition does not change the mixing time very much.

We explain how to identify the mixing time of continuous-time random walk on the hypercube and after this explain how to obtain that of NBW. The main advantage of working in continuous time is that the random walk becomes *aperiodic*.

Continuous-time random walk on $\{0, 1\}^d$ can be seen as a random walk on subsets of $[d]$. Indeed, we can identify $x \in \{0, 1\}^d$ with a subset $A(x) \subseteq [d]$ by letting $i \in A(x)$ precisely when $x_i = 1$. Then, continuous-time random walk has the interpretation of d independent rate $1/d$ clocks, and when the ith clock rings, element i is removed from A_t when $i \in A_t$ and added to A_t otherwise. We let $A_0 = \varnothing$, which corresponds to the random walk starting in $0 \in \{0, 1\}^d$. Then, the transition probability at time t, which we denote by s_t, is given by

$$s_t(A) = \mathbb{P}(A_t = A) = \prod_{i \in [d]} q_t(i)^{\mathbb{1}_{\{i \in A\}}} \left(1 - q_t(i)\right)^{\mathbb{1}_{\{i \notin A\}}}$$

$$= q_t(1)^{|A|} \left(1 - q_t(1)\right)^{d - |A|}, \tag{13.4.21}$$

where $q_t(i) = \mathbb{P}(i \in A_t)$ and we have used symmetry of the different elements. To determine $q_t(1)$, we note that

$$\frac{\mathrm{d}}{\mathrm{d}t} q_t(1) = \frac{1}{d}\left[-q_t(1) + \left(1 - q_t(1)\right)\right] = \frac{1}{d}[1 - 2q_t(1)], \tag{13.4.22}$$

so that

$$q_t(1) = \tfrac{1}{2}(1 - e^{-2t/d}). \tag{13.4.23}$$

This makes it easy to deduce the uniform mixing time in (13.4.3). The details of this proof are left as an exercise:

Exercise 13.10 (Uniform mixing time hypercube). Fill in the details of the proof that $T_{\mathrm{mix}} = (1 + o(1))(d/2) \log d$ for continuous-time simple random walk on the hypercube $\{0, 1\}^d$.

The proof of Thm. 13.18 is based on Fourier theory. The analysis is a bit harder due to the bipartite nature of the hypercube, meaning that, when started from $0 \in \{0, 1\}^d$, at time t the random walk (nonbacktracking or simple alike) has to be at an even location when t is even and at an odd location when t is odd. This is also reflected in the definition of the mixing time in (13.4.3), where the average between two consecutive times is taken. Let us now introduce Fourier theory on the hypercube. Let the Fourier dual of the hypercube $\{0, 1\}^d$ be given by $\{0, 1\}^d$, where, for a bounded function $f : \{0, 1\}^d \to \mathbb{R}$ and $k \in \{0, 1\}^d$,

$$\hat{f}(k) = \sum_{x \in \{0,1\}^d} (-1)^{x \cdot k} f(x). \tag{13.4.24}$$

Then, the Fourier inversion theorem states that

$$f(x) = \frac{1}{2^d} \sum_{k \in \{0,1\}^d} (-1)^{x \cdot k} \hat{f}(k). \tag{13.4.25}$$

Let the simple random walk step distribution be $D(x) = (1/d)\mathbb{1}_{\{\{0,x\}\in\mathcal{E}\}}$, so that

$$\widehat{D}(k) = \sum_{x\in\{0,1\}^d} (-1)^{x\cdot k} D(x) = \frac{1}{d}\sum_{i=1}^d (-1)^{k_i} = 1 - \frac{2a(k)}{d}, \qquad (13.4.26)$$

where $a(k) = \#\{i : k_i = 1\}$ is the number of ones in k.

The main technical estimate in the analysis of the mixing time for NBW on the hypercube shows that the Fourier transform decays exponentially fast toward zero for any Fourier variable $k \neq 0$, that is,

$$\mathbf{p}_t(k) \leq \left(\left|\widehat{D}(k)\right| \vee 1/\sqrt{d-1}\right)^{t-1} \qquad (13.4.27)$$

(keep in mind that $\left|\widehat{D}(k)\right| < 1$ for every $k \neq 0$). The proof that this implies Thm. 13.18 is left as an exercise:

Exercise 13.11 (Uniform mixing time NBW on hypercube). Fill in the details of the proof that, for any $\varepsilon > 0$, $T_{\mathrm{mix}}(1/\log d) \leq (1 + \varepsilon)(d/2)\log d$ for d large enough for discrete-time nonbacktracking random walk on the hypercube $\{0, 1\}^d$.

We continue with the necessary bound on p_c in order to show that $[(d-1)p_c]^{T_{\mathrm{mix}}} = 1 + o(1)$. Since T_{mix} is now known to be $\Theta(d \log d)$, this means that we need a bound of the form $(d-1)p_c \leq 1 + o((d \log d)^{-1})$. The following lemma can be used to prove that $(d-1)p_c \leq 1 + O(d^{-2})$:

Lemma 13.19 (Recursive bounds on $\mathbb{E}_p|\partial B(k)|$, [164]). *For any $c > 0$, there exists a $K > 0$ such that if*

$$p = \frac{1 + 5/(2d^2) + K/d^3}{d-1},$$

then, for $d = d(K)$ sufficiently large and for any $k \geq 1$ satisfying $\mathbb{E}_p|B(k)| \leq 2^{d/2}/d^3$,

$$\mathbb{E}_p|\partial B(k)| \geq [1 + c/d^3]\,\mathbb{E}_p|\partial B(k-1)|. \qquad (13.4.28)$$

The proof of Lem. 13.19 follows by induction combined with an appropriate split of the expectations to account for self-intersections. Lem. 13.19 can be used to show that for p with $(d-1)p = 1 + 5/(2d^2) + K/d^3$ it is true that $\mathbb{E}_p|B(k)| \geq 2^{d/2}/d^3$, so that $p > p_c$. This is left as an exercise:

Exercise 13.12 (Upper bound on p_c on hypercube). Prove that Lem. 13.19 implies that $\mathbb{E}_p|B(k)| \geq 2^{d/2}/d^3$ when $(d-1)p = 1 + 5/(2d^2) + K/d^3$ with K sufficiently large. Conclude that $(d-1)p_c < 1 + 5/(2d^2) + K/d^3$.

Open Problems for Hypercube Percolation. We close this section by discussing two of the main open problems for hypercube percolation. The first one is the scaling limit in the critical window:

Open Problem 13.3 (Scaling limit critical hypercube). *For the hypercube* $\{0, 1\}^d$ *and* $p = p_c(\{0, 1\}^d)(1 + \theta 2^{-d/3})$, *prove that*

$$V^{-2/3}\big(|\mathcal{C}_{(i)}|\big)_{i \geq 1} \xrightarrow{d} \big(\gamma_i(\theta)\big)_{i \geq 1} \qquad (13.4.29)$$

for some scaling limits $\big(\gamma_i(\theta)\big)_{i \geq 1}$.

We expect the scaling limit in Open Problem 13.3 to be closely related to the scaling limit on the Erdős and Rényi random graph, as discussed in more detail in Sect. 13.5. There, we also discuss recent progress on this problem for the *Hamming graph*.

We continue to discuss the *discrete duality principle* for hypercube percolation. Comparing (13.1.1) for subcritical Erdős–Rényi random graphs, and (13.1.3) for supercritical Erdős–Rényi random graphs shows that the *second* largest supercritical cluster for $p = p_c(1 + \varepsilon)$ is closely related to the *largest* subcritical cluster for $p = p_c(1 - \varepsilon)$. This is known as *discrete duality principle*, and it is a classical result for branching processes (which we discuss around (2.1.21)). Due to the "almost infinite-dimensional nature" of the hypercube, we believe that the discrete duality principle extends to hypercube percolation:

Open Problem 13.4 (Duality principle hypercube). *Prove that the hypercube satisfies the discrete duality principle by showing that the asymptotics of* $|\mathcal{C}_{(2)}|$ *for* $p = p_c(1 + \varepsilon)$ *with* $\varepsilon = o(1)$ *with* $\varepsilon \gg 2^{-d/3}$, *to leading order, agrees with that of* $|\mathcal{C}_{(1)}|$ *for* $p = p_c(1 - \varepsilon)$ *with* $\varepsilon = o(1)$ *with* $\varepsilon \gg 2^{-d/3}$.

A major difficulty in establishing Open Problem 13.4 is that we do not know how $|\mathcal{C}_{(1)}|$ for $p = p_c(1 - \varepsilon)$ with $\varepsilon = o(1)$ with $\varepsilon \gg 2^{-d/3}$ behaves. Indeed, (13.3.6) in Thm. 13.8 implies that whp $|\mathcal{C}_{(1)}| \leq 2\varepsilon^{-2} \log(\varepsilon^3 2^d)$, similarly to (13.1.1) for the subcritical Erdős–Rényi random graph. However, a matching lower bound is missing. Hulshof and Nachmias [178] recently showed that $|\mathcal{C}_{(1)}| = \Theta_{\mathbb{P}}\big(\varepsilon^{-2} \log(\varepsilon^3 2^d)\big)$, which, even though it is a major result, still does not identify the correct constant in the subcritical regime. Interestingly, Hulshof and Nachmias [178] *do* identify the correct constant for the related problem of the largest *diameter* of a subcritical cluster. Results on $|\mathcal{C}_{(2)}|$ in the supercritical regime, beyond the statement that it is $o_{\mathbb{P}}(|\mathcal{C}_{(1)}|)$, are completely lacking and appear to be quite hard to derive.

Bollobás, Kohayakawa, and Łuczak [57] establish the discrete duality principle for the hypercube in a certain range for the parameter p, which is a hopeful sign. For their range of p's, this problem is slightly easier, as comparisons to branching processes are much more powerful there. A version of the discrete duality principle for the two-dimensional Hamming graph can be found in [163].

13.5 Scaling Limits of Critical Random Graphs

In this section, we discuss scaling limits of cluster sizes in the critical window. We focus on the Erdős–Rényi random graph, where the scaling limit is identified in the beautiful work of David Aldous [17]. Aldous' theorem can be seen as concluding a long line of research, mainly in the probabilistic combinatorics community, on scaling behavior of the Erdős–Rényi random graph at, and close to, the critical point. This work was initiated by Erdős and Rényi [98, 99] and explored to full detail in works by Bollobás, Łuczak, Knuth, Pittel, and Wierman [184, 209]. Bollobás [54] and Łuczak [208] were the first to identify the *scaling window* of the Erdős–Rényi random graph. For results on the Erdős–Rényi random graph, we further refer to the monographs [20, 55, 155, 185] and the references therein.

Let us start by describing the main result. Take $p = (1 + \theta n^{-1/3})/n$ and recall that $\mathcal{C}_{(i)}$ denotes the ith largest component. It is convenient to make the dependence on θ explicit, so that from now on we write $\mathcal{C}_{(i)}(\theta)$. Then, Thm. 13.1 shows that $n^{-2/3}|\mathcal{C}_{(1)}(\theta)|$ is a tight sequence of random variables. In this section, we informally describe the scaling limit of $n^{-2/3}\big(|\mathcal{C}_{(i)}(\theta)|\big)_{i \geq 1}$.

To formulate the convergence result, define $\ell^2_{\searrow}$ to be the set of infinite sequences $x = (x_i)_{i \geq 1}$ with $x_1 \geq x_2 \geq \cdots \geq 0$ and $\sum_{i \geq 1} x_i^2 < \infty$, and define the $\ell^2_{\searrow}$ metric by

$$d(x, y) = \sqrt{\sum_{i \geq 1}(x_i - y_i)^2} \,. \tag{13.5.1}$$

For fixed $\theta \in \mathbb{R}$, consider the inhomogeneous Brownian motion, or *Brownian motion minus a parabola*, as

$$B^\theta(t) = B(t) + \theta t - \tfrac{1}{2}t^2 \,, \tag{13.5.2}$$

where B is standard Brownian motion. The reflected version of this process is denoted by R^θ and can be obtained as

$$R^\theta(t) = B^\theta(t) - \min_{0 \leq u \leq t} B^\theta(u) \,. \tag{13.5.3}$$

Aldous [17] shows that the excursions of R^θ from 0 can be ranked in increasing order as, say, $\gamma_1(\theta) > \gamma_2(\theta) > \cdots$. The main result for the cluster sizes of the Erdős–Rényi random graph within the scaling window is as follows:

Theorem 13.20 (Cluster sizes Erdős–Rényi random graph within the scaling window [17]). *For the Erdős–Rényi random graph with $p = (1 + \theta n^{-1/3})/n$, as $n \to \infty$,*

$$\big(n^{-2/3}|\mathcal{C}_{(i)}(\theta)|\big)_{i \geq 1} \xrightarrow{d} \big(\gamma_i(\theta)\big)_{i \geq 1} \,, \tag{13.5.4}$$

in distribution and with respect to the $\ell^2_{\searrow}$ topology.

While we do not fully prove Thm. 13.20, we do explain the ideas leading up to it. The proof relies on an *exploration* of the components of the Erdős–Rényi random graph. To set this exploration up, we successively explore clusters, starting from the cluster of a single vertex. In the exploration, we keep track of the number of active and neutral vertices. Here, we call a vertex *active* when it has been found to be part of the cluster that is currently explored, but its neighbors in the random graph have not yet been identified. We call a vertex *neutral* when it has not been active yet.

This exploration is described in terms of a stochastic process $(S_i)_{i \geq 0}$, which encodes the cluster sizes as well as their structure. To describe the exploration process, we let $S_0 = 0$, $N_0 = n - 1$, $R_0 = 1$, and let S_i for $i \geq 1$ satisfy the recursion

$$S_i = S_{i-1} + X_i - 1, \tag{13.5.5}$$

where $X_i \sim \mathsf{Bin}(N_{i-1}, p)$. Here, N_{i-1} denotes the number of neutral vertices after we have explored $i - 1$ vertices. The fact that N_i decreases as i increases plays a major role in determining the scaling limit of critical clusters and is caused by the so-called *depletion-of-points effect*, indicating that during the exploration, the number of potential additional neighbors of vertices decreases.

Let us highlight some of the properties of the above exploration.

The number of disjoint clusters that have been fully explored after the ith exploration is given by $-\inf_{j \in [i]} S_j$, while, for $i \geq 1$,

$$R_i = S_i - \inf_{j \in [i-1]} S_j + 1 \tag{13.5.6}$$

denotes the number of active vertices at time i in the cluster that we are currently exploring. We see that, for $i \geq 1$,

$$N_i = n - i - R_i \tag{13.5.7}$$

denotes the number of neutral vertices after the ith exploration.

We can view $(R_i)_{i \geq 0}$ as the reflection of the process $(S_i)_{i \geq 0}$. The exploration can perhaps be more easily understood in terms $(R_i)_{i \geq 0}$ When $R_i = 0$, we have fully explored a cluster. In particular, when we explore $\mathcal{C}(v_1)$, then

$$|\mathcal{C}(v_1)| = \inf\{i > 0 : R_i = 0\} . \tag{13.5.8}$$

After having explored $\mathcal{C}(v_1)$, we explore $\mathcal{C}(v_2)$ for some $v_2 \notin \mathcal{C}(v_1)$ and obtain that

$$|\mathcal{C}(v_2)| = \inf\{i > |\mathcal{C}(v_1)| : R_i = 0\} - |\mathcal{C}(v_1)| . \tag{13.5.9}$$

Iterating this procedure, we see that

$$|\mathcal{C}(v_j)| = \inf\{i > |\mathcal{C}(v_1)| + \cdots + |\mathcal{C}(v_{j-1})| : R_i = 0\} \\ - |\mathcal{C}(v_1)| - \cdots - |\mathcal{C}(v_{j-1})|. \tag{13.5.10}$$

Inspecting (13.5.6), we thus see that a cluster is fully explored when $S_i - \inf_{j \in [i-1]} S_j + 1 = 0$, which is the same as saying that $S_i = \inf_{j \in [i]} S_j$ for the first time. Thus, the total number of clusters that are fully explored up to time i indeed equals $-\inf_{j \in [i]} S_j$.

It turns out to be more convenient to deal with the process $(S_i)_{i \geq 0}$ than with $(R_i)_{i \geq 0}$, since the scaling limit of $(S_i)_{i \geq 0}$ can be more easily described:

Theorem 13.21 (Cluster exploration Erdős–Rényi random graph within the scaling window [17]). *For the Erdős–Rényi random graph with $p = (1 + \theta n^{-1/3})/n$, as $n \to \infty$,*

$$\left(n^{-1/3} S_{tn^{2/3}}\right)_{t \geq 0} \xrightarrow{d} \left(B^\theta(t)\right)_{t \geq 0} . \qquad (13.5.11)$$

Thm. 13.21, together with (13.5.6) and the continuity of the reflection mapping, implies that

$$\left(n^{-1/3} R_{tn^{2/3}}\right)_{t \geq 0} \xrightarrow{d} \left(R^\theta(t)\right)_{t \geq 0} , \quad \text{where } R^\theta(t) = B^\theta(t) - \inf_{s \in [0,t]} B^\theta(s) . \quad (13.5.12)$$

Since the excursions of $\left(R^\theta(t)\right)_{t \geq 0}$ describe the component sizes, (13.5.12) can be seen to imply Thm. 13.20 and describes the limiting law of the largest connected components.

Let us next informally describe the proof of Thm. 13.21. Since $n^{-1/3} S_{tn^{2/3}}$ will be seen to converge in distribution, and we are investigating the exploration process at times $tn^{2/3}$, we simplify our lives considerably and approximate (13.5.7) to $N_i \approx n - i$. This means that the random variables $(X_i)_{i \geq 1}$ in (13.5.5) are close to being independent with

$$X_i \approx \text{Bin}(n - i, p) = \text{Bin}\left(n - i, \left(1 + \theta n^{-1/3}\right)/n\right) . \qquad (13.5.13)$$

Here, and in what follows, $\approx$ denotes an uncontrolled approximation.

We now apply the Poisson limit theorem, which states that $\text{Bin}(\mu/n, n)$-distributed random variables are well approximated by a Poisson r.v. with parameter μ whenever n is large. In our case,

$$\mu = (n - i)p = (n - i)\left(1 + \theta n^{-1/3}\right)/n \approx 1 + \theta n^{-1/3} - i/n . \qquad (13.5.14)$$

Note that when $i = tn^{2/3}$, both correction terms are of the same order in n. Thus, we approximate

$$S_i \approx \sum_{j=1}^{i}(Y_j - 1) , \qquad (13.5.15)$$

where $Y_j \sim \text{Poi}(1 + \theta n^{-1/3} - j/n)$ are independent. Since sums of independent Poisson variables are Poisson again, we thus obtain that

$$S_i \approx S_i^* , \qquad (13.5.16)$$

where

$$S_i^* \sim \text{Poi}\left(\sum_{j=1}^{i}(1 + \theta n^{-1/3} - j/n)\right) - i = \text{Poi}\left(i + i\theta n^{-1/3} - \frac{i^2}{2n}\right) - i \,, \quad (13.5.17)$$

and $(S_i^* - S_{i-1}^*)_{i \geq 1}$ are independent. Now, we multiply by $n^{-1/3}$ and take $i = tn^{2/3}$ to obtain

$$n^{-1/3}S_{tn^{2/3}}^* \sim n^{-1/3}\left(\text{Poi}\left(tn^{2/3} + t\theta n^{1/3} - \tfrac{1}{2}t^2 n^{1/3}\right) - tn^{2/3}\right) \,, \quad (13.5.18)$$

Since the leading order term of a rescaled Poisson process is deterministic, we can approximate

$$n^{-1/3}S_{tn^{2/3}}^* \sim n^{-1/3}\left(\text{Poi}\left(tn^{2/3}\right) - tn^{2/3}\right) + t\theta - \tfrac{1}{2}t^2 \xrightarrow{d} B(t) + t\theta - \tfrac{1}{2}t^2 \,, \quad (13.5.19)$$

where $(B(t))_{t \geq 0}$ is standard Brownian motion and we use the fact that

$$m^{-1/2}(\text{Poi}(tm) - tm)_{t \geq 0} \xrightarrow{d} (B(t))_{t \geq 0} \quad \text{as } m \to \infty \,.$$

This informally explains the proof of Thm. 13.21. We also directly see that the quadratic term in (13.5.19) arises from the depletion-of-point effect discussed below (13.5.5). To make this proof rigorous, one typically resorts to Martingale Functional Central Limit Theorems, see, for example, how Aldous [17] does this nicely.

Additional Results on Critical Erdős–Rényi Random Graphs. There are many recent extensions to the above scaling limit results. For example, already in his original paper, Aldous also investigates the *surplus* of the largest connected components. The surplus of a graph is the minimal number of edges that need to be removed in order to obtain a tree and is related to the number of cycles in the graph. We leave this as an exercise:

Exercise 13.13 (Surplus critical Erdős–Rényi clusters). Prove, using Thm. 13.21 and (13.5.12), that the number of cycle edges between vertices found in the time interval $\left[t_1 n^{2/3}, t_2 n^{2/3}\right]$ has an approximate Poisson distribution with random parameter

$$\int_{t_1}^{t_2} R^\theta(t)\, dt \,. \quad (13.5.20)$$

Interestingly, Aldous gives a second interpretation of the scaling limit of clusters in terms of *multiplicative coalescents*. Indeed, when we interpret θ as a time variable and use the Harris coupling of percolation (see page 14), we see that within the time interval $(\theta, \theta + \varepsilon\theta)$, two clusters of size $|\mathcal{C}_{(i)}(\theta)| = xn^{2/3}$ and $|\mathcal{C}_{(j)}(\theta)| = yn^{2/3}$ merge with probability close to $(\varepsilon\theta/n^{4/3})(xn^{2/3})(yn^{2/3}) = \varepsilon\theta xy$ to create a cluster of size $(x + y)n^{2/3}$. When rescaling the cluster sizes by $n^{-2/3}$, this dynamics is called the *multiplicative coalescent*.

More recent work of Addario-Berry, Broutin, and Goldschmidt [2, 3] shows that the largest critical clusters, viewed as metric spaces, converge in distribution in an appropriate topology to some limiting graphs. The components $\mathcal{C}_{(i)}$ are considered as metric spaces using the graph distance rescaled by $n^{-1/3}$. The authors prove that the sequence of rescaled components converges to a sequence of continuous compact metric spaces. These limiting continuum graphs are *almost* trees, but they contain a finite number of cycles corresponding to the surpluses of the clusters (recall Exerc. 13.13). The convergence is in the Gromov–Hausdorff–Prokhorov topology on metric spaces. This result has several interesting consequences, such as the convergence in distribution of the diameter of $\mathcal{C}_{(i)}$, rescaled by $n^{-1/3}$ and of the array of distances between uniformly chosen vertices in the cluster.

Scaling Limit of Cluster Sizes in Inhomogeneous Random Graph. Scaling limits for critical random graphs have been obtained in many settings, in particular when the graphs themselves are quite *inhomogeneous*. Here, we can think of percolation on base graphs that have variable degrees, or inhomogeneous percolation, where the edge occupation probabilities are different from vertex to vertex. These edge probabilities are moderated by vertex weights $(w_i)_{i \in [n]}$, where w_i describes the propensity of edges for vertex i. Examples are rank-1 inhomogeneous random graphs, where the edge probability between vertex i and j is given by

$$p_{ij} = 1 - e^{-w_i w_j / \sum_{k \in [n]} w_k} . \tag{13.5.21}$$

This model is sometimes also called the *Norros–Reittu model* or the *Poisson random graph*. The interpretation of w_i is that it is (close to) the expected degree of vertex i, as investigated in more detail in the following exercise:

Exercise 13.14 (Degrees in rank-1 inhomogeneous random graphs). Prove that the degree of vertex i is converging to a $\mathsf{Poi}(w_i)$-distributed random variable as $n \to \infty$.

We assume that the weights are *sufficiently regular*, in the sense that their empirical distribution function F_n converges weakly to some limiting distribution function F, i.e.,

$$F_n(x) = \frac{1}{n} \sum_{i \in [n]} \mathbb{1}_{\{w_i \leq x\}} \to F(x) \quad \text{for all continuity points } x \in \mathbb{R} \text{ of } F . \tag{13.5.22}$$

Exercise 13.15 (Weak convergence degrees in rank-1 inhomogeneous random graphs). Let $U \in [n]$ denote a vertex that is chosen uniformly from $[n]$. Prove that (13.5.22) implies that the degree of U converges in distribution to a $\mathsf{Poi}(W)$ random variable, where W has distribution function F.

Assume that the second moment of the weights converges, i.e., $(1/n) \sum_{i \in [n]} w_i^2 \to \mathbb{E}[W^2]$. Then the random graph is critical when

$$\nu = \frac{\mathbb{E}[W^2]}{\mathbb{E}[W]} = 1 . \tag{13.5.23}$$

This can be seen by noting that, under these conditions, the number of other neighbors of a random neighbor of a random vertex (i.e., neighbors of neighbors not equal to the random vertex itself) has asymptotic distribution $\mathsf{Poi}(W^\star)$, where $W^\star$ is the *size-biased* version of W for which

$$\mathbb{P}(W^\star \leq w) = \frac{\mathbb{E}[W\mathbb{1}_{\{W \leq w\}}]}{\mathbb{E}[W]} . \tag{13.5.24}$$

It turns out that the local neighborhood of a uniformly picked individual is close to a branching process where the root has offspring distribution $\mathsf{Poi}(W)$, while all other vertices have offspring distribution $\mathsf{Poi}(W^\star)$. The expected value of $\mathsf{Poi}(W^\star)$ equals $\mathbb{E}[\mathsf{Poi}(W^\star)] = \nu$, so that for $\nu = 1$ the branching process is critical. This informally explains (13.5.23).

When we also assume that the third moment of the weights converges, i.e.,

$$\frac{1}{n}\sum_{i\in[n]} w_i^3 \to \mathbb{E}[W^3] < \infty , \tag{13.5.25}$$

where W has distribution function F, then also the scaling limits, at criticality, are given by rescaled versions of the multiplicative coalescent scaling limit in the critical Erdős–Rényi random graph. The third moment in (13.5.25) implies that $\mathsf{Poi}(W^\star)$ has finite variance, in which case one can expect Brownian limits in the cluster exploration process as in Thm. 13.21. See, for example, the work by Bhamidi et al. [50] and Turova [255].

Interestingly, when the third moment of the weights tends to infinity, different scalings and scaling limits arise. The simplest example arises when $w_i = (cn/i)^{1/(\tau-1)}$ for some constant $c > 0, \tau > 1$, which corresponds to asymptotic weight distribution $F(x) = 1 - cx^{-(\tau-1)}$ for $x > c^{1/(\tau-1)}$. We take $\tau \in (3, 4)$, so that $\int_0^\infty x^3\, dF(x) = \mathbb{E}[W^3] = \infty$, where W has distribution function F. By choosing c appropriately, we can make $\nu = \mathbb{E}[W^2]/\mathbb{E}[W] = 1$, so that the random graph is critical. In this case, $(1/n)\sum_{i\in[n]} w_i^3 \approx n^{3/(\tau-1)-1} \gg 1$. It turns out that now $n^{-(\tau-2)/(\tau-1)}(|\mathscr{C}_{(i)}|)_{i\geq 1}$ converges, interestingly again to a multiplicative coalescent. The scaling limit of the cluster exploration process in Thm. 13.21 changes dramatically and is now described in terms of so-called *thinned Lévy processes*. See [51] for more details.

Many more results have appeared about the critical behavior of such random graphs, and we refer the reader to [48, 49, 52, 88, 89, 188, 217, 227] and the references therein. The proofs all revolve around the appropriate exploration processes and their scaling limits.

Problems in Identifying the Scaling Limits for Critical High-Dimensional Tori. Now, we return to percolation on critical high-dimensional tori. As we have seen in the above description, scaling limits of critical random graphs are usually obtained by applying weak convergence results, such as Martingale Functional Central Limit Theorems, to the exploration processes that describe the exploration of such graphs. Here, we are helped by the fact that the models are sufficiently *mean field*, meaning that vertices interact with one another in roughly the same way. For example, for the Erdős–Rényi random graph, all vertices are the same, which makes that the depletion-of-points effect comes out in a really simple form as the $-t^2/2$ term in (13.5.2) and as the $N_i = n - i - S_i + \inf_{j\in[i-1]} S_j - 1 \approx n - i$ term in (13.5.7) (recall also the discussion below (13.5.5)).

Unfortunately, on lattices, the depletion-of-points effect is far less homogeneous. Some vertices that we explore still have many neighbors that are neutral, while others have far fewer than expected. On average, after a long time, most active vertices should have of the order 1 minus something small neighbors that become active after exploration. For example, on the hypercube, the scaling window is of width $2^{-d/3}/d$, so that, after of the order $2^{2d/3}$ explorations, we can expect vertices to have on average $[1 + O(2^{-d/3}/d)]/p_c$ neutral neighbors. This number is extremely close to d, and order 1 of them will be found to be neighbors of the vertex that is currently explored. One cannot expect to achieve this by the simple exploration process that was used for the Erdős–Rényi random graph. However, we *do* believe that the scaling limits are the same:

Open Problem 13.5 (Scaling limit critical high-dimensional tori). *Consider the* $\mathbb{T}_{n,d}$ *with* $d > 6$ *and fix* $p = p_c(\mathbb{Z}^d)(1 + \theta V^{-1/3})$. *Prove that*

$$V^{-2/3}\big(|\mathscr{C}_{(i)}|\big)_{i\geq 1} \xrightarrow{\text{d}} \big(\gamma_i(\theta)\big)_{i\geq 1} \quad \text{as } n \to \infty , \tag{13.5.26}$$

where $\big(\gamma_i(\theta)\big)_{i\geq 1}$ *are multiplicative coalescents. One would expect a translation of time and rescaling of* γ_i *as well as* θ *to be necessary to make* $\big(\gamma_i(\theta)\big)_{i\geq 1}$ *have the same law as the scaling limit on the Erdős–Rényi random graph* $\big(\gamma_i(\theta)\big)_{i\geq 1}$ *in Thm. 13.20.*

Open Problem 13.5 appears out of reach with current methodology. We believe that a rescaling in time in Open Problem 13.5 may be necessary, since we do not exactly know the location of $p_c(\mathbb{Z}^d)$ inside the scaling window. Mind that Open Problem 13.5 is much harder than Open Problem 13.3, since the hypercube $\{0, 1\}^d$ has a larger amount of symmetry, as shown by the bounds in terms of the nonbacktracking random walk.

At the moment, there is only one example of a high-dimensional torus where Open Problem 13.5 has been resolved. Indeed, for the Hamming graph, Federico, the second author, den Hollander, and Hulshof [104] have proved that the scaling limit in Open Problem 13.5 exists for $d = 2, 3, 4$, with the scaling limit indeed being equal to that of the Erdős–Rényi random graph. Interestingly, while the *cluster sizes* behave in a completely universal way, the *surplus edges* behave rather differently. Indeed, the number of surplus edges grows like $a(d)|\mathscr{C}_{(i)}|/n$ for some appropriate constant $a(d)$ depending on the dimension. It is conjectured that then the number of *long* cycles, however, *does* satisfy the same scaling as for the Erdős–Rényi random graph as in Exerc. 13.13. This would show that even within the Erdős–Rényi random graph differences can arise. We believe that in great generality the number of surplus edges on high-dimensional tori satisfies different scaling as on the Erdős–Rényi random graph.

The proof in [104] is by a *two-scale exploration*, which remedies the problem that the number of potential neighbors in the exploration process varies substantially. Additionally, Federico, the second author, den Hollander, and Hulshof [104] rely on the precise characterization of the critical value as identified by them in [103], so as to cancel out short cycles appearing in the exploration. This exact cancelation also partly explains why their methods fail in higher dimensions, as this would require a more refined analysis of the critical point.

We refer to Sect. 15.5 for more details on asymptotics of percolation critical values, where also the critical value of the Hamming graph is discussed.

13.6 The Role of Boundary Conditions

In this section, we describe some results that explore the role of boundary conditions in high-dimensional percolation. So far, we have studied *periodic* boundary conditions, leading to percolation on high-dimensional tori. The key advantage of periodic boundary conditions is that the resulting graph is *transitive*, meaning that every vertex plays the exact same role. Alternative boundary conditions that we consider in this section are the following:

Zero boundary conditions. Zero boundary conditions arise when we consider connections that only use edges in $[0, n-1]^d$. In other words, the clusters for zero boundary conditions are $\left(\mathcal{C}^{(z)}(x)\right)_{x \in [0, n-1]^d}$ with $\mathcal{C}^{(z)}(x) = \mathcal{C}(x; [0, n-1]^d)$, where $\mathcal{C}(x; A)$ denotes the cluster of x in the percolation configuration where only the edges between vertices in A are considered.

Bulk boundary conditions. Bulk boundary conditions arise when we declare two vertices $x, y \in [0, n-1]^d$ connected whenever $y \in \mathcal{C}(x)$. In other words, the clusters for bulk boundary conditions $\left(\mathcal{C}^{(b)}(x)\right)_{x \in [0, n-1]^d}$ are the original $\mathbb{Z}^d$-clusters intersected with the cube, $\mathcal{C}^{(b)}(x) = \mathcal{C}(x) \cap [0, n-1]^d$.

Wired boundary conditions. In wired boundary conditions, we identify all vertices on the boundary $\partial[0, n-1]^d = [0, n-1]^d \setminus [1, n-2]^d$, and say that x and y are connected when either $y \in \mathcal{C}(x; [0, n-1]^d)$ or x and y are both connected to the boundary. Denote the cluster of x under wired boundary conditions by $\mathcal{C}^{(w)}(x)$ for $x \in [0, n-1]^d$.

In each of the above boundary conditions, we let $|\mathcal{C}_{\max}|$ denote the maximal cluster size, so that, for example, $|\mathcal{C}^{(b)}_{\max}| = \max_{x \in [0, n-1]^d} |\mathcal{C}^{(b)}(x)|$. Clearly, $|\mathcal{C}^{(z)}_{\max}| \leq |\mathcal{C}^{(b)}_{\max}| \leq |\mathcal{C}^{(w)}_{\max}|$ a.s., since these three objects can all be defined in terms of a percolation configuration on $\mathbb{Z}^d$. Additionally, since $[0, n-1]^d \setminus [1, n-2]^d$ is identified to one point, the wired boundary conditions satisfy that $|\mathcal{C}^{(w)}_{\max}| \geq cn^{d-1}$ for some $c > 0$.

When $p \neq p_c$, the boundary conditions ought to play a minor role. See, for example, the work of the second author and Redig [166], where this is proved when $p \neq p_c$ is independent of n. We see that boundary conditions *are* relevant in the critical case. Theorem 13.9 shows that the largest connected components are of order $V^{2/3} = n^{2d/3}$ for periodic boundary conditions, and we give bounds on the largest connected components under the related boundary conditions in this section. Most of the results mentioned in this section are direct consequences of the work by Aizenman in [8], but we also extend some of his results:

Theorem 13.22 (The role of boundary conditions [8]). *For percolation on $\mathbb{Z}^d$ with $d \geq 11$, there exist constants C_1 and $\varepsilon > 0$ such that*

$$\lim_{n \to \infty} \mathbb{P}_{p_c}(\varepsilon n^4 \leq |\mathcal{C}^{(b)}_{\max}| \leq C_1 n^4 \log n) = 1 , \qquad (13.6.1)$$

and

$$\lim_{n\to\infty} \mathbb{P}_{p_c}(\varepsilon n^4 \leq |\mathcal{C}^{(z)}_{\max}| \leq C_1 n^4 \log n) = 1 \,. \tag{13.6.2}$$

The upper bound in (13.6.1) in Thm. 13.22 follows from Aizenman [8, (4.18)], together with the fact that $\eta = 0$ in the x-space sense by Thm. 11.4. The upper bound in (13.6.2) follows from that in (13.6.1). The lower bounds are proved below. Aizenman [8, (4.18)] proves a lower bound with $\varepsilon > 0$ replaced by ε_n for any $\varepsilon_n = o(1)$ and only for *bulk* boundary conditions. Our results improve [8, Thm. 5].

The proof of Aizenman [8, (4.18)] follows by using tree-graph inequalities like the ones used in the proof of Lem. 13.10, but now for bulk boundary conditions. We note that the largest cluster in a cube under bulk boundary conditions scales as $n^{4+o(1)}$, which suggests that large critical clusters are *four-dimensional*. We return to this issue in Sect. 15.1 below. Note that $n^4 \ll V^{2/3} = n^{2d/3}$ precisely when $d > d_c = 6$. This suggests that the maximal clusters under bulk and periodic boundary conditions grow at the same rate when $d < d_c$, but proving this seems out of reach at this moment.

Aizenman [8] proves several more interesting results related to spanning clusters, which are the clusters that connect two opposite sides of the cube (i.e., clusters connecting the hyperplanes $\{0\} \times [0, n-1]^{d-1}$ and $\{n-1\} \times [0, n-1]^{d-1}$). He proves that there are $n^{d-6+o(1)}$ of such spanning clusters and that the largest one has size $n^{4+o(1)}$, as for the largest cluster. Thus, in high dimensions, there are many large spanning clusters.

We now present the proof of the lower bound in (13.6.2), which is novel and implies the lower bound in (13.6.1):

Proof of the lower bound in (13.6.2). Fix $\varepsilon > 0$. Without loss of generality, assume that n is a multiple of 4. Let

$$Z_{\geq \varepsilon n^4} = \sum_{x\in[n/4, 3n/4]^d} \mathbb{1}_{\{|\mathcal{C}^{(z)}(x)| \geq \varepsilon n^4\}} \,. \tag{13.6.3}$$

We use the second moment method on $Z_{\geq \varepsilon n^4}$ to show that $Z_{\geq \varepsilon n^4} \geq 1$ with high probability. For this, it is enough to show that

$$\mathbb{E}_{p_c}[Z_{\geq \varepsilon n^4}] \to \infty \,, \quad \mathrm{Var}_{p_c}(Z_{\geq \varepsilon n^4}) \ll \mathbb{E}_{p_c}[Z_{\geq \varepsilon n^4}]^2 \,. \tag{13.6.4}$$

We start by proving a lower bound on $\mathbb{E}_{p_c}[Z_{\geq \varepsilon n^4}]$, for which we notice that

$$\mathbb{E}_{p_c}[Z_{\geq \varepsilon n^4}] = \sum_{x\in[n/4, 3n/4]^d} \mathbb{P}_{p_c}(|\mathcal{C}^{(z)}(x)| \geq \varepsilon n^4) \,. \tag{13.6.5}$$

Since $|\mathcal{C}(x)| \geq \varepsilon n^4$ implies that either $|\mathcal{C}^{(z)}(x)| \geq \varepsilon n^4$ or $x \longleftrightarrow \partial[0, n-1]^d$, we obtain that

$$\mathbb{P}_{p_c}(|\mathcal{C}^{(z)}(x)| \geq \varepsilon n^4) \geq \mathbb{P}_{p_c}(|\mathcal{C}(x)| \geq \varepsilon n^4) - \mathbb{P}_{p_c}(x \longleftrightarrow \partial[0, n-1]^d) \,. \tag{13.6.6}$$

We use Thm. 9.2 to bound the first term from below and Thm. 11.5 to bound the second term from above using $\mathbb{P}_{p_c}(x \leftrightarrow \partial[0, n-1]^d) \leq \mathbb{P}_{p_c}(0 \leftrightarrow \partial\Lambda_{n/4})$, leading to

$$\mathbb{P}_{p_c}\left(|\mathcal{C}^{(z)}(x)| \geq \varepsilon n^4\right) \geq \frac{c_\delta}{\sqrt{\varepsilon n^4}} - \frac{C_{\text{ex}}}{(n/4)^2} = \frac{c_\delta \varepsilon^{-1/2} - 16 C_{\text{ex}}}{n^2} \geq \frac{a}{n^2} \qquad (13.6.7)$$

with $a = c_\delta \varepsilon^{-1/2} - 16 C_{\text{ex}} > 0$ when $\varepsilon < c_\delta^2/(16 C_{\text{ex}}^2)$. We conclude that

$$\mathbb{E}_{p_c}[Z_{\geq \varepsilon n^4}] \geq \sum_{x \in [n/4, 3n/4]^d} \frac{a}{n^2} = \frac{a}{2^d} n^{d-2}. \qquad (13.6.8)$$

We proceed to investigate the variance of $Z_{\geq \varepsilon n^4}$, for which we note that

$$\text{Var}_{p_c}(Z_{\geq \varepsilon n^4}) = \sum_{x_1, x_2 \in [n/4, 3n/4]^d} \left[\mathbb{P}_{p_c}\left(|\mathcal{C}^{(z)}(x_1)|, |\mathcal{C}^{(z)}(x_2)| \geq \varepsilon n^4\right) \right.$$
$$\left. - \mathbb{P}_{p_c}\left(|\mathcal{C}^{(z)}(x_1)| \geq \varepsilon n^4\right) \mathbb{P}_{p_c}\left(|\mathcal{C}^{(z)}(x_2)| \geq \varepsilon n^4\right) \right]. \qquad (13.6.9)$$

We split depending on whether $x_1 \leftrightarrow x_2$ or not to obtain

$$\mathbb{P}_{p_c}\left(|\mathcal{C}^{(z)}(x_1)|, |\mathcal{C}^{(z)}(x_2)| \geq \varepsilon n^4\right)$$
$$= \mathbb{P}_{p_c}\left(|\mathcal{C}^{(z)}(x_1)|, |\mathcal{C}^{(z)}(x_2)| \geq \varepsilon n^4, x_1 \leftrightarrow x_2\right)$$
$$+ \mathbb{P}_{p_c}\left(|\mathcal{C}^{(z)}(x_1)|, |\mathcal{C}^{(z)}(x_2)| \geq \varepsilon n^4, x_1 \nleftrightarrow x_2\right), \qquad (13.6.10)$$

and bound both terms separately starting with the second. Note that, by the BK inequality,

$$\mathbb{P}_{p_c}\left(|\mathcal{C}^{(z)}(x_1)|, |\mathcal{C}^{(z)}(x_2)| \geq \varepsilon n^4, x_1 \nleftrightarrow x_2\right)$$
$$\leq \mathbb{P}_{p_c}\left(\{|\mathcal{C}^{(z)}(x_1)| \geq \varepsilon n^4\} \circ \{|\mathcal{C}^{(z)}(x_2)| \geq \varepsilon n^4\}\right)$$
$$\leq \mathbb{P}_{p_c}\left(|\mathcal{C}^{(z)}(x_1)| \geq \varepsilon n^4\right) \mathbb{P}_{p_c}\left(|\mathcal{C}^{(z)}(x_2)| \geq \varepsilon n^4\right). \qquad (13.6.11)$$

so that, by Thm. 11.4,

$$\text{Var}_{p_c}(Z_{\geq \varepsilon n^4}) \leq \sum_{x_1, x_2 \in [n/4, 3n/4]^d} \mathbb{P}_{p_c}\left(|\mathcal{C}^{(z)}(x_1)|, |\mathcal{C}^{(z)}(x_2)| \geq \varepsilon n^4, x_1 \leftrightarrow x_2\right)$$
$$\leq \sum_{x_1, x_2 \in [n/4, 3n/4]^d} \mathbb{P}_{p_c}(x_1 \leftrightarrow x_2)$$
$$\leq \sum_{x_1, x_2 \in [n/4, 3n/4]^d} \frac{A + O(|x_1 - x_2|^{-2/d})}{|x_1 - x_2|^{d-2}}, \qquad (13.6.12)$$

Computing the sum leads to

$$\text{Var}_{p_c}(Z_{\geq \varepsilon n^4}) \leq cn^{d+2} , \tag{13.6.13}$$

for some $c > 0$. Combining (13.6.8) and (13.6.13) shows that

$$\text{Var}_{p_c}(Z_{\geq \varepsilon n^4}) \leq cn^{d+2} \ll (an^{d-2})^2 \leq \mathbb{E}_{p_c}[Z_{\geq \varepsilon n^4}]^2 \tag{13.6.14}$$

precisely when $d > 6$. We conclude that

$$\frac{Z_{\geq \varepsilon n^4}}{\mathbb{E}_{p_c}[Z_{\geq \varepsilon n^4}]} \xrightarrow{\mathbb{P}} 1 , \tag{13.6.15}$$

and thus $Z_{\geq \varepsilon n^4} \geq 1$ whp since $\mathbb{E}_{p_c}[Z_{\geq \varepsilon n^4}] \to \infty$. This proves that $|\mathcal{C}_{\text{max}}^{(z)}| \geq \varepsilon n^4$ whp. Since $\mathbb{E}_{p_c}[Z_{\geq \varepsilon n^4}] \geq (a/2^d)n^{d-2}$, we even get that the number of vertices in components of size at least εn^4 exceeds $n^{d-2+o(1)}$ whp. □

While Aizenman's results and their extensions identify the critical exponents related to the cluster growth for different boundary conditions, they leave the exact scaling open:

Open Problem 13.6 (Scaling of critical high-dimensional clusters for different boundary conditions). *For critical percolation on the box* $[0, n-1]^d$*, identify the exact scaling of* $|\mathcal{C}_{\text{max}}^{(b)}|, |\mathcal{C}_{\text{max}}^{(z)}|$ *and* $|\mathcal{C}_{\text{max}}^{(w)}|$*. We believe that there exist constants* a_b, a_z, a_w *such that*

$$\frac{|\mathcal{C}_{\text{max}}^{(b)}|}{n^4 \log n} \xrightarrow{\mathbb{P}} a_b , \quad \frac{|\mathcal{C}_{\text{max}}^{(z)}|}{n^4 \log n} \xrightarrow{\mathbb{P}} a_z , \quad \frac{|\mathcal{C}_{\text{max}}^{(w)}|}{n^{d-1}} \xrightarrow{\mathbb{P}} a_w. \tag{13.6.16}$$

We continue by discussing results related to the boundary effects for periodic boundary conditions as proved by the second author and Sapozhnikov [169]. They prove that (all results *with high probability*) clusters have long cycles with positive probability, and the length of such cycles is of order $V^{1/3}$. Additionally, any cycle that is long (e.g., any cycle having displacement at least $n/4$ is considered to be long) contains at least $V^{1/3}$ edges. Finally, such long cycles go around the torus at least $V^{1/6}/n = n^{(d-6)/6}$ times. Again, this highlights the importance of the boundary conditions, as well as the special role that $d_c = 6$ plays. Finally, this means that when "unwrapping" large percolation clusters on the torus, their spatial extent becomes $V^{1/6}$, while their size is $V^{2/3}$. Since $V^{2/3} = (V^{1/6})^4$, this is again a sign of the four-dimensional nature of large percolation clusters.

We close this section by discussing the *lack* of a discrete duality principle for high-dimensional tori of fixed dimension, which can be viewed as an explanation why proving mean-field behavior of supercritical percolation on high-dimension tori is hard. While this result may appear highly surprising at first sight, the reason behind it is relatively simple. Indeed, if $p > p_c$, then the *second* largest cluster has size $c(\log V)^{d/(d-1)}(1 + o_{\mathbb{P}}(1))$ for some constant $c = c(p, d)$ [166]. This is quite different from the behavior of the largest cluster for $p < p_c$, where it is $c' \log V (1 + o_{\mathbb{P}}(1))$ for some $c' = c'(p,d)$. These results are independent of the precise boundary conditions (at least, when the boundary conditions are not wired). Thus, the discrete duality principle, as observed for the Erdős–Rényi random graph (recall (13.1.1) and (13.1.3)), and predicted for the hypercube in Open Problem 13.4, cannot be valid for high-dimensional tori. This leads us to the following open problem:

Open Problem 13.7 (Second largest component of high-dimensional supercritical percolation). *For supercritical percolation on the high-dimensional torus* $\mathbb{T}_{n,d}$, *identify the exact scaling of the second largest component for* $p = p_c(1 + \varepsilon)$ *with* $\varepsilon = o(1)$ *but* $\varepsilon \gg V^{-1/3}$. *How does this interpolate between the strictly supercritical value* $c(\log V)^{d/(d-1)}$ *for* $p > p_c$ *(which is unlike the Erdős–Rényi random graph), and the critical behavior (where it is* $V^{2/3}$ *as for the critical Erdős–Rényi random graph)?*

The occurrence of the exponent $d/(d-1) > 1$ in the strictly supercritical case hints at a role of geometry in the supercritical percolation phase that is absent in the mean-field model for percolation, the Erdős–Rényi random graph. We thus conclude that while the critical and subcritical phases of percolation on high-dimensional tori are, as far as we know, very well predicted by the mean-field model, the supercritical behavior, in general, is *not*.

Here is an explanation for the $(\log V)^{d/(d-1)}$ scaling of clusters in the strictly supercritical percolation setting. In this setting, it is not hard to see that there is a giant component of size $\theta(p)V(1 + o_{\mathbb{P}}(1))$. Thus, the second largest component needs to avoid this giant component that has a positive density. For Erdős–Rényi random graphs, avoiding a cluster of positive density effectively decreases the percolation value, and in the supercritical regime, this decreased percolation value becomes subcritical, so that the distribution of clusters avoiding the giant has an exponential tail. This explains why the second largest cluster has a logarithmic size in the strictly supercritical Erdős–Rényi random graph, where $p = \lambda/n$ with $\lambda > 1$ (independent of n).

For percolation on a finite-dimensional torus, however, it is much easier to create a large cluster that is not the giant. Indeed, when we fix m large, then making the boundary of a box of side length $m^{1/d}$ completely vacant has costs of order $e^{-am^{(d-1)/d}}$, where we observe a boundary versus volume effect that is absent in the complete graph. When making the boundary of a large cube vacant, the inside of that cube will still be supercritical, so the largest cluster inside the cube has size close to $\theta(p)(m^{1/d})^d$. Thus, we can create a cluster of size $\Theta(m)$ with probability at most $e^{-am^{(d-1)/d}}$. Assuming independence of the cluster sizes, extreme value theory suggests that the second largest cluster has size m if $e^{-am^{(d-1)/d}}$ is roughly $1/V$; therefore, it has size $m \approx (\log V)^{d/(d-1)}$. Of course, the cluster sizes are not independent, but the dependency is sufficiently weak to push the argument through. Indeed, the second author and Redig [166] provide a rigorous proof for this, and they even study the fluctuations around the leading order asymptotics.

In the analysis of the tails of large nongiant clusters in the supercritical setting, results on the percolation *Wulff shape* are crucial. The Wulff shape describes the geometry of a large finite cluster on $\mathbb{Z}^d$ in the supercritical regime and has attracted quite some attention. See Cerf [71] for a discussion of the Wulff shape in percolation and related models and [69, 70] for his results in two- and three-dimensional percolation. In particular, the results by Cerf show that there exists an $a = a(p, d)$ such that the probability that a supercritical cluster has size m behaves as $e^{-am^{(d-1)/d}\left(1+o(1)\right)}$. Wulff shape-type results on finite tori, for sequences of p that approach p_c, are crucial in order to resolve Open Problem 13.7 but have not yet been derived.

Part IV
Related and Open Problems

Chapter 14
Random Walks on Percolation Clusters

Random structures have an intricate relationship with the random walks defined on them. We are now focussing on random walks on percolation clusters, which is a prime example of the random conductance model with the charm of not being elliptic. We shall see that random walks on supercritical and critical percolation structures behave completely differently, underlining the remarkable features of critical structures.

We discuss random walks on supercritical percolation clusters in Sect. 14.1, on finite critical clusters in Sect. 14.2, and on the incipient infinite cluster in Sect. 14.3.

14.1 Random Walks on the Infinite Cluster

One of the most classical results in modern probability theory is Donsker's invariance principle: A simple symmetric random walk on the lattice $\mathbb{Z}^d$, diffusively rescaled, converges in distribution to a d-dimensional Brownian motion. The question to be addressed now is the following variation: If the lattice is not perfect, but a few edges are missing, will the scaling limit of a random walk still be Brownian motion? Indeed, the answer is affirmative.

Let us start with a formal description of the problem. To this end, let $D(\mathbb{R}^+, \mathbb{Z}^d)$ be the space of $\mathbb{Z}^d$-valued càdlàg functions equipped with the Skorokhod topology. For a given percolation configuration $\omega \in \{0, 1\}^{\mathcal{E}(\mathbb{Z}^d)}$, we consider the *simple symmetric random walk* $(X(n))_{n \geq 0}$ with law $P_0^\omega(X(0) = 0) = 1$ and transition probabilities, for $n \in \mathbb{N}$,

$$
P_0^\omega\big(X(n) = y \mid X(n-1) = x\big)
$$
$$
= \begin{cases} \frac{1}{\deg_\omega(x)} & \text{if } |x - y| = 1 \text{ and } \omega(\{x, y\}) = 1 \, ; \\ 0 & \text{otherwise} \, , \end{cases} \tag{14.1.1}
$$

© Springer International Publishing Switzerland 2017
M. Heydenreich and R. van der Hofstad, *Progress in High-Dimensional Percolation and Random Graphs*, CRM Short Courses, DOI 10.1007/978-3-319-62473-0_14

where $\deg_\omega(x)$ denotes the number of occupied bonds (in ω) that contain x. The walk is well defined as long as the cluster of 0 is not an isolated vertex without any connections. For $p > p_c$, we define the conditional measure

$$\mathbb{P}_p^*(\cdot) := \mathbb{P}_p(\cdot \mid 0 \leftrightarrow \infty), \qquad (14.1.2)$$

which guarantees us that the walk starts in the infinite component. Barlow [25] proves sharp heat kernel bounds for the random walk $(X(n))_{n\geq 0}$ under $\mathbb{P}_p^*$. This was extended to a full Donsker's invariance principle by Berger and Biskup [46], and independently by Mathieu and Piatnitski [213]:

Theorem 14.1 (Quenched invariance principle on the infinite percolation cluster [46, 213]). *Let $d \geq 2$ and $p > p_c$. For $\mathbb{P}_p^*$-almost all $\omega \in \{0, 1\}^{\mathcal{E}(\mathbb{Z}^d)}$, under P_0^ω, the process*

$$\left(X^\varepsilon(t)\right)_{t\in\mathbb{R}+}, \quad \text{where } X^\varepsilon(t) \equiv \varepsilon X(\lfloor t/\varepsilon^2 \rfloor),$$

converges in law as $\varepsilon \searrow 0$ to an isotropic d-dimensional Brownian motion, whose diffusion constant $\sigma^2(p)$ depends on p and d, but not on ω.

Theorem 14.1 is known as a *quenched invariance principle*, since it proves weak convergence for almost all realizations of the environment. An annealed invariance principle, where one averages over the environment, was proven earlier by De Masi et al. [83]. Sidoravicius and Sznitman [241] prove the quenched result for $d \geq 4$. Full generality has been obtained simultaneously by Berger and Biskup [46] and Mathieu and Piatnitsky [213], the latter for continuous-time random walk.

The central idea in the proofs, rooting back to Kipnis and Varadhan [199] and other work of the 1980s, is to find an harmonic embedding of the infinite cluster $\mathcal{C}_\infty$ into $\mathbb{R}^d$ such that the random walk on the deformed lattice is a *martingale*. The difference between this harmonic embedding and the original lattice is given by the *corrector map* $\chi^\omega: \mathbb{Z}^d \to \mathbb{R}^d$. In finite volume, this is obtained without further ado by solving an appropriate Dirichlet problem; the challenge is to construct the corrector in the infinite domain while keeping the distributional shift invariance. Berger and Biskup [46] point out that the corrector can also be obtained probabilistically through

$$\chi^\omega(x) = \lim_{n\to\infty} \left(E_x^\omega(X_n) - E_0^\omega(X_n)\right), \qquad (14.1.3)$$

although this representation does not appear to be very fruitful yet.

Once the corrector is established, the proof then consists of two parts, showing first that the rescaled martingale converges to Brownian motion, and second that the deformation of the path due to the corrector is asymptotically negligible.

We summarize that we have an invariance principle on the infinite cluster just as on the full lattice $\mathbb{Z}^d$, and we thus see hardly any difference between the full lattice and supercritical percolation in the scaling limit of random walks—the only dependence on p is through the diffusion constant $\sigma^2(p)$. It is highly plausible that $\lim_{p \searrow p_c} \sigma^2(p) = 0$, since we believe that $\theta(p_c) = 0$, so that the infinite component becomes more and more sparse and fractal-like as $p \searrow p_c$, thus slowing down the walk more and more. Indeed, it is suggested that there is another critical exponent characterizing the barely supercritical regime as

$$\sigma(p)^2 \approx (p - p_c)^\theta \quad \text{as } p \searrow p_c. \tag{14.1.4}$$

Heuristic arguments suggest that $\theta = 2$ in high dimensions, although no proof for this is known:

Open Problem 14.1 (Critical exponent for diffusion constant). *For random walk on the infinite percolation cluster as in Thm. 14.1, show that in sufficiently high dimension,*

$$\sigma(p)^2 \approx (p - p_c)^2 \text{ as } p \searrow p_c$$

for a suitable mode of convergence $\approx$.

Even monotonicity properties of the function $p \mapsto \sigma^2(p)$ are not known rigorously. We will give a heuristic explanation of the scaling in Open Problem 14.1 in Sect. 14.3.

14.2 Random Walks on Finite Critical Clusters

We shall now turn toward random walk on *critical* clusters, where much of the fun happens. A slight issue arises from the fact that all critical clusters are finite, thus a scaling limit in the sense of the previous section is no sensible object to study. Instead, we consider the *mixing time* of critical clusters on finite high-dimensional tori (as considered in Sect. 13.2). To this end, we call a *lazy simple random walk* on a finite graph $\mathcal{G} = (\mathcal{V}, \mathcal{E})$ a Markov chain on the vertices $\mathcal{V}$ with transition probabilities

$$\mathbf{p}(x, y) = \begin{cases} \frac{1}{2} & \text{if } x = y; \\ \frac{1}{2 \deg(x)} & \text{if } (x, y) \in \mathcal{E}; \\ 0 & \text{otherwise}, \end{cases} \tag{14.2.1}$$

where $\deg(x)$ denotes the degree of a vertex $x \in \mathcal{V}$. The attribute "lazy" refers to $\mathbf{p}(x, x) = \frac{1}{2}$, a term that is not present for (ordinary) random walk on the cluster as defined in (14.1.1).

The stationary distribution of this Markov chain π is given by $\pi(x) = \deg(x)/(2|\mathcal{E}|)$. The *mixing time* of lazy simple random walk on $\mathcal{G}$ is defined as

$$T_{\text{mix}}(\mathcal{G}) = \min \left\{ n : \|\mathbf{p}_n(x, \cdot) - \pi(\cdot)\|_{\text{TV}} \leq \tfrac{1}{4} \text{ for all } x \in \mathcal{V} \right\}, \tag{14.2.2}$$

with $\mathbf{p}_n$ being the distribution after n steps, and $\|\cdot\|_{\text{TV}}$ denoting the total variation distance. Mind that this definition differs from the uniform mixing time defined in (13.4.3).

Loosely speaking, the mixing time $T_{\text{mix}}(\mathcal{G})$ identifies the time scale at which the (lazy) random walker "forgets" its starting point. Working with the lazy walk instead of the usual simple walk has certain technical advantages, for example, it avoids any periodicity issues.

Motivation again is obtained from the Erdős–Rényi random graph models $\text{ER}_n(p)$ (cf. Sect. 13.1). When scaling $p = \lambda/n$, $\lambda \in \mathbb{R}$, we obtain the following results. For $\lambda > 1$, the mixing time of the giant component of $\text{ER}_n(\lambda/n)$ is of the order $\log^2(n)$, as proven by Fountoulakis and Reed [112] and Benjamini, Kozma, and Wormald [37]. For $\lambda < 1$, a similar bound holds, because clusters are simply too small, cf. (13.1.1). Interestingly, the fact that we are dealing with *worst-case starting points* is highly relevant. Indeed, Berestycki, Lubetzky, Peres, and Sly [42] show that starting from a *uniform* starting point in the giant component, the mixing time is of the order $\log(n)$ instead. This discrepancy can be understood by noting that the giant component in the Erdős–Rényi random graph has rare stretches of vertices of degree two of length of the order of $\log(n)$. The random walk is like a one-dimensional walk on these stretches, so it will take time $\log^2(n)$ to leave them. When started from a uniform vertex, on the other hand, these stretches are very far away, and thus the random walk is unlikely to even notice them. See also [36] where Ben-Hamou and Salez prove similar results for nonbacktracking walk. There, the worst possible mixing time is identified as being a specific constant times $\log(n)$, under the restriction that the minimal degree in the random graph is at least three, thus avoiding the long stretches of degree, two vertices that form traps.

Mixing Time of Large Critical Clusters. The interesting critical case is settled by Nachmias and Peres [216], who prove that the mixing time of $\text{ER}_n(1/n)$ is of the order n. Thus, we see that the mixing times blows up in the critical case.

Let us inspect the "engine room" of Nachmias's and Peres's argument a little more closely. The authors prove the following, rather general and highly fruitful, criterion for the mixing time of critical clusters. In its statement, we let $\mathbb{P}_{\mathcal{G},p}$ denote the percolation measure with percolation parameter p on the graph $\mathcal{G}$:

Theorem 14.2 (Nachmias–Peres [216]). *Consider bond percolation on the graph $\mathcal{G}$ with vertex set $\mathcal{V}$, $V = |\mathcal{V}| < \infty$, with percolation parameter $p \in (0, 1)$. Assume that for all subgraphs $\mathcal{G}' \subset \mathcal{G}$ with vertex set $\mathcal{V}'$,*

(a) $\mathbb{E}_{\mathscr{G}',p}|\mathscr{E}(\{u \in \mathscr{C}(v) : d_{\mathscr{C}(v)}(v,u) \le k\})| \le d_1 k, v \in \mathcal{V}'$,

(b) $\mathbb{P}_{\mathscr{G}',p}(\exists u \in \mathscr{C}(v) : d_{\mathscr{C}(v)}(v,u) = k) \le d_2/k, v \in \mathcal{V}'$,

where $\mathscr{E}(\mathscr{C})$ denotes the set of open edges with both endpoints in $\mathscr{C}$ and d_1, d_2 are uniform constants. If for some cluster $\mathscr{C}$

$$\mathbb{P}_{\mathscr{G},p}(|\mathscr{C}| \ge A^{-1}V^{2/3}) \ge 1 - \frac{b}{A}, A \ge 1 \qquad (14.2.3)$$

then there exists $c > 0$ such that for all $A \ge 1$,

$$\mathbb{P}_{\mathscr{G},p}(T_{\mathrm{mix}}(\mathscr{C}) > AV) \le \frac{c}{A^{1/6}}, \quad \mathbb{P}_{\mathscr{G},p}(T_{mix}(\mathscr{C}) < A^{-1}V) \le \frac{c}{A^{1/34}}. \qquad (14.2.4)$$

This version of the theorem is in the line of [216, Thm. 2.1], the precise adaptations are explained in [144, Sect. 4].

Theorem 14.2 works in fairly general setup, for example various cases of "sufficiently uniform" random graphs. A particularly beautiful result arises when we apply it to the case $\mathscr{G} = \mathbb{T}_{n,d}$ and $p = p_c(\mathbb{Z}^d)$, where Thm. 13.6 provides excellent control of the volume of the large clusters:

Corollary 14.3 (Mixing time of large critical percolation clusters on high-dimensional tori, [144]). There is $d_0 > 6$ such that for percolation on the torus $\mathbb{T}_{n,d}$ with $d \ge d_0$, every $m = 1, 2, \ldots$, there exist constants $c_1, \ldots, c_m > 0$, such that for all $A \ge 1$, $n \ge 1$, and all $i = 1, \ldots, m$,

$$\mathbb{P}_{p_c(\mathbb{Z}^d)}(A^{-1}V \le T_{\mathrm{mix}}(\mathscr{C}_{(i)}) \le AV) \ge 1 - \frac{c_i}{A^{1/34}}. \qquad (14.2.5)$$

Proof of Cor. 14.3. In order to prove Cor. 14.3, we need to verify the two conditions in Thm. 14.2(a) and (b) for critical percolation on the high-dimensional torus:

Verification of Thm. 14.2(a). Recall that $\mathscr{C}_{\mathbb{T}}(v)$ denotes the cluster of v on the torus, and $\mathscr{C}_{\mathbb{Z}}(v)$ denotes the cluster of v on $\mathbb{Z}^d$. The cluster $\mathscr{C}_{\mathbb{T}}(v)$ is a subgraph of the torus with degree at most $2d$, therefore we can replace the number of edges on the left-hand side by the number of vertices (and accommodate the factor $2d$ in the constant d_1). In Prop. 13.7, a coupling between $\mathscr{C}_{\mathbb{T}}(v)$ and $\mathscr{C}_{\mathbb{Z}}(v)$ was presented that shows that $\mathscr{C}_{\mathbb{T}}(v)$ can be obtained by identifying points that agree modulo n in a subset of $\mathscr{C}_{\mathbb{Z}}(v) \subseteq \mathbb{Z}^d$. A careful inspection of this construction shows that this coupling is such that it preserves graph distances. Since $|\{u \in \mathscr{C}(v) : d_{\mathscr{C}(v)}(v,u) \le k\}|$ is monotone in the number of edges of the underlying graph, the result in Thm. 14.2(a) for the torus follows from the bound $\mathbb{E}|\{u \in \mathscr{C}_{\mathbb{Z}}(v) : d_{\mathscr{C}_{\mathbb{Z}}(v)}(v,u) \le k\}| \le d_1 k$, which was established in Thm. 11.5.

Verification of Thm. 14.2(b). For percolation on $\mathbb{Z}^d$, this bound was proved by Kozma and Nachmias as [201, Thm. 1.2(ii)] and reproduced in Thm. 11.5. However, the event $\{\exists u \in \mathscr{C}_{\mathbb{T}}(v) : d_{\mathscr{C}_{\mathbb{T}}(v)}(v,u) = k\}$ is not a monotone event, and therefore this does not prove our claim. However, a close inspection of the proof of Thm. 11.5 shows that it only relies on the bound

$$\mathbb{P}_{p_c(\mathbb{Z}^d)}(|\mathcal{C}_{\mathbb{T}}(v)| \geq k) \leq C_1/k^{1/2} . \tag{14.2.6}$$

This, however, is implied by the corresponding $\mathbb{Z}^d$-result in Thm. 9.2 and the fact that $\mathbb{Z}^d$-clusters stochastically dominate $\mathbb{T}$-clusters in Prop. 13.7. This completes the verification of Thm. 14.2(b). $\qquad\square$

We discuss the example of random walks on large *finite* critical clusters on $\mathbb{Z}^d$ below, as it is more closely related to random walk on the IIC.

14.3 Random Walk on the Incipient Infinite Cluster

As we have seen in the previous section, random walk on *critical* clusters are quite charming, and very much different from the supercritical setting and everything we are used to. We are now targeting random walk on critical clusters on the infinite lattice $\mathbb{Z}^d$. It is clear that we need to condition on the cluster $\mathcal{C}(0)$ being big (since for small clusters, there is not much to explore for the random walk). Actually, it is most convenient to push the conditioning one step forward, and to condition on an *infinitely large* critical cluster $\mathcal{C}(0)$—which is nothing else than the *incipient infinite cluster* (IIC) that we have discussed in Chap. 12. It provides us with an infinite cluster that locally resembles a (large) critical cluster.

A well-known conjecture in the field has been made by the physicists S. Alexander and R. Orbach: Based on simulations in dimension $d = 2, \ldots, 5$, they conjectured that

$$P_0^\omega(X(2n) = 0) \approx n^{-2/3} \quad \mathbb{P}_{\text{IIC}} -a.s. \tag{14.3.1}$$

for random walk on the incipient infinite cluster in *all* dimension $d \geq 2$. This conjecture led to a wave of criticism in the literature. Indeed, large-scale simulation suggests that the exponent $\frac{2}{3}$ is *wrong* in dimension $d < 6$. In a seminal paper, Kozma and Nachmias [201] prove that the exponent is correct in high dimensions. Mind that for supercritical percolation, as well as on the ordinary lattice, (14.3.1) is valid with exponent $-d/2$, thus random walk on critical clusters is strongly subdiffusive. Indeed, the first study of random walk on the incipient infinite cluster is due to Kesten [196], who proved that random walk on the two-dimensional IIC is subdiffusive.

The reason behind this subdiffusivity lies in the specific structure of the IIC cluster. The vertices in the cluster can be classified into two groups: the *backbone* Bb is formed by those vertices that are part of a self-avoiding path from the origin to infinity,

$$\text{Bb} := \{x \colon \{0 \leftrightarrow x\} \circ \{x \leftrightarrow \infty\}\} ; \tag{14.3.2}$$

and all other vertices form large finite clusters (which are called *dangling ends* by some authors). The key point is that random walk on the IIC can only escape through the backbone, but spends most of its time on the dangling ends. Indeed, Cames van Batenburg [68] identified the *mass dimension* of the (entire) IIC cluster in high dimension as

$$d_{\mathrm{m}}(\mathrm{IIC}) := \lim_{n\to\infty} \frac{\log|\mathcal{C}(0) \cap \Lambda_n|}{\log n} = 4 \quad \mathbb{P}_{\mathrm{IIC}}\text{-a.s.} . \tag{14.3.3}$$

On the other hand, the backbone vertices with diffusive rescaling converge to a Brownian motion path (we present this result in Thm. 15.5), and Brownian motion has Hausdorff dimension 2. Sparsity of the backbone w.r.t. the entire IIC cluster in dimension $d > 4 + 2$ makes it very difficult for the random walk to *find* the backbone, and consequently the dangling ends function as *traps* for the random walk.

We summarize what is known for random walk on the high-dimensional IIC. Recall from Sect. 11.3.1 the notion $B(n)$ for the n-ball in the intrinsic distance $d_{\mathcal{C}(0)}$ centered at the origin and Λ_n the ℓ^∞-ball, write further $\tau'(A)^1$ for the *exit time* from the set $A \subset \mathbb{Z}^d$ and $W(n) = \{S(j) \mid j = 0,\dots,n\}$ for the *range* of the walk until time n.

Theorem 14.4 (Random walk on the incipient infinite cluster [146, 201]). *For random walk on the incipient infinite cluster, under the triangle condition (4.1.1), for $\mathbb{P}_{\mathrm{IIC}}$-almost all ω,*

$$\lim_{n\to\infty} \frac{P_0^\omega(X(2n) = 0)}{\log n} = -\frac{2}{3} , \qquad \lim_{n\to\infty} \frac{|W(n)|}{\log n} = \frac{2}{3} , \tag{14.3.4}$$

$$\lim_{n\to\infty} \frac{E_0^\omega \tau'\big(B(n)\big)}{\log n} = 3 , \qquad \lim_{n\to\infty} \frac{E_0^\omega \tau'(\Lambda_n)}{\log n} = 6 . \tag{14.3.5}$$

This behavior is very much in contrast to random walk on the ordinary lattice and supercritical clusters as in Sect. 14.1, where the exponents in (14.3.5) both equals 2. The fact that the latter exponent is twice as large as the former is, however, no surprise if we contrast the linear growth of $\mathbb{E}_{p_c}|B(n)|$ in Lem. 11.4 against the quadratic growth of $\mathbb{E}_{p_c}|\Lambda_n \cap \mathcal{C}(0)|$ (as follows from Thm. 11.4). The intuition is that the boundary of $B(n)$ is at Euclidean distance $\sqrt{n}$ from 0 under $\mathbb{P}_{p_c}$ as well as under $\mathbb{P}_{\mathrm{IIC}}$.

A proof of the statements in Thm. 14.4 follows a standard recipe. Indeed, it turns out that two properties of the medium characterize the random walk behavior. The first one is the ball growth rate. The second quantity is the *effective resistance* R_{eff}, and formally defined on a general graph $\mathcal{G} = (\mathcal{V}, \mathcal{E})$ as follows. Consider the quadratic form

$$\mathcal{Q}(f, g) = \frac{1}{2} \sum_{(x,y)\in\mathcal{E}} \big(f(x) - f(y)\big)\big(g(x) - g(y)\big) , \quad f, g \colon \mathcal{V} \to \mathbb{R} , \tag{14.3.6}$$

and let $H^2 := \{f : \mathcal{Q}(f, f) < \infty\}$. Note that the sum in (14.3.6) is over *directed* bonds, so in a symmetric setting, every bond contributes twice. Then

$$R_{\mathrm{eff}}(A, B)^{-1} := \inf_{f\in H^2} \big\{\mathcal{Q}(f, f) \colon f|_A = 1, f|_B = 0\big\} ; \quad A, B \subset \mathcal{V}. \tag{14.3.7}$$

Indeed, the effective resistance has a beautiful interpretation when we consider the graph (in our case: the cluster) as an electric network: For vertex sets A and B, $R_{\mathrm{eff}}(A, B)$ is the

---

[1] We do not use τ for risk of confusion with the two-point function that appears so prominently in this text.

effective resistance between A and B (in physics sense) when every edge in the graph is a resistor of 1 Ohm. In particular, R_{eff} obeys Kirchhoff's series and parallel law, which makes it easy to compute.

Exercise 14.1 (Lower bound on effective resistance along a bond). Let $\mathcal{G} = (V, \mathcal{E})$ and $\{a, b\} \in \mathcal{E}$. Use the definition in (14.3.7) to show that $R_{\mathrm{eff}}(\{a\}, \{b\}) \geq 1/d_a$, where d_a is the degree of a.

Exercise 14.2 (Upper bound on effective resistance along a bond). Let $\mathcal{G} = (V, \mathcal{E})$ and $\{a, b\} \in \mathcal{E}$. Use the definition in (14.3.7) to show that $R_{\mathrm{eff}}(\{a\}, \{b\}) \leq 1$.

Exercise 14.3 (Effective resistance along a pivotal bond). Let $\mathcal{G} = (V, \mathcal{E})$ and $\{a, b\} \in \mathcal{E}$ and there is no other path from a to b. Use the definition in (14.3.7) to show that $R_{\mathrm{eff}}(\{a\}, \{b\}) = 1$.

Exercise 14.4 (Series law for effective resistance). Let $A, B \subseteq \mathbb{Z}^d$. Let $x \in \mathbb{Z}^d \setminus (A \cap B)$ be such that every path from A to B passes through x. Use the definition in (14.3.7) to show that $R_{\mathrm{eff}}(A, B) = R_{\mathrm{eff}}(A, \{x\}) + R_{\mathrm{eff}}(\{x\}, B)$.

The basic line of arguments is as follows: Suppose that the volume of the n-ball grows like $|B(n)| \approx n^\alpha$, and the effective resistance between the origin and the boundary of the n-ball grows like $R_{\mathrm{eff}}(0, B(n)^c) \approx n^\beta$; and a number of technical conditions hold (to control that the medium is sufficiently regular), then the return probability $P_0^\omega(X(2n) = 0)$ scales like $n^{-\alpha/(\alpha+\beta)}$ and the expected exit time of the n-ball scales like $n^{\alpha+\beta}$. This has been proven for $\alpha = 2$ and $\beta = 1$, which gives the magic exponent $2/3$ in the Alexander–Orbach conjecture (14.3.1), by Barlow, Járai, Kumagai, and Slade [27] (who apply it to high-dimensional-oriented percolation, see Sect. 15.2). A generalized setup has been obtained by Kumagai and Misumi [203]. The correspondence between random walk properties on the one hand and volume + R_{eff} scaling on the other hand is very much in the folklore of random walk on graphs, see, e.g., the classic book by Doyle and Snell [91].

Since the volume growth of the incipient infinite cluster is fairly well understood, the main challenge is to control the effective resistance. With the interpretation of R_{eff} as a physical resistance, it is quite clear that the number of pivotals between 0 and $B(n)^c$ is a lower bound for $R_{\mathrm{eff}}(0, B(n)^c)$.

Exercise 14.5 (Effective resistance and pivotal bonds). Let $0 \in A \subseteq \mathbb{Z}^d$. Prove that $R_{\mathrm{eff}}(0, A^c)$ for random walk on the IIC is bounded below by the number of pivotal bonds for $0 \leftrightarrow A$ in the IIC configuration. [*Hint:* Use Exercs. 14.1–14.4.]

In high dimensions, this lower bound is of the correct order, and a corresponding upper bound exists. In other words, there are *lots* of pivotal bonds in the IIC cluster. Indeed, the triangle condition (4.1.1) is the central tool in establishing the upper bound. Details can be found in [146, 201].

In Sect. 14.1, we have considered the *scaling limit* of random walk on (infinite) percolation clusters. Does there exist a scaling limit of random walk on the incipient infinite cluster? The second limit in (14.3.4) suggests $n^{1/6}$ as the correct spatial scaling of random walk after n steps, since the exit time of a ball of radius n scales as n^6. Yet, the scaling limit has not been obtained. Only very recently, a promising candidate for a scaling limit has been found: in a seminal work, Ben Arous, Cabezas, and Fribergh [35] consider random walk on the high-dimensional *branching random walk* conditioned to have total population size n. This is a natural candidate for the scaling limit of random walks on critical high-dimensional percolation clusters of size n, since branching random walk is a mean-field model for percolation, as we have explained in detail in Sect. 2.2. It is believed that such critical clusters of finite size have the same scaling limit as for critical branching random walk conditioned on the population size, as we discuss in more detail in Sect. 15.1. This scaling limit is called Integrated Super-Brownian Excursion (ISE). In [35], the authors prove that random walk on critical branching random walk conditioned on the population size being n converges to *Brownian motion on Integrated Super-Brownian Excursion*, the latter object being introduced by Croydon [80]. More formally, they prove that for $d \geq 14$,

$$\left(n^{-1/4} X^{\omega_n}_{tn^{3/2}}\right)_{t \geq 0} \longrightarrow \left(\sigma B^{\text{ISE}}_t\right)_{t \geq 0}, \qquad (14.3.8)$$

where $\sigma > 0$ is a constant, ω_n is the range of critical branching random walk conditioned to be of size n, B^{ISE} is Brownian motion on the ISE, convergence is annealed and occurs in the topology of uniform convergence over compact sets. To make the connection to the order $n^{1/6}$ spatial scaling for random walk on the IIC after n steps, we note that in (14.3.8), the spatial displacement is $n^{1/4}$ at time $n^{3/2}$. Denoting $m = n^{3/2}$, we indeed have that $n^{1/4} = m^{1/6}$. Extending (14.3.8) to high-dimensional percolation is a major challenge, as one would in particular need that the scaling limit of critical high-dimensional percolation clusters of fixed size is ISE. We discuss progress in this direction in Sect. 15.1 below.

Mind that in (14.3.8), the (spatial and temporal) rescaling of the random walk on the one hand and the conditioning on the size of the critical structure on the other hand are done simultaneously. When studying the scaling limit on the percolation IIC in high dimension, the IIC itself is an infinite structure, and therefore we are only left to rescale the random walk accordingly. However, this leads to a different limit; Ben Arous et al. [34] suggest that the scaling limit might be Brownian motion on the infinite canonical super-Brownian motion (connected to Open Problem 15.2 below). Croydon [79] used regular resistance forms to identify the scaling limit of random walk on (scaling limits) of various random objects provided that the underlying spaces are converging in the Gromov-Hausdorff-vague topology. Indeed, Croydon [79, Sect. 8.1] conjectures that the properly rescaled IIC does converge to a limiting object in the required sense and therefore the scaling limit of the processes associated to certain resistance forms converge, too.

The scaling limit of random walk on the high-dimensional IIC remains a challenging Open Problem:

Open Problem 14.2 (Scaling limit of random walk in IIC [34, 79]). *Prove that random walk on the high-dimensional incipient infinite cluster has a nondegenerate scaling limit.*

This Open Problem is closely linked with the scaling limit of the critical clusters itself, which we elaborate on in the forthcoming chapter.

We close this chapter by giving a heuristic explanation of the scaling of the slightly supercritical diffusion constant.

Heuristics for Scaling Diffusion Constant in Open Problem. 14.1 We use a multiscale analysis. We fix $p > p_c$ and partition $\mathbb{Z}^d$ into cubes whose side length is the percolation correlation length $\xi(p)$. The correlation length is intuitively the length scale at which we cannot see the difference between the percolation configuration for $p > p_c$ and the *critical* system. Thus, heuristically, the random walk on a cube of width $\xi(p)$ behaves similarly as the random walk on the IIC. As a result, by Thm. 14.4 (in particular (14.3.4)), it will take roughly $\xi(p)^6$ steps for the random walker to leave a cube of width $\xi(p)$. Further, the correlation length also has the interpretation that clusters further away than $\xi(p)$ are approximately independent. We reinterpret this by assuming that the random walk hops between the different cubes of width $\xi(p)$ as a *simple random walk*. Together, these two heuristic assumptions imply that $\mathrm{Var}(S_n) = \mathrm{Var}(S_{\xi(p)^6(n/\xi(p)^6)}) \approx \xi(p)^2 \mathrm{Var}(S'_{n/\xi(p)^6})$, where $(S'_k)_{k \geq 0}$ is simple random walk on $\mathbb{Z}^d$. Since $\mathrm{Var}(S'_k) = k$, we thus arrive at

$$\mathrm{Var}(S_n) \approx \xi(p)^2 \mathrm{Var}(S'_{n/\xi(p)^6}) = \xi(p)^2 n/\xi(p)^6 = n\xi(p)^{-4} . \qquad (14.3.9)$$

By Thm. 11.2, together with the prediction that $\nu' = \nu$, we obtain that $\xi(p) \sim (p - p_c)^{-1/2}$, so that

$$\mathrm{Var}(S_n) \approx n(p - p_c)^2 , \qquad (14.3.10)$$

which is the reason for the conjectured behavior in Open Problem 14.1.

Chapter 15
Related Results

In this chapter, we discuss related problems in high-dimensional percolation. These topics have attracted tremendous attention in the past two decades, and their investigation is far from complete. We discuss the relation between critical high-dimensional percolation clusters and super-processes in Sect. 15.1, oriented percolation in Sect. 15.2, and scaling limits of percolation backbones in Sect. 15.3. We continue to discuss long-range percolation in Sect. 15.4 and the asymptotic expansion of the percolation threshold in Sect. 15.5. We close this chapter by describing percolation on nonamenable graphs in Sect. 15.6.

15.1 Super-Process Limits of Percolation Clusters

Super-processes or measure-valued diffusions are continuous-time and continuous-space processes that describe the random distribution of mass undergoing simultaneous *branching* and *motion*. In this section, we describe three examples of them and conjecture that each of them arises as an appropriate scaling limit of critical percolation clusters.

Super-processes can be understood in terms of scaling limits of branching random walk (BRW). This text aims to argue that branching random walk is the mean-field model for percolation on $\mathbb{Z}^d$, and obtaining super-process scaling limits would be one of the strongest confirmations of this heuristic. Unfortunately, not many results exist that link high-dimensional percolation to super-processes.

Informal descriptions of super-processes can be found in two nice introductory papers by Gordon Slade [244, 245]. The super-processes that we discuss here can all be seen as variations on *super-Brownian motion*, which is a measure-valued process arising as the scaling limit of critical branching random walk. See the work by Dawson [82] or Perkins [224], or one of the books [96, 101, 206] for general expositions on super-processes.

We explain three different super-process constructions. The first super-process is integrated super-Brownian excursion (ISE), which can be seen as the scaling limit of the random distribution of all mass in a critical BRW conditioned on having a large and fixed total population and where we ignore the generations. ISE already made its appearance in Sect. 14.3, where we discussed random walk on critical clusters of a given size. The second super-process is the

M. Heydenreich and R. van der Hofstad, *Progress in High-Dimensional Percolation and Random Graphs*, CRM Short Courses, DOI 10.1007/978-3-319-62473-0_15

canonical measure of super-Brownian motion (CSBM), which can be seen as the scaling limit of critical branching random walk when we rescale time by n and space by $\sqrt{n}$, and multiply probabilities by n to obtain a nondegenerate limiting measure. The third super-process is obtained in the scaling limit where we *first* condition the critical BRW to survive forever, and *then* rescale time by n and space by $\sqrt{n}$, so as to obtain incipient infinite canonical super-Brownian motion or IICSBM.

These three super-processes are tightly connected to one another. Indeed, when we condition SBM on having total mass equal to 1, and then integrate the random measure out over time (so as to forget the generations of the particles involved), then we retrieve ISE. When we condition SBM to survive forever (a construction that can be made sense of, even though, as for critical BRW, SBM dies out in finite time), then we obtain IICSBM. We next informally describe how these three limits can be obtained by suitable rescalings of space and time of critical percolation clusters.

Integrated super-Brownian excursion (ISE). Integrated super-Brownian excursion (ISE) arises as the scaling limit of the random distribution of mass in a critical BRW conditioned on having a large and fixed total population where we ignore the generation structure of the population involved. Thus, we take a critical BRW, condition on a fixed total progeny $T = n$, rescale space by $n^{-1/4}$, and take the scaling limit. The limit, which can be seen to exist, is ISE. One way to describe ISE is through its r-point functions, as we explain now. For $\vec{x} \in (\mathbb{Z}^d)^{r-1}$, let

$$p^{(r)}(\vec{x}; n) = \mathbb{E}\left[\prod_{i=1}^{r-1} N(x_i) \mathbb{1}_{\{T=n\}}\right] ,$$

$$A_n^{(r)}(\vec{x}) = \frac{p^{(r)}(\vec{x}; n)}{\mathbb{P}(T = n)} = \mathbb{E}\left[\prod_{i=1}^{r-1} N(x_i) \,\bigg|\, T = n\right] , \tag{15.1.1}$$

where T is the total progeny of a critical infinite-variance branching process, and $N(x)$ is the total number of particles ever present at $x \in \mathbb{Z}^d$, i.e., $N(x) = \sum_{n=0}^{\infty} N_n(x)$, where $N_n(x)$ is the number of particles at x at time n. We see that time is *integrated out*, which explains the name *Integrated* Super-Brownian Excursion. The ISE total mass functions describe the scaling limit $A^{(r)}(\vec{x})$ of $A_n^{(r)}(\vec{x})$, with space rescaled by a factor $n^{-1/4}$ and mass by $1/n$ so that the total mass becomes 1. We do not enter this topic fully, but rather only describe $A^{(r)}(\vec{x})$ for $r = 2$, for which

$$A^{(2)}(x) = \int_0^{\infty} t e^{-t^2/2} p_t(x)\, dt , \quad \text{where } p_t(y) = \frac{1}{(2\pi t)^{d/2}} e^{-|y|^2/(2t)} \tag{15.1.2}$$

denotes the standard Brownian transition density. It is the case that

$$\lim_{n \to \infty} n^{d/4} A_n^{(2)}(x/n^{1/4}) = A^{(2)}(x) . \tag{15.1.3}$$

The intuition behind the form in (15.1.2) and the convergence in (15.1.3) is as follows. In $A_n^{(2)}(x)$, we need to have a particle at $x \in \mathbb{Z}^d$ when the tree has total progeny n. Condition on the length of the path from 0 at time 0 to x, and denote this length by m. We call the simple random walk path between 0 and x of length m the *backbone*. Then, given n, the $(m + 1)$ trees that are branching off the backbone path are themselves critical branching processes. Their total sizes need to add up to n. When we let T_i denote the total population of the tree branching off the ith point along the backbone (including the backbone vertex itself), then $T_0 + \cdots + T_m = T = n$, where we recall that T denotes the total size of the tree, and we condition on $T = n$.

Now, $\mathbb{P}(T_i = \ell) = C_\delta \ell^{-3/2}(1 + o(1))$ by Thm. 2.1, see in particular (2.1.28). By the random walk hitting time theorem (recall [162]), and similarly to (2.1.24) (recall also Exercs. 2.3–2.4),

$$\mathbb{P}(T_0 + \cdots + T_m = n) = \frac{m+1}{n} \mathbb{P}(X_1 + \cdots + X_n = n - m - 1) , \qquad (15.1.4)$$

where $(X_i)_{i \geq 1}$ are i.i.d. random variables having the offspring distribution of the critical BRW. Therefore, using that $\mathbb{E}[X_i] = 1$,

$$\frac{\mathbb{P}(T_0 + \cdots + T_m = n)}{\mathbb{P}(T = n)} = (m+1)\frac{\mathbb{P}(X_1 + \cdots + X_n = n - m - 1)}{\mathbb{P}(X_1 + \cdots + X_n = n - 1)} \qquad (15.1.5)$$
$$\sim (m+1)e^{-m^2/(2\gamma n)} ,$$

by the local central limit theorem, and where $\gamma = \text{Var}(X_1)$ is the variance of the offspring distribution. We obtain that

$$A_n^{(2)}(x) = \sum_{m=0}^{\infty} \frac{\mathbb{P}(T_0 + \cdots + T_m = n)}{\mathbb{P}(T = n)} D^{\star m}(x)$$
$$\sim \sum_{m=0}^{\infty} (m+1)e^{-m^2/(2\gamma n)} D^{\star m}(x) . \qquad (15.1.6)$$

The main contribution comes from m of the order $\sqrt{n}$. Take $m = tn^{1/2}$ and note that

$$D^{\star(tn^{1/2})}(x/n^{1/4}) \sim n^{-d/4} p_t(x). \qquad (15.1.7)$$

Combining 15.1.6–15.1.7 leads to (15.1.3) with $A^{(2)}(x)$ given by (15.1.2).

Partial results linking high-dimensional percolation and ISE can be found in the papers of Hara and Slade [139, 140]. This is a fairly difficult problem. For example, since we condition on the cluster size being equal to n, we need to know how this probability behaves as $n \to \infty$. Let us start by introducing some notation. Let

$$\tau_{p_c}(x; n) = \mathbb{P}_{p_c}(x \in \mathscr{C}(0), |\mathscr{C}(0)| = n) ,$$
$$q_n(x) = \frac{\tau_{p_c}(x; n)}{\mathbb{P}_{p_c}(|\mathscr{C}(0)| = n)} . \qquad (15.1.8)$$

Hara and Slade [139, 140] prove the following links between high-dimensional critical percolation and ISE:

Theorem 15.1 (Critical percolation and ISE [139, 140]). *For percolation in dimension $d \geq d_0$ with $d_0 > 6$ sufficiently large, for every $\varepsilon < \frac{1}{2}$, there exists $\upsilon > 0$ such that, for every $k \in \mathbb{R}^d$,*

$$\hat{q}_n\big(k/(\upsilon n^{1/4})\big) = \hat{A}^{(2)}(k) + O(n^{-\varepsilon}) \,. \tag{15.1.9}$$

The fact that k is rescaled by $n^{1/4}$ shows that the majority of mass of the percolation cluster is at distance $n^{1/4}$ from the origin. Thus, we have an object of size n whose radius is $n^{1/4}$, suggesting that the dimension of large critical clusters is 4. This was also observed when discussing the role of boundary conditions in Sect. 13.6 and in Eq. (14.3.3). In general, super-processes related to super-Brownian motion are four-dimensional structures (see, e.g., Perkins [224]). As already noted in Thm. 9.4, the papers of Hara and Slade [139, 140] also prove that $\delta = 2$ in the strongest possible form.

For lattice tree and lattice animals, a statistical mechanical model like percolation, Derbez and Slade [86, 87] proved that the finite-dimensional distributions converge to those of ISE. It is natural to believe that this is also the case for high-dimensional percolation:

Open Problem 15.1 (ISE scaling limit of critical clusters of fixed size). *Show that critical percolation on $\mathbb{Z}^d$ with $d > 6$ satisfies that the random measure*

$$\mu_n(E) = \frac{1}{n} \sum_{x \in \mathscr{C}(0)} \mathbb{1}_{\{x/\sqrt{n} \in E\}} \tag{15.1.10}$$

conditionally on $|\mathscr{C}(0)| = n$ converges in distribution to ISE. This follows by showing that there exist constants γ and υ such that $\hat{q}_n^{(r)}(n^{-1/4}\vec{k}/\upsilon) \to \gamma^{r-2}\hat{A}^{(r)}(\vec{k})$, where

$$q_n^{(r)}(x) = \mathbb{P}_{p_c}\big(0 \leftrightarrow x_i \ \forall i = 1, \ldots, r-1 \ \big| \ |\mathscr{C}(0)| = n\big) \tag{15.1.11}$$

denote the percolation r-point functions conditioned on its cluster size, and $\hat{A}^{(r)}(\vec{k})$ is the Fourier transform of $A^{(r)}(\vec{x})$.

For critical branching random walk, γ is equal to the variance of the offspring distribution. In particular, for Poisson branching random walk, $\gamma = 1$. For interacting models, the role of the so-called *vertex factor* γ is less obvious, and we return to this matter when discussing the canonical measure of super-Brownian motion below. The parameter υ should be thought of as the standard deviation of the Brownian motion describing percolation paths. While υ is probably close to 1 in high dimensions, we expect that it is different from the random walk standard deviation 1 in any dimension.

Canonical Measure of Super-Brownian Motion. The so-called canonical measure of super-Brownian motion (CSBM) can be seen as the scaling limit of critical branching random walk when we rescale time by n and space by $\sqrt{n}$ and multiply probabilities by $n\gamma$. Since critical

BRW dies out a.s., this limit puts all mass on finite structures, a problem that is resolved by multiplying probabilities by a factor n so as to still give sizeable mass to particles in generations proportional to n. A downside is that the limiting measure no longer is a probability measure, as it arises by multiplying a probability measure by a factor that tends to infinity.

We next describe these objects for BRW. Define the measure-valued process $(X_t^{(n)})_{t \geq 0}$ by

$$X_t^{(n)}(f) = \frac{1}{\gamma n} \sum_{x \in A_{nt}} f(x/\sqrt{\upsilon n}) \quad \text{and} \quad \mu_n(\cdot) = n\gamma \, \mathbb{P}(\cdot) \, , \tag{15.1.12}$$

where A_m is the set of locations of all particles alive at time m so that $|A_m| = N_m$. Further, for $k \in (-\pi, \pi]^d$, let $\widehat{X}_t^{(n)}(k) = X_t^{(n)}(\phi_k)$, where $\phi_k(x) = e^{ik \cdot x}$ is the complex exponential. Thus, $k \mapsto \widehat{X}_t^{(n)}(k)$ is the (random) Fourier transform of the random measure $X_t^{(n)}$. The distribution of the process $(X_t^{(n)})_{t \geq 0}$ can again be described in terms of r-point functions. Similar to (15.1.1), let

$$p_{\vec{n}}^{(r)}(\vec{x}) = \mathbb{E}\left[\prod_{i=1}^{r-1} N_{n_i}(x_i) \right], \quad \vec{n} \in \mathbb{N}^{r-1} \, , \quad \vec{x} \in (\mathbb{Z}^d)^{r-1} \, , \tag{15.1.13}$$

denote the BRW r-point functions. Here, again, $N_n(x)$ denotes the number of particles present at x at time n. Then,

$$\mu_n\left(\prod_{i=1}^{r-1} \widehat{X}_{t_i}^{(n)}(k_i) \right) = (n\gamma)^{r-2} \hat{p}_{n\vec{t}}^{(r)}(\vec{k}/\sqrt{\upsilon n}) \, , \tag{15.1.14}$$

so that the rescaled r-point functions correspond to the moments of the Fourier transform of $(X_t^{(n)})_{t \geq 0}$. We refer to [151] for an extensive discussion of the CSBM and its moment measures describing the limits of the right-hand side of (15.1.14). Again, it is natural to conjecture that critical percolation has CSBM as a scaling limit, where now time is interpreted in terms of the graph distance:

Open Problem 15.2 (CSBM scaling limit of critical clusters). *Show that critical percolation on $\mathbb{Z}^d$ with $d > 6$ satisfies that there exist γ, υ such that the random measure-valued process*

$$X_{nt}^{(n)}(\cdot) = \frac{1}{n\gamma} \sum_{x \in \partial B(\lfloor nt \rfloor)} \mathbb{1}_{\{x/\sqrt{\upsilon n} \in \cdot\}} \, , \tag{15.1.15}$$

under the measure $\mu_n(\cdot) = n\gamma \, \mathbb{P}_{p_c}(\cdot)$, converges in distribution to the canonical measure of SBM.

In Open Problem 15.2, we recall that $\partial B(n)$ denotes the vertices in $\mathscr{C}(0)$ at graph distance exactly n from 0. Let us explain how υ and γ arise. The variable υ is the variance of the underlying motion, which should be equal to

$$v = \lim_{n\to\infty} \frac{1}{n\hat{t}_n(0)} \sum_{x\in\mathbb{Z}^d} |x|^2 t_n(x) , \tag{15.1.16}$$

assuming that this limit exists (which is highly nontrivial), and where $t_n(x) = \mathbb{P}_{p_c}(x \in \partial B(n))$ denotes the critical two-point function at "time" n. The parameter γ is the branching ratio of the CSBM, which in the percolation case can be computed as

$$\gamma = AV , \quad \text{where } A = \lim_{n\to\infty} \mathbb{E}_{p_c}[|\partial B(n)|] , A^3 V = \lim_{n\to\infty} \frac{1}{n} \mathbb{E}_{p_c}\big[|\partial B(n)|^2\big] , \tag{15.1.17}$$

assuming that these limits exist (which is not yet known and technically quite a challenging problem, see below for more details). For BRW, $A = 1$ since the random process $(N_m)_{m\geq 0}$ is a nonnegative martingale with expectation 1, whereas γ is the variance of the offspring distribution. For percolation, we expect that $A > 1$ and $V < 1$. The parameter V is sometimes called the *vertex factor*.

Holmes and Perkins [175] prove that convergence of the r-point functions

$$\mu_n\left(\prod_{i=1}^{r-1} \widehat{X}_{t_i}^{(n)}(k_i)\right) = (n\gamma)^{r-2}\hat{t}_{n\vec{t}}^{(r)}(\vec{k}/\sqrt{vn}) \longrightarrow \widehat{M}_{\vec{t}}^{(r-1)}(\vec{k}) , \tag{15.1.18}$$

where $\widehat{M}_{\vec{t}}^{(r-1)}(\vec{k})$ are the Fourier transforms of the so-called moment measures of CSBM, together with the convergence of the survival probability $\gamma n \, \mathbb{P}_{p_c}(\partial B(n) \neq \varnothing)$ $\to 2$, imply convergence in finite-dimensional distributions. This is weaker than Open Problem 15.2. In turn, the conditions in (15.1.18) and the convergence of the intrinsic one-arm are much stronger than the identification of the intrinsic one-arm probability by Kozma and Nachmias in [201] as discussed in Sect. 11.3. In particular, Open Problem 15.2 implies Open Problem 11.2, which underlines the relevance of Open Problem 15.2. The fact that Open Problem 15.2 involves *intrinsic* distances certainly makes the problem significantly harder.

The CSBM moment measure $\widehat{M}_{\vec{t}}^{(r-1)}(\vec{k})$ can be seen to involve a sum over $(2r - 3)!!$ trees, where $(2r - 3)!!$ arises as the number of shapes of trees with r labeled leaves. See also Lem. 13.10, where this number appeared in the tree-graph inequalities for percolation.

There is some work in this direction for interacting models. The second author and Slade prove the convergence of the oriented percolation r-point functions in (15.1.18) in [171], and, jointly with den Hollander, the convergence of the oriented percolation survival probability in [157, 158] (see also [159]). This is discussed in more detail in Sect. 15.2 below. Holmes [174] proves the convergence of the lattice tree r-point functions, and the convergence of the survival probability follows again from the work of Holmes and the second author [159]. Holmes, Perkins, and the second author provide a tightness criterion in [160] based on r-point functions. While they do not verify it for percolation in high dimensions, they do so for (spread-out) lattice trees for $d > 8$, suggesting once more that similar results ought to be true for percolation in $d > 6$.

Incipient Infinite Super-Brownian Motion. Incipient infinite CSBM is obtained by taking the scaling limit of critical BRW and conditioning it to survive forever, and can be thought of as the mean-field model for the percolation IIC as discussed in Chap. 12. Of course, since critical BRW dies out a.s., we are conditioning on a null event, and a proper definition needs to rely on a limiting argument. For details on such limiting arguments, we refer to Chap. 12. Here, we follow the presentation of the second author in [152].

As discussed in Chap. 12, the BRW IIC measure can be explicitly computed. This implies that also the BRW IIC r-point functions, which we denote as $s_{\bar{n}}^{(r)}$, can be determined as

$$s_{\bar{n}}^{(r)}(\vec{x}) = \mathbb{E}_{\mathrm{IIC}}\left[\prod_{i=1}^{r-1} N_{n_i}(x_i)\right] = \mathbb{E}_{p_c}\left[N_{\bar{n}} \prod_{i=1}^{r-1} N_{n_i}(x_i)\right], \qquad (15.1.19)$$

where $\bar{n} = \max_{i=1}^{r} n_i$, and $N_n = \sum_x N_n(x)$ denotes the total number of individuals in generation n.

Exercise 15.1 (Proof IIC BRW higher-point function). Prove (15.1.19) using (12.2.11).

Using (15.1.19), one can identify the scaling limits of the higher-point functions of the BRW IIC, which correspond to the higher-point functions of SBM with one *immortal particle* in the formulation of Evans [102]. This immortal particle performs a Brownian motion, and there are critical SBMs hanging off it. We call this *incipient infinite canonical measure of super-Brownian motion* (IICSBM). This leads to the following natural conjecture stating that the IIC, when properly rescaled, converges to this same measure-valued process:

Open Problem 15.3 (SBM with immortal particle limit of the IIC). *Show that critical percolation on $\mathbb{Z}^d$ with $d > 6$ satisfies that there exist γ, υ such that the random measure-valued process*

$$X_{nt}^{(n)}(\cdot) = \frac{1}{n\gamma} \sum_{x \in \partial B(\lfloor nt \rfloor)} \mathbb{1}_{\{x/\sqrt{\upsilon n} \in \cdot\}} \qquad (15.1.20)$$

under the measure $\mu_n(\cdot) = \mathbb{P}_{\mathrm{IIC}}(\cdot)$, converges in distribution to incipient infinite SBM.

Toward a Proof of Super-Process Limits. In Open Problems 15.1, 15.2, and 15.3, we propose to show that various critical high-dimensional percolation clusters converge to related super-processes. These results are intimately related. Here, we discuss how such proofs can be approached. This approach is similar in spirit to the three settings, yet there are also substantial differences. Let us focus on the similarities and remark upon the differences along the way.

The first step to identify the super-process scaling limit of critical high-dimensional percolation clusters is to identify the scaling limits of the r-point functions. See, e.g., (15.1.11) for the r-point function in the ISE setting and (15.1.18) for what such a result could look like in the context of the canonical measure of SBM. Such scaling again should be proved using the lace expansion. Let us illustrate this by discussing the first step, which is the convergence of the three-point function, in more detail. The lace expansion for the two-point function, as discussed in detail in Chap. 6, is an expansion for a *linear structure*. See Fig. 6.1 for a linear picture of percolation paths that summarize the event $\{0 \leftrightarrow x\}$, and that is made up of doubly connected pieces between the ordered set of pivotals. For the three-point function, however, we are interested in the event $\{0 \leftrightarrow x_1, 0 \leftrightarrow x_2\}$. Each of the two connections $\{0 \leftrightarrow x_1\}$ and $\{0 \leftrightarrow x_2\}$ has their own sets of ordered pivotals. This picture thus gives rise to a *branching structure* where the sets of ordered pivotals are split into the pivotals that are common to both events $\{0 \leftrightarrow x_1\}$ and $\{0 \leftrightarrow x_2\}$, and the ones that are pivotal only for $\{0 \leftrightarrow x_1\}$ or $\{0 \leftrightarrow x_2\}$, respectively. In the lace expansion for the three-point function, we would like to factor the three-point function $\tau_p(x_1, x_2) = \mathbb{P}_p(0 \leftrightarrow x_1, 0 \leftrightarrow x_2)$ as

$$\tau_p(x_1, x_2) = \sum_{w_0, w_1, w_2} \tau_p(w_0) \psi_p(w_1 - w_0, w_2 - w_0) \tau_p(x_1 - w_1) \tau_p(x_2 - w_2)$$
$$+ R_p(x_1, x_2). \qquad (15.1.21)$$

Here, we think of $R_p(x_1, x_2)$ as being some kind of "error" term and ψ_p as a lace-expansion coefficient that plays a similar role for the three-point function as Π_p for the two-point function. This means that we need to extract three two-point functions, giving rise to a "vertex factor" that we denote by ψ_p. In order to achieve (15.1.21), one typically starts in a rather similar way as for the lace expansion for the two-point function. Indeed, we successively look for the first pivotal bond, use the Factorization Lemma, etc. However, now there are two cases: (a) The first pivotal can be pivotal for *both* $\{0 \leftrightarrow x_1\}$ and $\{0 \leftrightarrow x_2\}$; or (b) $\{0 \leftrightarrow x_1\}$ and $\{0 \leftrightarrow x_2\}$ have two *different* first pivotals. (The case where one of the events $\{0 \leftrightarrow x_1\}$ or $\{0 \leftrightarrow x_2\}$ does not have a pivotal turns out to be an error term.) Case (a) can be treated very similarly as for the expansion of the two-point function and (ignoring the error term that arises) eventually gives rise to a contribution of the form

$$\sum_{w_0} (J_p \star \Pi_p)(w_0) \tau_p(x_1 - w_0, x_2 - w_0), \qquad (15.1.22)$$

where we recall that $J_p(x) = 2dp D(x)$. Case (b) (which also arises each time we find a "first" pivotal in the treatment of Case (a)) is more involved and leads to a contribution after expansion (again ignoring the error term) of the form

$$\sum_{w_1, w_2} \varphi_p(w_1, w_2) \tau_p(x_1 - w_1) \tau_p(x_2 - w_2). \qquad (15.1.23)$$

Here, we note that we need to perform a *double* expansion, the first extracts the factor $\tau_p(x_1 - w_1)$, while the second extracts a factor $\tau_p(x_2 - w_2)$. Such a double expansion is highly nontrivial and quite a bit more difficult than the expansion for the two-point function. In the end, this three-point function expansion gives rise to

$$\tau_p(x_1, x_2) = \sum_{w_0} (J_p \star \Pi_p)(w_0) \tau_p(x_1 - w_0, x_2 - w_0)$$
$$+ \sum_{w_1, w_2} \varphi_p(w_1, w_2) \tau_p(x_1 - w_1) \tau_p(x_2 - w_2) + Q_p(x_1, x_2), \qquad (15.1.24)$$

where $Q_p(x_1, x_2)$ is the combined error term that arises. Equation 15.1.24 is not quite equal to (15.1.21), but (15.1.21) can be obtained from it by iterating indefinitely. Indeed, we can apply (15.1.24) again to the three-point function $\tau_p(x_1 - w_0, x_2 - w_0)$, etc. This way, we obtain

$$\tau_p(x_1, x_2) = \sum_{w_0} \sum_{m=0}^{\infty} (J_p \star \Pi_p)^{\star m}(w_0) \varphi_p(w_1, w_2) \tau_p(x_1 - w_0) \tau_p(x_2 - w_0)$$
$$+ \sum_{w_0} \sum_{m=0}^{\infty} (J_p \star \Pi_p)^{\star m}(w_0) Q_p(x_1 - w_0, x_2 - w_0) . \quad (15.1.25)$$

Realizing that

$$\sum_{m=0}^{\infty} (J_p \star \Pi_p)^{\star m}(w) = (J_p \star \tau_p)(w) , \quad (15.1.26)$$

we arrive at (15.1.21) with

$$\psi_p(w_1, w_2) = \sum_{u} J_p(u) \varphi_p(w_1 - u, w_2 - u) , \quad (15.1.27)$$

$$R_p(x_1, x_2) = \sum_{m=0}^{\infty} (J_p \star \Pi_p)^{\star m}(w_0) Q_p(x_1 - w_0, x_2 - w_0) . \quad (15.1.28)$$

The above argument applies to the "regular" three-point function $\tau_p(x_1, x_2) = \mathbb{P}_p(0 \leftrightarrow x_1, 0 \leftrightarrow x_2)$. In Open Problems 15.1, 15.2, and 15.3, we need to deal with several complications. Indeed, the expansion for the three-point function in Open Problem 15.1 needs a conditioning on the cluster size. This analysis was performed by Hara and Slade in [140], where they proved the convergence for the three-point function. This is a major result in the connection between critical percolation clusters conditioned on their size and ISE. Convergence results for higher-point functions are still missing. The expansion for the three-point function in Open Problem 15.2 is phrased in terms of the *intrinsic distance* between 0 and x_1 and x_2, respectively. Such an expansion is far more elaborate, and partial results (though not for the three-point function, and phrased in terms of the number of pivotals rather than the intrinsic distance) are derived in [147]. We present this result in Sect. 15.3. A similar comment applies to the IIC setting in Open Problem 15.3. Here, one should bear in mind that an IIC two-point function, loosely speaking, corresponds to a regular three-point function, due to the "extra point" at infinity. In a similar vein, an IIC r-point function, loosely speaking, corresponds to a regular $(r + 1)$-point function.

The expansion for the three-point function is only the first step toward resolving Open Problems 15.1, 15.2, and 15.3. The next step involves an extension to the r-point function for general r. This looks quite frightening, but luckily the expansion for an r-point function is closely related to the expansion for the three-point function. Intuitively, this is because the limiting tree-like objects have *binary* branching, so that the major contributions come from a single split as in (15.1.24). In fact, an equation as in (15.1.24) is expected to hold for the r-point function for *any* $r \geq 3$, where only the error changes dramatically, while the vertex factor $\varphi_p(w_1, w_2)$ is identical, while the two-point functions $\tau_p(x_1 - w_1)$ and $\tau_p(x_2 - w_2)$ are replaced by s-point functions with $s < r$. This sets the stage for an inductive analysis in r. Such an analysis was performed in the context of *oriented* percolation by the second author and Slade in [170] (see also Sect. 15.2 below). Such an inductive analysis would identify the scaling limit of the r-point functions in each of the different settings. For Open Problem 15.1, this is enough, as convergence of the r-point functions in the ISE setting proves weak convergence of the mass measure. See Derbez and Slade [87] for an example where such a result is proved in the context of lattice trees.

For Open Problems 15.2 and 15.3, however, an additional tightness argument is needed to prove convergence in path space. A criterion for such tightness in the context of Open Problem 15.2, phrased in the language of lace expansions, was recently formulated by the second author together with Holmes and Perkins [160]. They applied this criterion to prove the convergence in path space in Open Problem 15.2 to lattice trees. However, we are quite distant from such results for percolation. A tightness criterion in the context of Open Problem 15.3 has not yet been formulated.

We see that even though we understand much about critical high-dimensional percolation clusters, the results linking such clusters to super-processes are not as strong as we would like them to be, and much more work is needed to bring such questions substantially forward.

15.2 Oriented Percolation

We now discuss so-called *oriented* percolation, in which vertices have a time variable, and edges are oriented in time. This is closely related to, yet slightly different from, *directed* percolation, where edges always move in positive direction coordinate-wise. For example, the edges of directed percolation in d dimensions are of the form $\{x, y\}$, where $y_i = x_i + 1$ for some i, while $x_j = y_j$ for all $j \neq i$. In the high-dimensional literature, the focus has been on oriented percolation, so we stick to this setting here.

For oriented percolation, the base graph $\mathcal{G} = (V, \mathcal{E})$ has vertex set $V = \mathbb{Z}^d \times \mathbb{Z}_+$, and bond set $\mathcal{E} = \{((x, n), (y, n + 1)) : |x - y| = 1\}$, and $\mathcal{G}$ is considered as a *directed* graph. This means that every directed bond $b \in \mathcal{E}$ is open with probability p, and we say that (x, n) is (forward) connected to (y, m) (denoted as: $(x, n) \to (y, m)$) whenever a directed path of occupied bonds exists starting in (x, n) and ending at (y, m). Mind that the direction of the bonds implies that $(x, n) \to (y, m)$ is only possible when $n \leq m$. Thus, we can only traverse edges in the direction of increasing last coordinate, and this last coordinate has the convenient interpretation of *time*. We define the *forward cluster* $\mathscr{C}(x, n)$ of $(x, n) \in \mathbb{Z}^d \times \mathbb{Z}_+$ to be

$$\mathscr{C}(x, n) = \{(y, m) : (x, n) \to (y, m)\}, \qquad (15.2.1)$$

so that, in particular, $\mathscr{C}(x, n) \subset \mathbb{Z}^d \times \{n, n + 1, \dots\}$.

While one might expect that percolation on oriented lattices poses mathematicians similar difficulties as percolation on unoriented lattices, this turns out not to be the case, since, for example, we *do* know that the oriented percolation function is continuous:

Theorem 15.2 (Continuity of oriented percolation [47, 124]). *For oriented percolation on $\mathbb{Z}^d \times \mathbb{Z}_+$, for $d \geq 1$, there is no infinite cluster at $p = p_c(\mathbb{Z}^d \times \mathbb{Z}_+)$, i.e., $\theta(p_c(\mathbb{Z}^d \times \mathbb{Z}_+)) = 0$.*

Theorem 15.2 was first proved by Bezuidenhout and Grimmett in [47] for directed percolation. The results were extended to the oriented percolation setting described above by Grimmett and Hiemer in [124].

The proof of Thm. 15.2 makes use of a block renormalization that was also used by Barsky, Grimmett, and Newman [29, 30] to prove that percolation does not occur in half-spaces. The proof in [47] also applies to the *contact process*, a continuous-time adaptation of oriented percolation. The deep relation between the contact process and oriented percolation has proved to be quite useful, and results in one model can typically also be proved for the other.

Durrett [95] investigates the one-dimensional contact process and oriented percolation models, focussing on the growth of the vertices in the cluster of the origin $(0, 0) \in \mathbb{Z}^d \times \mathbb{Z}_+$ at time n. These results basically show that when the cluster of the origin is infinite, then the part of it at time n grows linearly in n with a specific growth constant. Sakai [233] investigates the hyperscaling inequalities for oriented percolation and the contact process, indicating that mean-field critical exponents can only occur for $d > 4$, thus suggesting that the upper critical dimension of oriented percolation equals $d_c = 4$. Indeed, the orientation of the percolation problem implies that mean-field behavior already occurs for $d > 4$:

Theorem 15.3 (Mean-field critical exponents for oriented percolation [221, 222]). *For oriented percolation on $\mathbb{Z}^d \times \mathbb{Z}_+$, there exists $d_0 > 4$ such that for all $d \geq d_0$, $\beta = \gamma = 1$, and $\delta = \Delta = 2$ in the bounded-ratio sense, while $\eta = 0$ in the Fourier-asymptotic sense.*

The statement of the theorem is also valid for a *spread-out* version of oriented percolation (similar to the spread-out model for ordinary percolation considered in Sect. 5.2) in dimension $d > 4$ and spread-out parameter L sufficiently large.

Nguyen and Yang [221, 222] prove Thm. 15.3 by following a similar strategy as that for unoriented percolation as described in this text, see also the work of the second author and Slade [170]. Sakai [232] proves similar results for the contact process, which can be viewed as a continuous-time version of oriented percolation. Indeed, the theorem follows by employing the Aizenman–Barsky results in [9, 13] assuming the triangle condition, and using the lace expansion as in [132]. In [157, 158], the second author, den Hollander, and Slade prove that in the spread-out setting, for $d > 4$, the probability that there is an occupied path at criticality connecting $(0, 0)$ to $\mathbb{Z}^d \times \{n\}$ equals $1/(Bn)(1 + o(1))$ an $n \to \infty$ for some positive constant B. This can be seen as a version of the statement that the critical exponent ρ in the intrinsic distance exists and takes the mean-field value $\rho = 1$. This result was reproved in a much simpler way by the second author and Holmes [159] at the expense of a weaker error estimate.

The main results of the second author and Slade in [170] make a connection between clusters at criticality for the spread-out oriented percolation model above 4 dimensions and the canonical measure of SBM (recall Sect. 15.1). The proof in [170] follows the outline presented at the end of Sect. 15.1, by deriving an expansion of the r-point function followed by an induction argument in r to identify the scaling limit of these r-point functions. Of course, the time variable n in Open Problem 15.2 has a highly natural meaning as "time" in the context of oriented percolation, which certainly simplifies the analysis. Together with the results in [157, 158] about the oriented survival probability, this proves convergence in finite-dimensional distribution (see [175], where this statement is proved). A tightness argument, for example by verifying the conditions in [160], is still missing. Related results that apply to oriented percolation as well as the contact process have been proved by the second author and Sakai in [167, 168], while the results for the survival probability of the contact process follow from [159]. The proof for the contact process in [167, 168] follows by noting that a time-discretization of the contact process is an oriented percolation model, and proving that all such oriented percolation models have the SBM scaling limits, uniformly in the discretization parameter.

We close this section by stating an open problem concerning *nearest-neighbor* oriented percolation:

Open Problem 15.4 (Upper critical dimension nearest-neighbor oriented percolation). *Show that nearest-neighbor percolation exhibits mean-field behavior for all* $d > 4$.

15.3 Scaling Limit of Percolation Backbones

In this section, we discuss a novel lace expansion derived with Hulshof and Miermont in [147] for the two-point function with a fixed number of pivotals. Recall that $\mathrm{Piv}(x, y)$ denotes the set of pivotal bonds for the event $\{x \leftrightarrow y\}$, and $|\mathrm{Piv}(x, y)|$ the number of such pivotal bonds. We adopt the convention that $|\mathrm{Piv}(x, y)| = \infty$ whenever x and y are not connected, and define

$$\tau_n(x, y) = \mathbb{P}_{p_c}(|\mathrm{Piv}(x, y)| = n), \quad n = 0, 1, 2, \dots, \tag{15.3.1}$$

to be the probability that x and y are connected and there are precisely n pivotal edges in between them for the critical value $p = p_c$. Mind that the number of pivotal bonds forms a random pseudometric on $\mathbb{Z}^d$. We write $\tau_n(x, y) = \tau_n(x - y)$. We also study the backbone two-point function of the IIC, denoted by

$$\rho_n(x) = \mathbb{P}_{\mathrm{IIC}}(S_n = x), \tag{15.3.2}$$

where S_n is the top of the nth pivotal bond for the connection from 0 to ∞ in the IIC backbone, with $S_0 = 0$. Since the IIC is single ended, it has infinitely many backbone pivotals, so that S_n (and thus $\rho_n(x)$) is well defined. The main result in this section is the following theorem:

Theorem 15.4 (Gaussian asymptotics for two-point function with fixed number of pivotals). *For percolation in dimension* $d \geq d_0 > 6$, *there exist constants* $\sigma^2 = \sigma^2(d)$ *and* A *such that*

$$\lim_{n\to\infty} \hat{\tau}_n(k/\sqrt{n}) = Ae^{-\sigma^2|k|^2/(2d)} , \qquad (15.3.3)$$

and

$$\lim_{n\to\infty} \hat{\rho}_n(k/\sqrt{n}) = e^{-\sigma^2|k|^2/(2d)}. \qquad (15.3.4)$$

In words, Thm. 15.4 states that the endpoint of a path consisting of n pivotals has an asymptotic Gaussian distribution. The result about the endpoint can be strengthened to obtain convergence of the entire paths to a Brownian motion path. This is best explained for the IIC backbone, whose definition we recall from (14.3.3). Further, recall that S_n is the top of the nth pivotal bond of the IIC cluster for the event $\{0 \leftrightarrow \infty\}$.

Theorem 15.5 (Backbone scaling limit). *Consider percolation on $\mathbb{Z}^d$ with $d \geq d_0$, with $d_0 > 14$ sufficiently large. There exists a constant $\sigma^2 = \sigma^2(d)$ such that the following convergence in distribution is valid $\mathbb{P}_{\mathrm{IIC}}$-a.s. as $n \to \infty$ in the space $D([0, \infty), \mathbb{R}^d)$ of right-continuous functions with left limits endowed with the Skorokhod J_1 topology:*

$$\left(n^{-1/2} S_{\lceil nt\rceil}\right)_{t\geq 0} \implies (\sigma B)_{t\geq 0} , \qquad (15.3.5)$$

where $(B_t)_{t\geq 0}$ is standard Brownian motion on $\mathbb{R}^d$.

Several variants of this result may be stated. For example, a corresponding result holds for convergence of the entire backbone (in the sense of (14.3.3)) on the space of nonempty compact subsets of $\mathbb{R}^d$ endowed with the Hausdorff distance. Similarly, one may phrase a version of the theorem for large critical clusters conditioned to be sufficiently large. For details, we refer to [147].

Theorem 15.5 gives a clear mathematical interpretation to the phrase "faraway pieces of critical percolation clusters are close to being independent" in high dimensions. Of course, the different pieces are *not* independent, as the spatial position of one part of the cluster has implications on other parts. However, if we pretend that they were independent (as we did, e.g., in (4.2.27)), then the resulting quantitative error is very small. Theorem 15.5 takes this intuition a little further by identifying Brownian motion paths as scaling limits of the backbone of critical clusters. Since the path of Brownian motion for disjoint time intervals is (fully) independent, we conclude that all dependencies of the cluster backbone vanish in the scaling limit.

A remark is due about the condition $d_0 > 14$ in Thm. 15.5. Indeed, we expect the theorem to hold for any $d > 6$, just as Thm. 15.4. However, the tightness criterion that is currently applied in the proof requires the *heptagon diagram* to be finite rather than the triangle diagram only, and for this we need $d > 14$ (cf. Prop. 5.5 with $l = 0$ and $n = 7$). The challenge is to find a more suitable tightness criterion in order to avoid this technical nuisance.

The proof of Thm. 15.4 makes use of a novel lace expansion, that we now explain. It is based on the existence of a family $\pi_l(x, y)$ (where $l \in \mathbb{N}$, $x, y \in \mathbb{Z}^d$) such that the lace-expansion equation

$$\tau_m(x, y) = \pi_m(x, y) + \sum_{l=0}^{m-1}\sum_b \pi_l(x, b)\, p\, D(b)\tau_{m-l-1}(\bar{b}, y) \qquad (15.3.6)$$

is valid for all $m \in \mathbb{N}$. We achieve this goal by a lace expansion. The lace expansion below is *novel* for percolation.

We can take the Fourier transform of (15.3.6) to get

$$\hat{\tau}_m(k) = \hat{\pi}_m(k) + \sum_{l=0}^{m-1} \hat{\pi}_l(k) \, p \, \widehat{D}(k) \, \hat{\tau}_{m-l-1}(k). \tag{15.3.7}$$

Equation (15.3.7) is the starting point of the analysis for the fixed-pivotals two-point function.

Exercise 15.2 (Uniqueness lace expansion). (a) Use (15.3.7) to prove that $\hat{\tau}_p(k)$ satisfies

$$\hat{\tau}_p(k) = \frac{1 + \widehat{\Pi}'_p(k)}{1 - 2dp\widehat{D}(k)[1 + \widehat{\Pi}'_p(k)]}, \tag{15.3.8}$$

where $\widehat{\Pi}'_p(k) = \sum_{l=0}^{\infty} \hat{\pi}_l(k)$.

(b) Prove that $\widehat{\Pi}'_p(k) = \widehat{\Pi}_p(k)$ for every p and k for which both are well defined. Thus, the lace-expansion coefficients are *unique*.

In order to understand the meaning of (15.3.6), it is most instructive to compare it to the classical lace expansion in Chap. 6. There we were expanding the probability of the increasing event $\{x \leftrightarrow y\}$, but now we consider events of the form $|\mathrm{Piv}(x, y)| = n$, which are intersections of increasing and decreasing events. A most notable difference is that the basic expansion identity in (6.2.6) is replaced by a more involved identity. In order to describe it, we need two more definitions: For any (deterministic) $A \subseteq \mathbb{Z}^d$ and vertices $x, y \in \mathbb{Z}^d$, we write $\mathrm{Piv}^A(x, y)$ for the collection of pivotal bonds for the event $\{v \leftrightarrow y \text{ in } \mathbb{Z}^d \setminus A\}$ and define

$$\tau_n^A(x, y) := \mathbb{P}_{p_c}\left(\left|\mathrm{Piv}^A(x, y)\right| = n\right). \tag{15.3.9}$$

Since $\left|\mathrm{Piv}^A(x, y)\right| \geq |\mathrm{Piv}(x, y)|$ holds on the event $x \leftrightarrow y$ in $\mathbb{Z}^d \setminus A$, we can write for all $A \subseteq \mathbb{Z}^d$,

$$\tau_n(x, y) - \tau_n^A(x, y) = \mathbb{P}_{p_c}\left(|\mathrm{Piv}(x, y)| = n, \mathrm{Piv}^A(x, y) = \varnothing\right)$$
$$+ \mathbb{P}_{p_c}\left(|\mathrm{Piv}(x, y)| = n, \left|\mathrm{Piv}^A(x, y)\right| > n\right)$$
$$- \mathbb{P}_{p_c}\left(|\mathrm{Piv}(x, y)| < n, \left|\mathrm{Piv}^A(x, y)\right| = n\right). \tag{15.3.10}$$

Here is an intuitive explanation for this identity: Let us suppose that x and y are connected, and this connection has precisely n pivotal bonds. If we "close off" a set of vertices A, then there are three possible scenarios: We can break the connection from x to y entirely (the first summand), we can make it too long (the second summand), or a connection that was previously too short now has the right length (the third summand). The last two summands were not present in (6.2.6). Hence, in every expansion step, we get both a positive as well as a negative term. In turn, this gives a more complicated structure for the coefficients $\pi_l(x, y)$.

Instead of working with the number of pivotal bonds, it would also be highly interesting to investigate the two-point function with fixed intrinsic distance. This is formulated more precisely in the following open problem:

> **Open Problem 15.5 (Intrinsic distance two-point function).** *Extend Thm. 15.4 to the* intrinsic distance *two-point function*
>
> $$\tau_n^{(\mathrm{in})}(x) = \mathbb{P}_{p_c}\big(d_{\mathscr{C}(0)}(0, x) = n\big) , \tag{15.3.11}$$
>
> *where $d_{\mathscr{C}(0)}(0, x)$ is the intrinsic or graph distance between 0 and x in the (random) graph $\mathscr{C}(0)$.*

15.4   Long-Range Percolation

Long-range percolation is a variation of percolation where edges of arbitrary length can be occupied, and edge occupation depends on the Euclidean distance between the vertices in the edge. It has been demonstrated that the lace expansion is a particularly useful method to study long-range percolation in high dimensions. The big picture is the following: By allowing (very) long edges to be occupied with sufficiently high probability, critical clusters become spatially more stretched out, and different parts of critical clusters are "more independent." This triggers mean-field behavior even in dimensions below 6 (which is impossible for ordinary percolation by Cor. 11.7). Some authors express this phenomenon by saying that the slow decay of edge occupation densities increases the "effective dimension" of the model.

We formally introduce the model by defining edge weights for arbitrary $L, \alpha > 0$ by

$$D_L(x, y) = \mathcal{N}_L \max\left\{\frac{|x - y|}{L}, 1\right\}^{-(d+\alpha)} , \quad x, y \in \mathbb{Z}^d , \tag{15.4.1}$$

where $\mathcal{N}_L$ is a normalizing constant, that is, $\mathcal{N}_L^{-1} = \sum_{x \in \mathbb{Z}^d} \max\{|x|/L, 1\}^{-(d+\alpha)}$ is chosen such that

$$\sum_{y \in \mathbb{Z}^d} D_L(x, y) = 1$$

for all x. For a parameter $\lambda \geq 0$, we make *any* edge $\{x, y\} \in \mathbb{Z}^d \times \mathbb{Z}^d$ occupied with probability

$$p_{xy} = \lambda D_L(x, y) \wedge 1 \tag{15.4.2}$$

independently of each other, and otherwise vacant.

We have now introduced the three parameters in long-range percolation, namely α, L and λ. Let us explain the role of all three of them. The *power-law exponent* parameter α controls the tail behavior of the edge-occupation probabilities. Loosely speaking, the smaller α, the more likely very long edges in percolation clusters are. The *spread-out parameter L* is of a purely technical nature. The role of L here is indeed the same as in Thm. 5.4; it "spreads out" the weights in order to make the triangle diagram small enough such that the bootstrap argument, as discussed in Chap. 8, works. The *percolation parameter* $\lambda \geq 0$ scales the edge-occupation probabilities, and thus has the same role as p in the original percolation model, but bear in mind that

> λ is **not a probability anymore**. Rather, λ is the expected degree of a percolation configuration ω at $x \in \mathbb{Z}^d$.

We denote $\tau_\lambda(x) = \mathbb{P}_\lambda(0 \leftrightarrow x) = \mathbb{P}_\lambda(v \leftrightarrow v + x)$ for $v, x \in \mathbb{Z}^d$, and further use the same notation as in Sect. 1.1 for this model with p replaced by λ; in particular, we define the critical threshold as $\lambda_c = \inf\{\lambda \geq 0 : \theta(\lambda) > 0\}$, where $\theta(\lambda) = \mathbb{P}_\lambda(|\mathscr{C}(0)| = \infty)$.

Interestingly, Berger [45] shows that the percolation function $\theta(\lambda)$ is *continuous* at $\lambda = \lambda_c$ in general dimension whenever $\alpha \in (0, d)$. Again this shows that the problems encountered for regular percolation can sometimes be resolved by slightly adapting the model, recall Open Problem 1.1.

The main result concerning mean-field behavior for long-range percolation states that the *upper critical dimension* of the model changes from 6 (in the original model) to $3(\alpha \wedge 2)$ in long-range percolation:

Theorem 15.6 (Infrared bound [145, 148]). *For long-range percolation in dimension $d > 3(\alpha \wedge 2)$, there is $L_0 > 0$ such that for every $L > L_0$ there exists a constant $A_2 = A_2(d, \alpha, L)$ and $\varepsilon > 0$ such that for $\lambda_c = \lambda_c(d, \alpha, L)$ and $k \in (-\pi, \pi]^d$,*

$$\hat{\tau}_{\lambda_c}(k) = \frac{A_2(d, \alpha, L)}{|k|^{\alpha \wedge 2}} \left(1 + O(|k|^\varepsilon)\right) . \tag{15.4.3}$$

Consequently, the triangle condition is satisfied in this regime, and the critical exponents $\beta, \gamma, \delta, \nu, \nu_2$, and Δ exist in bounded-ratio sense and take on their mean-field values.

The infrared bound in Thm. 15.6 was proved in our joint paper with Sakai [148], and the extension to sharp asymptotics follows from joint work with Hulshof [145], where strong spatial bounds are obtained for $\Pi_{\lambda_c}(x)$.

Notice that the right-hand side of (15.4.3) is comparable to $[1 - \hat{D}_L(k)]^{-1}$, which is the Fourier transform of the Green's function of a random walk with step distribution D_L. This observation is also the key to the proof of the theorem: We follow the same strategy as for ordinary percolation (outlined in Chaps. 5–8), but instead of comparing τ_λ to the Green's function of simple random walk, we compare it to the Green's function of long-range walk. In fact, Thm. 15.6 holds not only for the specific form of D_L in (15.4.1), but for a much more general class, which we disregard in this discussion; we refer the interested reader to our paper with Sakai [148].

A natural next step in the analysis is the x-space asymptotics of $\tau_{\lambda_c}(x)$ as $|x| \to \infty$. This requires an excellent control of the convolution bounds in (11.2.4), which is notoriously difficult for infinite-range models. Chen and Sakai [77] nevertheless manage to identify the decay of $\tau_{\lambda_c}(x)$ under some extra assumptions on D_L. One example of D_L that satisfies all these extra assumptions is

$$D_L(x) = \sum_{n \in \mathbb{N}} U_L^{*n}(x) T_\alpha(n) , \qquad (15.4.4)$$

where U_L is the discrete uniform distribution on $[-L, L]^d \cap \mathbb{Z}^d$, and T_α is the stable distribution on $\mathbb{N}$ with parameter $\alpha/2 \neq 1$ (cf. the appendix of [77]). The main result by Chen and Sakai in this setting is the following:

Theorem 15.7 ($\eta = 0$ in x-space [77]). *For long-range percolation with edge weights given by* (15.4.4) *in dimension* $d > 3(\alpha \wedge 2)$, $\alpha \in (0, 2) \cup (2, \infty)$, *there is* $L_0 > 0$ *such that for every* $L > L_0$ *there exist constants* $A_2 = A_2(d, \alpha, L)$ *and* $\mu \in (0, \alpha \wedge 2)$ *such that*

$$\tau_{\lambda_c}(x) = \frac{A_2(d, \alpha, L)}{|x|^{d-(\alpha \wedge 2)}} + \frac{O(L^{-(\alpha \wedge 2)+\mu})}{|x|^{d-(\alpha \wedge 2)+\mu}} . \qquad (15.4.5)$$

The conclusion of Thm. 15.6 holds for a large class of bond weights D_L with the appropriate decay rates, and universality suggests that the rather restricted result of Thm. 15.7 generalizes in the same way. It is an open problem to prove this generalization of Thm. 15.7:

Open Problem 15.6 (Universality in x-space asymptotics long-range percolation).
Extend Thm. 15.7 to more general long-range settings, for example, that in (15.4.1).

In view of the results presented so far, it is tempting to conjecture that virtually all the results for ordinary percolation in high dimensions hold for long-range percolation with 2 replaced by $\alpha \wedge 2$ at appropriate places. This is also quite natural, as we are comparing the two-point function $\tau_{\lambda_c}(x)$ to the Green's function of a random walk with step distribution D_L. It is well known that the Green's function asymptotics $|x|^{-(d-2)}$ is restricted to random walks with finite variance step distributions, which holds for (15.4.4) precisely when $\alpha > 2$. It turns out, however, that $\alpha > 2$ is not enough for all critical exponents to take their finite-range mean-field values. As an illustration, we consider the extrinsic one-arm exponent of Thm. 11.5. The crossover, for which long-range percolation differs from ordinary percolation, is not for $\alpha = 2$ as before, but here it is for $\alpha = 4$:

Theorem 15.8 (Long-range one-arm exponent [146, 177]). *For long-range percolation in dimension* $d > 3(\alpha \wedge 2)$ *with* $\alpha \in (0, 2) \cup (2, \infty)$ *for which* (15.4.5) *holds in a bounded-ratio sense, there is* $L_0 > 0$ *such that for every* $L > L_0$ *there exist constants* $C_{\text{ex}}, c_{\text{ex}} > 0$ *such that*

$$\frac{c_{\text{ex}}}{n^{(\alpha \wedge 4)/2}} \leq \mathbb{P}_{\lambda_c}(0 \leftrightarrow \mathbb{Z}^d \setminus \Lambda_n) \leq \frac{C_{\text{ex}}}{n^{(\alpha \wedge 4)/2}} . \qquad (15.4.6)$$

The lower bound in Thm. 15.8 does not rely on (15.4.5), but the upper bound does.
We summarize that the critical exponent $\rho_{\text{ex}} = \frac{1}{2}$ of ordinary percolation changes to $\rho_{\text{ex}} = 2/(\alpha \wedge 4)$ in long-range percolation. This may look surprising at first, but indeed critical

branching random walk shows the same behavior, as proven by Hulshof [177] (and hinted at by Janson and Marckert [186]).

We conclude this discussion by a remark on the continuity of the percolation function. In the mean-field regime discussed so far, the critical exponent β equals 1, and thus the percolation function $\lambda \mapsto \theta(\lambda)$ is continuous at $\lambda = \lambda_c$. This is the case in particular for $d = 1$ and $\alpha < \frac{1}{3}$. In a celebrated result, Aizenman and Newman [14] show that for $d = 1$ and $\alpha = 1$ (under some minor extra conditions), indeed $\theta(\lambda_c) > 0$! This shows that Open Problem 1.1 does not hold for *all* percolation models and arguably explains why proving continuity of $p \mapsto \theta(p)$ at $p = p_c$ for general percolation models has remained unsolved for such a long time.

15.5 The Asymptotic Expansion of the Critical Value

In this section, we investigate the asymptotics of $p_c(\mathbb{Z}^d)$ as $d \to \infty$. The main result is as follows:

Theorem 15.9 (The asymptotic expansion of $p_c(\mathbb{Z}^d)$ in powers of $2d$ [136, 138, 172, 173]). *Consider bond percolation on $\mathbb{Z}^d$. There are rational numbers a_i such that for all $M \geq 1$,*

$$p_c(\mathbb{Z}^d) = \sum_{i=1}^{M} a_i (2d)^{-i} + O((2d)^{-M-1}) \quad \text{as } d \to \infty . \tag{15.5.1}$$

Further, $a_1 = a_2 = 1$, $a_3 = \frac{7}{2}$.

The fact that there is an asymptotic expansion to all order was proved by the second author and Slade in [173]. The first 3 coefficients were first identified by Hara and Slade in [138] (see also [136]), and they were recomputed in [172]. In the latter paper, the asymptotic expansion was extended to the critical value on the hypercube $p_c(\{0, 1\}^d)$, whose first three coefficients agree with those on $\mathbb{Z}^d$. The proof uses the lace expansion, and a careful analysis of how the lace-expansion coefficients depend on the dimension. The crucial identity is given by (6.1.6), namely

$$2dp_c = \frac{1}{1 + \widehat{\Pi}_{p_c}(0)} . \tag{15.5.2}$$

The asymptotic expansion is then proved to exist by induction in M and an analysis of $\widehat{\Pi}_{p_c}(0)$ in terms of p and d.

The expansion

$$p_c(\mathbb{Z}^d) = \frac{1}{2d} + \frac{1}{(2d)^2} + \frac{7}{2(2d)^3} + \frac{16}{(2d)^4} + \frac{103}{(2d)^5} + \cdots \tag{15.5.3}$$

was reported by Gaunt and Ruskin in [116], but with no rigorous bound on the remainder. The fact that $p_c(\mathbb{Z}^d) = (1 + o(1))/(2d)$ has a long history [19, 56, 118, 132, 197], with various error estimates. It is believed that the full asymptotic expansion $\sum_{i=1}^{\infty} a_i (2d)^{-i}$ does not converge, the reason being that the absolute values $|a_i|$ grow too quickly as $i \to \infty$. However, it appears plausible that the sequence might be Borel summable, which we explain next.

For convenience, we abbreviate $p_c(d) = p_c(\mathbb{Z}^d)$. The main question is, in view of the (presumed) divergence of the formal series $\sum_i a_i (2d)^{-i}$, how can we reconstruct $p_c(d)$ from the sequence $(a_i)_{i \geq 1}$? To this end, we consider the Borel transform

$$B(t) = \sum_{i=1}^{\infty} \frac{a_i t^i}{i!}, \tag{15.5.4}$$

and assume existence for all $t > 0$. Suppose that p_c (as a function of $d \in \mathbb{N}$) can be extended to an analytic function on an open neighborhood U around p_c in the complex plane $\mathbb{C}$ for some specific value d, and assume that

$$\left| p_c(d) - \sum_{i=1}^{M} a_i (2d)^{-i} \right| \leq \frac{C^M M!}{(2d)^M} \tag{15.5.5}$$

for some fixed constant $C > 0$. Then,

$$p_c(d) = 2d \int_0^{\infty} e^{-2dt} B(t) \, dt \tag{15.5.6}$$

and the right-hand side, known as the *Borel sum*, converges. See Graham [119] for details in the slightly different context of self-avoiding walk.

Open Problem 15.7 (Borel summability of the asymptotic expansion $p_c(\mathbb{Z}^d)$).
Prove that the sequence $(a_i)_{i \geq 1}$ in (15.5.1) is Borel summable, that is, prove (15.5.6).

It is instructive to compare the expansion of p_c for percolation with another model called self-avoiding walk. A principle challenge for the study of self-avoiding walk is to understand the *connective constant*, which is a parameter for that model similar to p_c for percolation. Indeed, similar expansions to the one in (15.5.1) have been obtained for the connective constant. Clisby, Liang, and Slade [78] have found rigorously that (for self-avoiding walk) the first sign change for the sequence $(a_i)_{i \geq 1}$ appears in the 10th digit. This supports the picture that $(a_i)_{i \geq 1}$ is a sequence with fast growing absolute values and sign changes after blocks of varying length. Graham [119] proved a bound similar to (15.5.5) in the context of self-avoiding walk.

Computing the coefficients a_i becomes increasingly more difficult when i grows large, since more and more lace-expansion coefficients $\widehat{\Pi}_{p_c}^{(N)}(0)$ need to be taken into account, and these coefficients become substantially more involved as N grows. Finding an efficient way to compute the coefficients is the content of the next open problem:

Open Problem 15.8 (Computation coefficients asymptotic expansion $p_c(\mathbb{Z}^d)$).
Find an argument that proves the existence of the asymptotic expansion of $p_c(\mathbb{Z}^d)$ in Thm. 15.9 that avoids the lace expansion, and use it to compute $a_4, a_5, \ldots$

As noted earlier, Thm. 15.9 also applies to hypercube percolation with $2d$ replaced with d and with the same coefficients $a_1 = a_2 = 1$, $a_3 = 7/2$. The proof in [172, 173] hints at the fact that a_i for $\mathbb{Z}^d$ and the hypercube are different for *some* $i \geq 4$.

Open Problem 15.9 (Computation   coefficients   asymptotic   expansion $p_c(\{0, 1\}^d; 1)$). *Extend Open Problem 15.8 to the hypercube and find the first i for which a_i on $\mathbb{Z}^d$ and the hypercube are distinct.*

Probably a slightly easier open problem involves the Hamming graph. Indeed, Thm. 13.13 applies to the Hamming graph:

Exercise 15.3 (Hamming graph and its critical behavior). Prove that Thm. 13.13 applies to the Hamming graph.

Open Problem 15.10 (Computation coefficients asymptotic expansion $p_c(K_n^{d}; \lambda)$).
Show that the critical value on the Hamming graph K_n^d satisfies

$$p_c(K_n^d; \lambda) = \sum_{i=1}^{d/3} a_i(d)n^{-i} + O(n^{-1-d/3}) . \tag{15.5.7}$$

It is not hard to see that $a_1(d) = 1/d$, but the higher-order contributions are less obvious to determine. For $d = 1, 2, 3$, however, this completes the proof. Recently, progress was made in this direction by identifying the second-order correction in (15.5.7) by Federico, the second author, den Hollander, and Hulshof [103], with $a_2(d) = (2d^2 - 1)/[2(d - 1)^2]$. This proves Open Problem 15.10 also for $d = 4, 5, 6$. For $d \geq 7$, also the third term becomes relevant.

The major distinction between the situation on the hypercube described in Open Problem 15.9 and that on the Hamming graph in Open Problem 15.10 is that on the hypercube, any truncation of the asymptotic expansion is not inside the critical window (assuming that $a_s \neq 0$ for infinitely many s, as the scaling window has width $\Theta(2^{-d/3})$, and any inverse power of d is much larger than $\Theta(2^{-d/3})$). On the other hand, the width of the scaling window on the Hamming graph is an inverse power of n, and therefore, eventually the asymptotic expansion is inside the scaling window.

15.6 Percolation on Nonamenable Graphs

We next focus on percolation on a class of infinite transitive graphs $\mathscr{G} = (\mathcal{V}, \mathcal{E})$ known as *nonamenable graphs*. A comprehensive reference for percolation on transitive graphs is the book of Lyons and Peres [210].

We start by defining what a nonamenable graph is. For a finite set of vertices $S \subseteq \mathcal{V}$, we recall from Sect. 3.2 that its *edge boundary* ΔS is defined by

$$\Delta S = \{(u, v) \in \mathcal{E} : u \in S, v \notin S\} . \tag{15.6.1}$$

The notion of amenability is all about whether the size of ΔS is of equal order as that of S, or is much smaller. To formalize this, we denote the *Cheeger constant* of a graph $\mathscr{G}$ by

$$\mathsf{Ch}(\mathscr{G}) = \inf_{S \subset \mathcal{V} : |S| < \infty} \frac{|\Delta S|}{|S|} . \tag{15.6.2}$$

A graph is called *amenable* when $\mathsf{Ch}(\mathscr{G}) = 0$, and it is called nonamenable otherwise. Key examples of amenable graphs are finite-range translation invariant graphs $\mathscr{G}$ with vertex set $\mathbb{Z}^d$. The simplest example of a nonamenable graph is the regular tree $\mathbb{T}_r$ with $r \geq 3$. For the regular tree $\mathbb{T}_r$ with $r \geq 3$, it is not hard to see that $\mathsf{Ch}(\mathbb{T}_r) = r - 2$. Benjamini and Schramm [40] prove many preliminary results of percolation on nonamenable graphs and state many open questions, some of which have been settled in the meantime. For example, [40, Theorem 1] proves that $p_c(\mathscr{G}) \leq 1/(\mathsf{Ch}(\mathscr{G}) + 1)$, so that $p_c(\mathscr{G}) < 1$ for every nonamenable graph.

A related definition of nonamenability can be given in terms of the *spectral radius* of a graph. Let $\mathbf{p}_n(u, v)$ be the probability that simple random walk on $\mathscr{G}$ starting at $u \in \mathcal{V}$ is at time n at $v \in \mathcal{V}$. The *spectral radius* of $\mathscr{G}$ is defined as

$$\rho(\mathscr{G}) = \lim_{n \to \infty} \mathbf{p}_{2n}(u, u)^{1/(2n)} . \tag{15.6.3}$$

By Kesten's Theorem [191, 192] (see also [90]), when $\mathscr{G}$ has bounded degree, then $\rho(\mathscr{G}) < 1$ if and only if $\mathsf{Ch}(\mathscr{G}) > 0$. This exemplifies the fact that there is a close relationship between graph theoretic properties on the one hand and the behavior of stochastic processes on the graph on the other. A similar relation between the existence of invariant site percolation and amenability of Cayley graphs is proved in [39, Thm. 1.1].

For percolation on $\mathbb{Z}^d$, in the supercritical regime, the Burton–Keane argument proves that the infinite cluster is unique, cf. [67]. It turns out (see e.g., the discussion following [198, Thm. 4]) that the uniqueness of the infinite cluster is valid for all amenable graphs. As the Burton–Keane argument shows, there is a close relation between the surface/volume ratio of the underlying graph on the one side and the uniqueness of the infinite cluster on the other side, which helps to explain the uniqueness for all amenable graphs. The situation for trees is very different, as the number of infinite components N equals $N = \infty$ a.s. in the supercritical phase, which can be attributed to the fact that if we remove one edge, then a tree falls apart into two infinite graphs, and each of these has at least one infinite component a.s., so that in total there are infinitely many infinite clusters. Thus, this phenomenon is more related to there not being any cycles rather than the boundary being large.

In order to investigate the number of infinite clusters, we define the *uniqueness critical value* by

$$p_u = p_u(\mathscr{G}) = \inf\{p: \mathbb{P}_p \text{-a.s. there is a unique infinite cluster}\}. \qquad (15.6.4)$$

For the regular tree with $r \geq 3$, $p_u = 1$, while for $\mathbb{Z}^d$, $p_u = p_c$. Below, we give examples where $p_c < p_u < 1$.

While the existence of an infinite cluster is clearly an increasing event, the uniqueness of the infinite cluster is *not*. Therefore, it is a priori not at all obvious that for all $p > p_u$, the infinite cluster is unique. This is the main content of the following theorem by Schonmann [237]. In its statement, we write $N(p)$ for the number of infinite percolation clusters under $\mathbb{P}_p$.

Theorem 15.10 (Uniqueness transition [237]). *For percolation on a connected, quasi-transitive, infinite graph of bounded degree, a.s.,*

$$N(p) = \begin{cases} 0 & for\ p \in [0, p_c)\ ; \\ \infty & for\ p \in (p_c, p_u)\ ; \\ 1 & for\ p \in (p_u, 1]\ . \end{cases} \qquad (15.6.5)$$

This theorem is due to Schonmann [237], see also related results by Häggström, Peres, and Schonmann [181, 182]. Note that, in general, not much is known for the critical cases $p = p_c$ and $p = p_u$.

Classical examples of graphs for which $0 < p_c < p_u < 1$ are the Cartesian product $\mathbb{Z} \times \mathbb{T}_r$, as shown by Grimmett and Newman [126] and, much more general, any transitive, non-amenable planar graph with one end (e.g., tilings of the hyperbolic plane) as shown by Benjamini and Schramm [41].

We continue by studying the nature of the phase transition on nonamenable graphs. Remarkably, we know that $p \mapsto \theta(p)$ is continuous for a number of special cases, in particular for *Cayley graphs of nonamenable groups* and for *quasi-transitive graphs of exponential growth*. We start by defining what a Cayley graph is. Let Γ be a group, and let $S = \{g_1, \ldots, g_n\} \cup \{g_1^{-1}, \ldots, g_n^{-1}\}$ be a finite set of generators. The Cayley graph $\mathscr{G} = \mathscr{G}(\Gamma)$ with generators S has vertex set $\mathcal{V} = \Gamma$, and edge set $\mathcal{E} = \{\{g, h\}: g^{-1}h \in S\}$. Furthermore, we call a graph $\mathscr{G}$ *quasi-transitive* (or almost transitive) if there is a finite set of vertices $\mathcal{V}_0 \subset \mathcal{V}$ such that any $v \in \mathcal{V}$ is taken into $\mathcal{V}_0$ by some automorphism of $\mathscr{G}$. The graph is said to have exponential growth whenever

$$\text{gr}(\mathscr{G}) := \liminf_{n \to \infty} |\{y \in \mathcal{V}: d_{\mathscr{G}}(x, y) \leq n\}|^{1/n} > 1, \qquad (15.6.6)$$

where x is some vertex of $\mathscr{G}$, and $d_{\mathscr{G}}$ is the graph distance on $\mathscr{G}$. Mind that the property (15.6.6) is independent of the initial vertex x if $\mathscr{G}$ is quasi-transitive.

Theorem 15.11 (Continuity on nonamenable graphs [38, 39, 179]). *Let $\mathscr{G}$ be a Cayley graph of a finitely generated nonamenable group or a quasi-transitive graph of exponential growth. For percolation on $\mathscr{G}$, there is no infinite cluster at $p = p_c(\mathscr{G})$, that is, $\theta(p_c(\mathscr{G})) = \mathbb{P}_{p_c}(|\mathscr{C}(v)| = \infty) = 0$.*

This result was proved by Benjamini, Lyons, Peres, and Schramm [38, 39], and generalized earlier work by Wu [257] on $\mathbb{Z} \times \mathbb{T}_r$ with $r \geq 7$. The proof makes use of the *mass-transport technique* with a clever choice of the mass-transport function. Hutchcroft's proof [179] for quasi-transitive graph of exponential growth is more recent and presented below. As we have discussed earlier for the case of $\mathbb{Z}^d$, continuity of the percolation function at the critical point is a question of high importance in percolation theory. Here is a stronger form of an earlier formulated open problem:

> **Open Problem 15.11 (Critical percolation dies out; general form [40]).** *For any quasi-transitive graph $\mathscr{G}$ with $p_c(\mathscr{G}) < 1$, there is no infinite cluster at criticality, that is $\theta(p_c) = 0$.*

This generalizes Open Problem 1.1 and was first formulated as Conjecture 4 in Benjamini and Schramm [40], see also [210, Conjecture 8.15]. Mind that one can construct certain trees for which there are infinitely many infinite open clusters at criticality, which shows that quasi-transitivity is essential.

Proof of Thm. 15.11 for quasi-transitive graphs of exp. growth. We present the argument of Hutchcroft [179]. First we note that on any quasi-transitive graph, the number $N(p)$ of infinite percolation clusters satisfies $N(p) \in \{0, 1, \infty\}$ a.s. for every $p \in [0, 1]$, as shown by Newman and Schulman [218]. Furthermore, the case $N(p_c) = \infty$ can be ruled out, as proven by Burton and Keane [67] and Gandolfi, Keane, and Newman [115] for amenable quasi-transitive graphs, by Benjamini, Lyons, Peres, and Schramm [38, 39] for nonamenable, quasi-transitive graphs with an extra assumption named unimodularity, and finally, Timár [254] removed the unimodularity assumption. It remains to show that $N(p_c) = 1$ cannot occur. To this end, we claim that

$$\kappa_{p_c}(n) \leq \mathrm{gr}(\mathscr{G})^{-n}, \tag{15.6.7}$$

where $\mathrm{gr}(\mathscr{G}) > 1$ was defined in (15.6.6) and

$$\kappa_p(n) := \inf\{\tau_p(x, y) : x, y \in \mathcal{V}, d_{\mathscr{G}}(x, y) \leq n\}. \tag{15.6.8}$$

To see that this implies the desired claim, assume that $N(p_c) = 1$. By the Harris inequality (1.3.1),

$$\tau_{p_c}(x, y) \geq \mathbb{P}_{p_c}(x \text{ is in the unique infinite cluster})^2,$$

and hence, $\lim_{n \to \infty} \kappa_{p_c}(n) > 0$, thus contradicting (15.6.7).

In order to show (15.6.7), we observe that $\kappa_p(n)$ is super-multiplicative in n for any p, because again the Harris inequality implies for $u, v \in V$ with $d_{\mathscr{G}}(u, v) < n + m$ that

$$\tau_p(u, v) \geq \tau_p(u, w) \, \tau_p(w, v) \geq \kappa_p(m) \, \kappa_p(n)$$

for any $w \in V$ with $d_{\mathscr{G}}(u, w) < n$ and $d_{\mathscr{G}}(w, v) < m$. Taking the infimum yields $\kappa_p(n + m) \geq \kappa_p(m) \, \kappa_p(n)$.

For any $p \in [0, 1]$ and every $n \geq 1$ and any fixed vertex $x \in V$,

$$\kappa_p(n) \cdot \left| \{ y \in V : d_{\mathscr{G}}(x, y) \leq n \} \right| \leq \sum_{y \in V: d_{\mathscr{G}}(x,y) \leq n} \tau_p(x, y) \leq \sum_{y \in V} \tau_p(x, y) ,$$

so that by Fekete's lemma (cf. [211, Lemma 1.2.2]),

$$\sup_{n \geq 1}(\kappa_p(n))^{1/n} = \lim_{n \to \infty}(\kappa_p(n))^{1/n} \leq \limsup_{n \to \infty} \left(\frac{\sum_{y \in V} \tau_p(x, y)}{\left| \{ y \in V : d_{\mathscr{G}}(x, y) \leq n \} \right|} \right)^{1/n} .$$

We next use that $\sum_{y \in V} \tau_p(x, y) = \mathbb{E}_p |\mathscr{C}(x)| < \infty$ whenever $p < p_c$ by Thm. 3.1 (the quasi-transitive case being treated in Antunović and Veselic [22]) and obtain

$$\sup_{n \geq 1} (\kappa_p(n))^{1/n} \leq \mathrm{gr}(\mathscr{G})^{-1}$$

for every $p < p_c$. Finally, we remark that the left-hand side is a supremum of lower semicontinuous functions and is therefore lower semicontinuous itself, which implies (15.6.7) and thus finishes the proof. $\square$

Since nonamenable graphs can be thought of as being *infinite* dimensional, one would expect that all nonamenable graphs should display mean-field behavior. While no counterexamples exist, also this result is not known in general. We complete this section by describing the available results on critical exponents. For this, we need to introduce the notions of *planar* and *unimodular* graphs. A graph is called *planar* when it can be embedded into $\mathbb{R}^2$ with vertices being represented by points in $\mathbb{R}^2$ and edges by lines between the respective vertices such that the edges only intersect at their endpoints. For $x \in V$, let the *stabilizer* $S(x)$ of x be the set of automorphisms of $\mathscr{G}$ that keep x fixed, i.e., $S(x) = \{\gamma : \gamma(x) = x\}$. The graph $\mathscr{G}$ is called *unimodular* if $|\{\gamma(y) : \gamma \in S(x)\}| = |\{\gamma(x) : \gamma \in S(y)\}|$ for every $x, y \in V$.

Finally, the *number of ends* of a graph $\mathscr{G}$ is

$$\mathsf{E}(\mathscr{G}) = \sup_{S \subset V: |S| < \infty} \{\text{number of infinite connected components of } \mathscr{G} \setminus S\} . \qquad (15.6.9)$$

Then, Schonmann [238, 239] proves that percolation has mean-field critical exponents in the following cases:

Theorem 15.12 (Mean-field critical exponent on nonamenable graphs). *For percolation on a locally finite, connected, transitive, nonamenable graph $\mathscr{G}$, $\beta = \gamma = 1$, $\delta = \Delta = 2$ in the bounded-ratio sense, in the following cases:* clearpage

(a) *Unimodular graphs $\mathcal{G}$ for which $\mathsf{Ch}(\mathcal{G}) > (\sqrt{2r^2 - 1} - 1)/2$, where r is the degree of the graph*;
(b) *Graphs $\mathcal{G}$ that are planar and have one end*;
(c) *Graphs $\mathcal{G}$ that are unimodular and have infinitely many ends*.

The proof of case (b) in Thm. 15.12 above proceeds via an embedding into the hyperbolic plane $\mathbb{H}^2$. To this end, let $\mathcal{G}$ be an infinite, locally finite, connected transitive nonamenable planar single-ended graph. Then, Babai [24] proves that $\mathcal{G}$ can be embedded as a graph $\mathcal{G}'$ in the hyperbolic plane $\mathbb{H}^2$ in such a way that the group of automorphisms Γ on $\mathcal{G}'$ extends to an isometric action on $\mathbb{H}^2$. Moreover, the embedding can be chosen such that the edges of $\mathcal{G}'$ are hyperbolic line segments; such embeddings are called "nice embeddings." Schonmann proceeds by employing results about percolation on hyperbolic tessellations due to Lalley [204] and Benjamini-Schramm [41] in order to derive a mean-field criterion. The proof for graphs with infinitely many ends (case (c) above) makes use of the tree-like structure in this situation. It is worthwhile to note that Schonmann's second paper on the matter [239] derives specific mean-field criteria suitable for nonamenable graphs (they would simply fail for amenable graphs), which are apparently simpler to verify in the nonamenable situation.

Since percolation in high dimensions is known to have mean-field critical exponents, which coincide with the critical exponents on the tree (which is a key example of a nonamenable graph), one would expect that, in great generality, percolation on nonamenable graphs do so, too:

Open Problem 15.12 (Mean-field behavior of nonamenable graphs). *Show that the percolation phase transition on any transitive nonamenable graph has mean-field critical exponents.*

Chapter 16
Further Open Problems

In this chapter, we investigate some topics related to percolation where one would hope that lace-expansion ideas could prove useful, but they have not been fully explored. We hope that this text sparks new interest in these topics.

16.1 Invasion Percolation

Invasion percolation is a dynamical percolation model that displays self-organized critical behavior. In invasion percolation, each edge is equipped with an i.i.d. random weight $(U_b)_{b \in \mathcal{E}}$ uniformly distributed on $(0, 1)$. These variables play a similar role as the i.i.d. uniform random variables present in the Harris coupling of all percolation models for $p \in [0, 1]$. In order to construct the invasion percolation cluster (IPC), we start at time zero from $\mathcal{G}_0 = (\{0\}, \varnothing)$ and iteratively grow a random graph in discrete time. At time $t \in \mathbb{N}$, given the graph $\mathcal{G}_{t-1} = (\mathcal{V}_{t-1}, \mathcal{E}_{t-1})$, we consider all the edges in the edge boundary $\Delta \mathcal{V}_{t-1} = \{(x, y) : x \in \mathcal{V}_{t-1}, y \notin \mathcal{V}_{t-1}\}$ and pick the one with the smallest weight:

$$b_t := \arg\min_{b \in \Delta \mathcal{V}_{t-1}} U_b . \qquad (16.1.1)$$

Denoting $b_t = (x_t, y_t)$, we take $\mathcal{V}_t = \mathcal{V}_t \cup \{y_t\}$ and $\mathcal{E}_t = \mathcal{E}_t \cup \{b_t\}$. Continue indefinitely, and let IPC denote the limiting graph (which is an infinite tree embedded into $\mathbb{Z}^d$ since we do not allow for cycles).[1]

Invasion percolation is closely related to *critical bond percolation*. Indeed, color those bonds whose weight is at most p_c red. Once a red bond is invaded, all other red bonds in its p_c-cluster are invaded before the invasion process leaves the cluster. For $\mathbb{Z}^d$, where critical clusters appear on all scales, we expect larger and larger critical clusters to be invaded, so that the invasion process spends a large proportion of its time in large critical clusters. When

---

[1] In the literature, there are versions of this construction that do allow for cycles; this leads to a different invasion percolation cluster with the same set of vertices as our construction, but more edges.

© Springer International Publishing Switzerland 2017
M. Heydenreich and R. van der Hofstad, *Progress in High-Dimensional Percolation and Random Graphs*, CRM Short Courses, DOI 10.1007/978-3-319-62473-0_16

$\theta(p_c) = 0$, all critical clusters are finite, so that we have to leave the critical cluster again, having to accept an edge b_t with weight $U_{b_t} > p_c$ infinitely often.

A reflection of this is the fact that the number of bonds in IPC with weight above $p_c + \varepsilon$ is almost surely finite for all $\varepsilon > 0$, as proved for $\mathbb{Z}^d$ by Chayes, Chayes, and Newman [75] and extended to much more general graphs by Häggström, Peres, and Schonmann [182]:

Theorem 16.1 (Limsup of edge weights in invasion percolation [75]). *Fix $d \geq 1$ and recall that U_{b_t} is the edge weight of the t-th invaded edge. Then, a.s.,*

$$\limsup_{t \to \infty} U_{b_t} = p_c . \tag{16.1.2}$$

The fact that invasion percolation is driven by the critical parameter p_c, even though there is no parameter specification in its definition, makes invasion percolation a prime example of *self-organized criticality*. Intuitively, the above statement is quite obvious. There is a unique infinite component at $p > p_c$ for any p. As soon as we hit the infinite component for $p = p_c + \varepsilon$, we never leave it, so that $U_{e_t} \leq p_c + \varepsilon$ forever after that. Letting $\varepsilon \searrow 0$ shows the claim.

Interestingly, due to the close connection to critical percolation, the statement that $\theta(p_c) = 0$ is equivalent to the fact that

$$\limsup_{n \to \infty} \frac{|\text{IPC} \cap \Lambda_n|}{|\Lambda_n|} = 0 . \tag{16.1.3}$$

However, also establishing (16.1.3) has proved to be difficult. Alternatively, the statement $\theta(p_c) > 0$ is equivalent to the fact that $\sup_{s \geq t} U_{b_s} = p_c$ a.s. for all t sufficiently large.

Another reflection of the relation to critical percolation has been established by Járai [189], who shows for $\mathbb{Z}^2$ that the probability of an event E under the *incipient infinite cluster* IIC measure (recall Chap. 12) is identical to the probability of the translation of E to $x \in \mathbb{Z}^2$ under the IPC measure, conditionally on x being invaded and in the limit as $|x| \to \infty$. It is tempting to take this a step further and conjecture that the scaling limit of invasion percolation on $\mathbb{Z}^d$ when $d > 6$ is the canonical measure of super-Brownian motion conditioned to survive forever (see [152, Conjecture 6.1]). Indeed, such a result was proved by the second author, den Hollander and Slade [152, 156] for the IIC of spread-out oriented percolation on $\mathbb{Z}^d \times \mathbb{Z}_+$ when $d > 4$, and presumably it holds for the IIC of unoriented percolation on $\mathbb{Z}^d$ when $d > 6$ as well. However, in Angel, Goodman, den Hollander, and Slade [21], this conjecture was shown to be *false* on the tree: the IIC and IPC have different scaling limits. In fact, their scaling limits even turn out to be *mutually singular*. Hints of a discrepancy between the IIC and IPC had been noted earlier by Nickel and Wilkinson [223]. Remarkably, all critical exponents that we are aware of *do* agree for the IIC and IPC on the tree, yet the scaling limits differ.

In high dimensions, not much is known about invasion percolation. It appears quite plausible that, like for the IIC, the two-point function is asymptotic to $|x|^{-(d-4)}$:

Open Problem 16.1 (Invasion percolation two-point function). *Show that there is a constant A such that, for $d > 6$, as $|x| \to \infty$,*

$$\mathbb{P}(x \in \text{IPC}) = \frac{A}{|x|^{d-4}}(1 + o(1)) \,. \tag{16.1.4}$$

A solution to Open Problem 16.1 is also crucial for insight into the phase structure of the Edwards–Anderson model of a spin glass, as proved by Newman and Stein [219].

It is tempting to believe that also the IPC, in high dimensions, has a measure-valued diffusion scaling limit (recall Sect. 15.1). One would expect this limit to be closely related to the scaling limit of invasion percolation on the tree:

Open Problem 16.2 (Scaling limit of invasion percolation). *Identify the scaling limit of the invasion percolation cluster for percolation for $d > 6$. In more detail, let the measure $Z_{nt}^{(n)}$ be given by*

$$Z_{nt}^{(n)}(\cdot) = \frac{1}{n\gamma} \sum_{x \in \partial B^{(\text{IP})}(nt)} \mathbb{1}_{\{x/\sqrt{n} \in \cdot\}} \,, \tag{16.1.5}$$

where $\partial B^{(\text{IP})}(n)$ consists of those vertices in IPC at graph distance precisely n from the origin. Show that $\left(Z_{nt}^{(n)}\right)_{t \geq 0}$ converges to some limiting measure-valued process for $d > 6$. Show that this process is different from the scaling limit of the IIC in Open Problem 15.3 and relate it to the scaling limit of invasion percolation on the tree as identified in [21].

A related, possibly easier, problem is to show that a certain equivalence relation between IPC and IIC proven by Járai [189] for $d = 2$ is true in high dimensions as well. This relation states that the probability of the translation of an event E by $x \in \mathbb{Z}^d$ under the IPC measure, conditional on $x \in \text{IPC}$ and in the limit as $|x| \to \infty$, equals the IIC measure. While this is intuitively obvious, since all the edges invaded close to x have weight at most $p_c + o(1)$ by Thm. 16.1, the difficulty in dealing with invasion percolation is that it is slightly *supercritical*, and we do not know a lot about supercritical percolation (recall also Open Problem 11.1).

Minimal Spanning Trees. When growing the IPC on $\mathbb{Z}^d$, we do not get the entire graph since the vertices that are surrounded by high weights are never found. On a *finite* graph $\mathcal{G}$, this is an entirely different matter. Invasion percolation can be defined in exactly the same way as for an infinite graph, and we see that we construct a spanning tree of the graph $\mathcal{G}$, which is a cycle-free connected subgraph of $\mathcal{G}$ with the same vertex set. This tree is called the *minimal spanning tree* (MST). Given weights $(U_b)_{b \in \mathcal{E}(\mathcal{G})}$, the MST is the spanning tree that minimizes the sum of the weights of all the edges in the tree $\sum_{b \in \mathcal{T}} U_b$. The invasion percolation dynamics is *Prim's algorithm* for the construction of the minimal spanning tree.

The MST can be investigated on any finite graph and for any edge-weight distribution. In fact, whenever the edge-weight distribution does not have atoms, the distribution of the edges chosen in the MST is the same. Of course, for the actual *weight* of the MST, this is not the case. It turns out to be particularly convenient to assume that the edge weights $(U_b)_{b \in \mathcal{E}(\mathcal{G})}$ are i.i.d. exponential random variables with parameter 1, and we denote this setting by writing E_b instead of U_b for the edge weight of edge b. For the time being, we focus on the mean-field situation, where $\mathcal{G}$ is the complete graph.

Let us start by introducing some notation to formalize the notion of the MST. For a spanning tree $\mathcal{T}$, we let $W(\mathcal{T})$ denote its weight, i.e.,

$$W(\mathcal{T}) = \sum_{b \in \mathcal{T}} E_b \ . \tag{16.1.6}$$

Here we assume that $(E_b)_{b \in \mathcal{E}(\mathcal{G})}$ are i.i.d. exponential random variables with parameter 1. Then, the MST $\mathcal{T}_n$ is the minimizer of $W(\mathcal{T})$ over all spanning trees. When the edge weights are all different, this minimizer is unique. Since we assume that the edge-weight distribution E_b is *continuous*, the MST is a.s. unique.

We discuss two important results on the MST on the complete graph. We start with Frieze's result [113] showing that the weight of the MST is asymptotically equal to $\zeta(3)$, where $\zeta(s)$ is the zeta-function given by $\zeta(s) = \sum_{n \geq 1} n^{-s}$:

Theorem 16.2 (The weight of the MST [113]). *Let* $(E_b)_{b \in \mathcal{E}(\mathcal{G})}$ *be i.i.d. exponential random variables with parameter 1 and let* $\mathcal{T}_n$ *denote the MST on the complete graph* K_n, *i.e., the tree that minimizes* $\sum_{b \in \mathcal{T}} E_b$ *over all spanning trees of* K_n. *Then,*

$$\lim_{n \to \infty} \mathbb{E}[W(\mathcal{T}_n)] = \zeta(3) \ . \tag{16.1.7}$$

Proof. We follow Addario-Berry [1]. Rather than Prim's algorithm for computing the MST, we rely on *Kruskal's algorithm*. In Kruskal's algorithm, we grow the MST by going through the edges in increasing edge weight, adding each edge as long as it does not create a cycle. Thus, we grow a forest, and, after having added the $(n-1)$th edge, we have obtained a spanning tree. The fact that this is the MST is not hard, and a nice exercise:

Exercise 16.1 (Kruskal's algorithm). Show that the spanning tree obtained in Kruskal's algorithm equals the MST.

We write $N = \binom{n}{2}$ for the total number of edges in the complete graph, order the exponential edge weights as $(E_{(m)})_{m \in [N]}$ with $E_{(1)}$ being the smallest weight, and call this the *weight sequence*. Further, let $b_{(m)}$ denote the edge that corresponds to the edge weight $E_{(m)}$. The resulting *edge sequence* $(b_{(m)})_{m \in [N]}$ is a sequence of uniform draws without replacement from $[N]$. Since weights have been assigned in an i.i.d. manner, the resulting edge sequence is *independent* from the weight sequence $(E_{(m)})_{m \in [N]}$. As a result, for any $M \leq N$, $(b_{(m)})_{m \in [M]}$ is a uniform draw from the edges in the complete graph, so that $(b_{(m)})_{m \in [M]}$ is the Erdős–Rényi random graph with a fixed number of edges. By Kruskal's algorithm,

$$W(\mathcal{T}_n) = \sum_{m \in [N]} E_{(m)} \mathbb{1}_{\{b_{(m)} \text{ does not create a cycle}\}} \ . \tag{16.1.8}$$

We can now take the expectation and use the independence between edge sequence and weight sequence to obtain

$$\mathbb{E}[W(\mathcal{T}_n)] = \sum_{m \in [N]} \mathbb{E}[E_{(m)}] \, \mathbb{P}\left(b_{(m)} \text{ does not create a cycle}\right). \tag{16.1.9}$$

The problem has decoupled, and we can investigate each of the terms separately. By the memoryless property of the exponential distribution,

$$\mathbb{E}[E_{(m)}] = \sum_{s=1}^{m} \frac{1}{N-s+1}. \tag{16.1.10}$$

This was the easier part of the two. For the second term, let

$$\chi_n(\lambda) = \mathbb{E}_\lambda[|\mathcal{C}(1)|] \tag{16.1.11}$$

be the expected cluster size after adding $m = \lambda N/(n-1) = \lambda n/2$ edges, where we recall that $N = \binom{n}{2}$ denotes the total number of edges. This is the susceptibility of the Erdős–Rényi random graph when adding m edges, which we call the *combinatorial* Erdős–Rényi random graph model (in contrast to the earlier introduced *binomial* Erdős–Rényi random graph model, where edges are inserted independently with probability $p = \lambda/n$). The binomial model can be obtained from the combinatorial model by taking $m \sim \mathsf{Bin}(N, \lambda/n)$. Morally, they should be the same, since a $\mathsf{Bin}(N, \lambda/n)$ distribution is very close to its mean $\lambda(n-1)/2 \approx \lambda n/2 = m$ for n large.

It is not too hard to show that

$$\frac{\chi_n(\lambda)}{n} \to \theta(\lambda)^2, \tag{16.1.12}$$

where $\theta(\lambda)$ is the survival probability of a branching process with a Poisson offspring distribution with parameter λ. In particular, $\theta(\lambda) = 0$ when $\lambda \le 1$, while, for $\lambda > 1$, $\theta(\lambda)$ is the largest solution to

$$\theta(\lambda) = 1 - e^{-\lambda\theta(\lambda)}. \tag{16.1.13}$$

The proof of (16.1.12) is left as an exercise:

Exercise 16.2 (Susceptibility of Erdős–Rényi random graph). Consider the binomial Erdős–Rényi random graph with edge probability $p = \lambda/n$. Prove (16.1.12).

Then, the important fact is that, with $\lambda = nm/N$ and using (16.1.12) for the combinatorial model,

$$\mathbb{P}\left(b_{(m+1)} \text{ does not create a cycle}\right) = 1 - \frac{\chi_n(\lambda) - 1}{n - 1 - \lambda}. \tag{16.1.14}$$

To see (16.1.14), we condition on the cluster when we have added m edges. Let $\mathcal{F}_m$ denote the σ-field describing the first m choices of the edges. Then,

$$\mathbb{P}(b_{(m+1)} \text{ creates a cycle} \mid \mathcal{F}_m) = \sum_{i \geq 1} \frac{|\mathcal{C}_{(i)}|(|\mathcal{C}_{(i)}| - 1)}{2(N - m)} . \tag{16.1.15}$$

Indeed, we have already chosen m edges, so there are $N - m$ edges left. Conditionally on the structure of the clusters formed up to the addition of the mth edge, we create a cycle precisely when we choose both end points of the next uniform edge inside the *same* connected component. There are precisely $|\mathcal{C}_{(i)}|(|\mathcal{C}_{(i)}| - 1)/2$ different ways to choose an edge in the ith largest cluster. We complete the proof of (16.1.14) by noting that

$$\sum_{i \geq 1} |\mathcal{C}_{(i)}|(|\mathcal{C}_{(i)}| - 1) = \sum_{v \in [n]} |\mathcal{C}(v)| - n , \tag{16.1.16}$$

so we arrive at

$$\mathbb{P}(b_{(m+1)} \text{ does not create a cycle}) = 1 - \sum_{v \in [n]} \frac{\mathbb{E}[|\mathcal{C}(v)|] - n}{n(n - 1) - 2m}$$

$$= 1 - \frac{\chi_n(\lambda) - 1}{n - 1 - \lambda} . \tag{16.1.17}$$

Since the Erdős–Rényi random graph with $\lambda = (1 + \varepsilon) \log n$ is whp connected for any $\varepsilon > 0$ (see the original paper by Rényi [231] or [155, Sect. 5.3]), we can restrict to $\lambda = \Theta(\log n)$. Thus, we can approximate

$$\mathbb{P}(b_{(m+1)} \text{ does not create a cycle}) \approx 1 - \frac{\chi_n(\lambda)}{n} \approx 1 - \theta(\lambda)^2 . \tag{16.1.18}$$

We see that all the action happens when $\lambda \approx 1$. Further, when $\lambda \leq 1$, almost *all* edges are accepted, since then $\theta(\lambda) = 0$, while as soon as $\lambda > 1$, a larger and larger positive proportion of the edges are rejected. For λ fixed and $m = \lambda n/2$, we obtain that $\mathbb{E}[E_{(m)}] \approx m/N = 2m/(n(n-1)) = \lambda/(n-1) \approx \lambda/n$, so that, with the convention that $m = \lambda n/2$,

$$\mathbb{E}[W(\mathcal{T}_n)] \approx \sum_{m \in [N]} \frac{\lambda}{n}[1 - \theta(\lambda)^2] \approx \frac{1}{2} \int_0^\infty \lambda[1 - \theta(\lambda)^2] \, d\lambda . \tag{16.1.19}$$

We are left to show that the integral equals $2\zeta(3)$. This was first done by Aldous and Steele [18]. It is remarkable that even though we do not have a nice description of $\lambda \mapsto \theta(\lambda)$, we can still compute the integral $\int_0^\infty \lambda[1 - \theta(\lambda)^2] \, d\lambda$. Let us do this cute computation here.

For this, we first use partial integration to write

$$\int_0^\infty \lambda[1 - \theta(\lambda)^2] \, d\lambda = \int_0^\infty \lambda^2 \theta'(\lambda)\theta(\lambda) \, d\lambda = \int_1^\infty \lambda^2 \theta'(\lambda)\theta(\lambda) \, d\lambda , \tag{16.1.20}$$

since $\theta(\lambda) = 0$ for $\lambda \in [0, 1]$. Then, we use (16.1.13) to write

$$\lambda = -\frac{\log[1 - \theta(\lambda)]}{\theta(\lambda)}, \tag{16.1.21}$$

and rewrite the integral as

$$\begin{aligned}\int_0^\infty \lambda[1 - \theta(\lambda)^2]\,d\lambda &= \int_1^\infty \frac{(\log[1 - \theta(\lambda)])^2}{\theta(\lambda)^2}\theta'(\lambda)\theta(\lambda)\,d\lambda \\ &= \int_1^\infty \frac{(\log[1 - \theta(\lambda)])^2}{\theta(\lambda)}\theta'(\lambda)\,d\lambda \\ &= \int_0^1 \frac{(\log[1 - \theta])^2}{\theta}\,d\theta,\end{aligned} \tag{16.1.22}$$

using that $\theta: [1, \infty) \to [0, 1)$ is a bijection. Now using the final change of variables $u = \log(1 - \theta)$, for which $\theta = e^{-u}/(1 - e^{-u})$, we arrive at

$$\begin{aligned}\int_0^\infty \lambda[1 - \theta(\lambda)^2]\,d\lambda &= \int_0^\infty u^2\frac{e^{-u}}{1 - e^{-u}}\,du \\ &= \sum_{k=1}^\infty \int_0^\infty u^2 e^{-ku}\,du = \sum_{k=1}^\infty \frac{2}{k^3},\end{aligned} \tag{16.1.23}$$

since $\int_0^\infty u^2 e^{-ku}\,du = 2/k^3$. This completes the proof. $\square$

We continue to discuss the scaling of the MST as a tree. It turns out that, when properly rescaled, the MST converges toward a scaling limit, which is a so-called *real tree*. A real tree is a continuum object that has many of the properties that we know for finite trees. As we have already seen above, there are deep relations between the MST and Erdős–Rényi random graphs. This connection is particularly tight for the scaling limit, as shown in a sequence of papers by Addario-Berry, Broutin, and Goldschmidt [2, 3], the final work and 'pièce de résistance' being jointly with Miermont [4]. In turn, the latter paper is a follow-up on earlier work of Addario-Berry, Broutin, and Reed [5]. We focus on the results in [4]. We interpret a discrete tree $\mathcal{T}$ along with the graph metric on $\mathcal{T}$ as a metric space. Further, for a tree $\mathcal{T}$, we write $a\mathcal{T}$ for the tree where graph distances are rescaled by a factor a. In particular, this changes the unit length of an edge in $\mathcal{T}$ to length a in $a\mathcal{T}$. Rescaled discrete trees are real trees themselves. The main result proved by Addario-Berry, Broutin, Goldschmidt, and Miermont [4] is the following:

Theorem 16.3 (The scaling limit of the MST [4]). *Let $\mathcal{T}_n$ denote the MST on the complete graph K_n. Then, the real tree $n^{-1/3}\mathcal{T}_n$ converges in distribution to a nondegenerate limiting real tree $\mathcal{T}_\infty$. In particular, this implies that the diameter of MST is of order $n^{1/3}$, i.e.,*

$$n^{-1/3}\,\mathrm{diam}(\mathcal{T}_n) \xrightarrow{\ d\ } \mathrm{diam}(\mathcal{T}_\infty). \tag{16.1.24}$$

The topology is crucial for weak convergence. The weak convergence in Thm. 16.3 is weak convergence in the *Gromov–Hausdorff–Prokhorov* topology. This is a graph and measure theoretic notion that we do not explain in more detail. The proof of Thm. 16.3 makes strong use of the scaling limit of critical clusters as identified by Addario-Berry, Broutin, and Goldschmidt in [2, 3]. Loosely speaking, the shape of the scaling limit of the MST is determined within the scaling window of the Erdős–Rényi random graph. This is remarkable, since these critical clusters have size $n^{2/3}$, while the MST has size n. This discrepancy appears prominently in the fact that the Minkowski dimension of the scaling limit of large critical clusters is 2, while the Minkowski dimension of the scaling limit of the MST is 3 [4].

Since we believe that the scaling limit of critical percolation on high-dimensional tori is closely related to that of critical Erdős–Rényi random graphs, it is natural to believe that the minimal spanning trees on high-dimensional tori are closely related to that on the complete graph. This is the content of the following two open problems:

> **Open Problem 16.3 (Diameter of MST on high-dimensional tori).** *Show that the diameter of the MST on high-dimensional tori, as well as on the hypercube, is of order $V^{1/3}$, where V denotes the volume of the torus.*

> **Open Problem 16.4 (Scaling limit of MST on high-dimensional tori).** *Show that the scaling limit of the MST on high-dimensional tori, as well as on the hypercube, with distances rescaled by $V^{-1/3}$, converges in distribution to a multiple of the scaling limit on the complete graph as identified in Thm.* 16.3.

One difficulty arising in these open problems is that the MST is determined by slightly *supercritical* edge weights, and thus by slightly *supercritical* percolation. In high dimensions, we have a reasonably good control over the subcritical and critical regimes, but we lack control in the supercritical regime.

16.2   Random Walk Percolation and Interlacements

Throughout this text, we have covered bond percolation, where the bonds (or edges) of the lattice are either occupied or vacant. Another option is to put all the randomness on the vertices, the corresponding model is called *(Bernoulli) site percolation*. In fact, all the results proven for bond percolation so far carry over to site percolation; the only (yet remarkable) difference is that the critical value p_c changes when switching from bond to site percolation. The prenomial "Bernoulli" is put here to remind us of the independence of site (resp., bond) occupation probabilities, which is in contrast to another percolation model that we discuss next.

We are now concentrating on *random interlacements*, which—after being established by Sznitman in 2010 [251]—gave rise to a vast amount of interesting and surprising results. We only touch upon some of the new developments and refer for further discussions and details to the recent monographs by Drewitz, Ráth, and Sapozhnikov [92], Sznitman [252], and Černý and Teixera [72].

Throughout this section, we assume that $d \geq 3$, such that random walks on $\mathbb{Z}^d$ are transient. Random interlacement is a name of a random set $\mathcal{I} \subset \mathbb{Z}^d$ with corresponding law $\mathcal{P}^u$, $u \in (0, \infty)$ being a parameter of the model. It shares many properties with the (unique) infinite component $\mathcal{C}_\infty$ of supercritical site percolation with corresponding law $\mathbb{P}_p$, $p \in (p_c, 1)$: Both are infinite, connected, transient, and measure preserving and ergodic with respect to the spatial shift. Furthermore, both measures $\mathcal{P}^u$ and $\mathbb{P}_p$ satisfy positive association in the sense that the FKG inequality is satisfied.

There are several ways to characterize the law $\mathcal{P}^u$. Arguably the easiest way to characterize random interlacement is via

$$\mathcal{P}^u(\mathcal{I} \cap K) = e^{-\operatorname{cap}(K)} \qquad (16.2.1)$$

for any finite subset $K \subset \mathbb{Z}^d$, where $\operatorname{cap}(K)$ is the (random walk) capacity of K defined by

$$\operatorname{cap}(K) := \sum_{x \in K} \mathbb{P}(\text{simple random walk started in } x \text{ never returns to } K) .$$

Uniqueness of the measure $\mathcal{P}^u$ satisfying (16.2.1) is straightforward, because sets of the form $\{\mathcal{I} \cap K\}$ form a π-stable generator of the corresponding σ-algebra of events. Existence of $\mathcal{P}^u$ requires deeper insight and has been established by Sznitman [251].

Another, more constructive approach to random interlacement is via bi-infinite paths of random walk trajectories. Let W^* be the set of bi-infinite paths that intersect any finite set $K \subset \mathbb{Z}^d$ only finitely often. It is possible to construct a sigma-finite measure ν on W^* that assigns weight to the paths in W^* as if they were obtained as simple random walk trajectories "starting and ending at infinity." With this measure ν at hand, we can construct a Poisson point process on the product space $W^* \times (0, \infty)$ with intensity measure $\nu \otimes \lambda$ (where λ is the Lebesgue measure). The random set $\mathcal{I}^u$ (which we write for $\mathcal{I}$ under $\mathcal{P}^u$) is then obtained from a realization $\{(w_h^*, u_h) : h \in \mathbb{N}\}$ of this point process through

$$\mathcal{I}^u = \bigcup_{u_h \leq u} \operatorname{range}(w_h^*) . \qquad (16.2.2)$$

One advantage of this construction via Poisson processes is that it provides us with a coupling that immediately yields a monotonicity result in u: The law of $\mathcal{I}^u$ is stochastically dominated by the law of $\mathcal{I}^{\bar{u}}$ if $u < \bar{u}$. This construction of the random interlacement via a Poisson process of random walk trajectories inspired the naming "random walk percolation."

Yet another approach has been worked out by Sznitman [252], where $\mathcal{I}$ has been constructed via Poisson gases of Markovian loops, which are in turn connected to level sets of Gaussian free fields.

A highly remarkable feature of random interlacement is the polynomially decaying correlation function for *all* values of u. To this end, recall from (2.2.8) the definition of the random walk Green's function $C_\mu(x)$ with critical value $\mu = 1$, and in particular the polynomial bound (2.2.9). Denoting $\Psi_x = \mathbb{1}_{\{x \in \mathcal{J}\}}$, one can show [251] that

$$\text{Cov}_{\mathcal{P}^u}(\Psi_x, \Psi_y) \sim \frac{2u}{C_1(0)^2} C_1(y - x) e^{-2u/C_1(0)} \quad \text{as } |x - y| \to \infty . \tag{16.2.3}$$

This polynomial behavior is quite remarkable, because for many models of statistical mechanics (e.g., for percolation), polynomial decay of correlation is a clear sign of criticality. Here, we observe it for the *entire range* of the parameter u.

It might be tempting to suspect that the measures $\mathcal{P}^u$ and $\mathbb{P}_p$ are stochastically monotone with respect to each other, but this turns out to be false for *all* values of $u \in (0, \infty)$ and $p \in (p_c, 1)$, cf. [92, Sect. 2.3].

As remarked earlier, a random interlacement is concentrated on infinite components by construction. A more interesting behavior occurs when we consider the *vacant set of random interlacement* $\mathcal{V} := \mathbb{Z}^d \setminus \mathcal{J}$, for which a percolation phase transition *does* occur. Indeed, there exists $u^* = u^*(d) \in (0, \infty)$ such that

$$\mathcal{P}^u(\mathcal{V} \text{ has an infinite component}) = \begin{cases} 1 & \text{if } u < u^* , \\ 0 & \text{if } u > u^* . \end{cases} \tag{16.2.4}$$

The above should be compared to the fact that the probability that there exists an infinite component in Bernoulli percolation equals 1 for $p > p_c$, while it equals 0 if $p < p_c$. We can also define a random interlacement percolation function by considering the function $u \mapsto \mathcal{P}^u(0$ is in an infinite component of $\mathcal{V})$. Very little is known about the behavior of the vacant set in the critical setting, i.e., for $u = u^*$. This leads us to the following two open problems:

Open Problem 16.5 (Continuity of random walk percolation). *Prove that $u \mapsto \mathcal{P}^u(0$ is in the infinite component) is a continuous function. In particular, this would imply that $\mathcal{P}^{u^*}(\mathcal{V}$ has an infinite component) $= 0$.*

Open Problem 16.6 (Upper critical dimension random walk percolation). *Does random walk percolation have an upper critical dimension, and, if so, what is it? What is the scaling limit of a large finite connected component in the vacant set above the upper critical dimension?*

For the rest of this discussion of random interlacements, we focus on the behavior of random walk on it, as investigated by Sapozhnikov [236]:

Theorem 16.4 (Quenched heat kernel bounds for random walk on random interlacement [236]). *For random walk on a realization of the random interlacement $\mathcal{I}$ (for any parameter u), the return probability $\mathbf{p}_n(x) = P_0^{\mathcal{I}}(X(n) = x)$ satisfies the following bound $\mathcal{P}^u(\cdot \mid 0 \in \mathcal{I})$-almost surely:*

$$C_1 \, n^{-d/2} \, e^{-C_2 |x|^2/2} \leq \mathbf{p}_n(x) + \mathbf{p}_{n+1}(x) \leq C_3 \, n^{-d/2} \, e^{-C_4 |x|^2/2} \quad \text{if } n > |x|^{3/2}$$

for certain constants $C_1, C_2, C_3, C_4 > 0$.

This heat kernel bound is very much in flavor of Barlow's results on supercritical percolation clusters [25]. The arguments of Sapozhnikov are fairly robust and extend also to infinite components of the vacant set of random interlacement (assuming $u < u^*$) and level sets of the Gaussian free field. The author also proves parabolic Harnack inequalities in the style of Barlow and Hambly [26] for the correlated percolation models. It is expected that the heat kernel bounds in Thm. 16.4 hold in great generality for dependent percolation models with polynomially decaying correlations, for example, for percolation of the stationary measures of the voter model in dimension $d \geq 5$ as identified by Ráth and Valesin [230].

The quenched heat kernel bounds in Thm. 16.4 suggest that, just as for the infinite percolation cluster $\mathcal{C}_\infty$, a quenched invariance principle as in Thm. 14.1 holds. We leave this as an open problem:

> **Open Problem 16.7 (Scaling limit of random walk on random interlacement).**
> *Prove a quenched invariance principle for random walk on random interlacement or on an infinite component of the vacant set of random interlacement.*

16.3 Scale-Free Percolation

In this section, we define a percolation model that interpolates between long-range percolation, as defined in Sect. 15.4, and the scale-free rank-1 inhomogeneous random graphs as discussed in Sect. 13.5. This model, termed *scale-free percolation* in the work of Deijfen, the second author and Hooghiemstra [84], provides a percolation model in which the degree of a vertex can have finite mean but infinite variance. Mind that this phenomenon is impossible for independent percolation models, since the independence of the edge variables implies that the variance of the degrees is always bounded by their mean:

Exercise 16.3 (Variance of degree is bounded by its mean in percolation). Let D_x denote the degree of $x \in \mathbb{Z}^d$ in percolation models where the edge statuses are independent random variables. Show that $\mathrm{Var}(D_x) \leq \mathbb{E}[D_x]$.

Scale-free percolation is defined on the lattice $\mathbb{Z}^d$. Let each vertex $x \in \mathbb{Z}^d$ be equipped with an i.i.d. weight W_x. Conditionally on the weights $(W_x)_{x \in \mathbb{Z}^d}$, the edges in the graph are independent, and the probability that there is an edge between x and y is defined by

$$p_{xy} = 1 - e^{-\lambda W_x W_y / |x-y|^\alpha} , \tag{16.3.1}$$

for $\alpha, \lambda \in (0, \infty)$. We say that the edge $\{x, y\}$ is *occupied* with probability p_{xy} and *vacant* otherwise.

Let us discuss the role of the different parameters in the scale-free percolation model. The parameter $\alpha > 0$ describes the *long-range nature* of the model, while we think of $\lambda > 0$ as the *percolation parameter*. The *weight distribution* is the last parameter that describes the model. We are mainly interested in settings where the W_x have unbounded support in $[0, \infty)$, and then particularly when they vary substantially.

Naturally, the model for fixed $\lambda > 0$ and weights $(W_x)_{x \in \mathbb{Z}^d}$ is the same as the one for $\lambda = 1$ and weights $(\sqrt{\lambda} W_x)_{x \in \mathbb{Z}^d}$, so there is some redundancy in the parameters of the model. However, we view the weights $(W_x)_{x \in \mathbb{Z}^d}$ as creating a *random environment* in which we study the percolative properties of the model. Thus, we think of the random variables $(W_x)_{x \in \mathbb{Z}^d}$ as fixed once and for all and we change the percolation configuration by varying λ. We can thus view our model as percolation in a random environment given by the weights $(W_x)_{x \in \mathbb{Z}^d}$. The downside of scale-free percolation is that the edge statuses are no longer independent random variables, but are rather positively correlated:

Exercise 16.4 (Positive correlation between edge statuses in scale-free percolation). Show that, for scale-free percolation, and for all x, y, z distinct and $\lambda > 0$,

$$\mathbb{P}(\{x, y\} \text{ and } \{x, z\} \text{ occupied})$$
$$\geq \mathbb{P}(\{x, y\} \text{ occupied}) \, \mathbb{P}(\{x, z\} \text{ occupied}) , \tag{16.3.2}$$

the inequality being strict when $\mathbb{P}(W_0 = 0) < 1$. In other words, the edge statuses are positively correlated.

Scale-free percolation interpolates between long-range percolation and rank-1 inhomogeneous random graphs. Indeed, we retrieve long-range percolation when we take $W_x \equiv 1$. We retrieve the Norros–Reittu model in (13.5.21) with i.i.d. edge weights when we take $\alpha = 0$, $\lambda = 1/\sum_{i \in [n]} W_i$ and consider the model on $[n]$ instead of $\mathbb{Z}^d$. Thus, this model can be considered to be an interpolation between long-range percolation and rank-1 inhomogeneous random graphs.

Choice of Edge Weights. We assume that the distribution F_W of the weights $(W_x)_{x \in \mathbb{Z}^d}$ has a regularly-varying tail with exponent $\tau - 1$, that is, denoting by W a random variable with the same distribution as W_0 and by F_W its distribution function, we assume that

$$1 - F_W(w) = \mathbb{P}(W > w) = w^{-(\tau-1)} L(w) , \tag{16.3.3}$$

where $w \mapsto L(w)$ is a function that varies slowly at infinity. Here we recall that a function L varies slowly at infinity when, for every $x > 0$,

$$\lim_{t \to \infty} \frac{L(tx)}{L(x)} = 1. \tag{16.3.4}$$

Examples of slowly-varying functions are powers of logarithms. See the classical work by Bingham, Goldie, and Teugels [53] for more information about slowly-varying functions. We interpret $\tau > 1$ as the final parameter of our model, next to α, λ (and the dimension d). Of course, there may be many vertex-weight distributions having the asymptotics in (16.3.3) with the same τ, but the role of τ is so important in the sequel that we separate it out.

Write D_x for the degree of $x \in \mathbb{Z}^d$ and note that, by translation invariance, D_x has the same distribution as D_0. The name *scale-free* percolation is justified by the following theorem:

Theorem 16.5 (Power-law degrees for power-law weights [84]). *Fix $d \geq 1$.*

(a) *Assume that the weight distribution satisfies* (16.3.3) *with* $\alpha \leq d$ *or* $\gamma = \alpha(\tau - 1)/d \leq 1$. *Then* $\mathbb{P}(D_0 = \infty \mid W_0 > 0) = 1$.
(b) *Assume that the weight distribution satisfies* (16.3.3) *with* $\alpha > d$ *and* $\gamma = \alpha(\tau - 1)/d > 1$. *Then there exists* $s \mapsto \ell(s)$ *that is slowly varying at infinity such that*

$$\mathbb{P}(D_0 > s) = s^{-\gamma} \ell(s). \tag{16.3.5}$$

The fact that the degrees have a power-law distribution is why this model is called **scale-free percolation**. The parameter γ measures how many moments of the degree distribution are finite:

Exercise 16.5 (Degree moments in scale-free percolation [84]). Show that $\mathbb{E}[D_0^p] < \infty$ when $p < \gamma$ and $\mathbb{E}[D_0^p] = \infty$ when $p > \gamma$. In particular, the variance of the degrees is finite precisely when $\gamma > 2$.

We continue by studying the percolative properties of scale-free percolation. As before, we denote by λ_c the infimum of all $\lambda \geq 0$ with the property $\mathbb{P}(|\mathcal{C}(0)| = \infty) > 0$. It is a priori unclear whether $\lambda_c < \infty$ or not. Deijfen et al. [84, Thm. 3.1] prove that $\lambda_c < \infty$ holds in most cases. Indeed, if $\mathbb{P}(W = 0) < 1$, then $\lambda_c < \infty$ in all $d \geq 2$. Naturally, $d = 1$ again is special and the results in [84, Thm. 3.1] are not optimal. It is shown that if $\alpha \in (1, 2]$ and $\mathbb{P}(W = 0) < 1$, then $\lambda_c < \infty$ in $d = 1$, while if $\alpha > 2$ and $\tau > 1$ is such that $\gamma = \alpha(\tau - 1)/d > 2$, then $\lambda_c = \infty$ in $d = 1$.

More interesting is whether $\lambda_c = 0$ or not. The following theorem shows that this depends on whether the degrees have infinite variance or not:

Theorem 16.6 (Positivity of the critical value [84]). *Assume that the weight distribution satisfies* (16.3.3) *with* $\tau > 1$ *and that* $\alpha > d$.

(a) *Assume that* $\gamma = \alpha(\tau - 1)/d > 2$. *Then,* $\theta(\lambda) = 0$ *for small* $\lambda > 0$, *that is,* $\lambda_c > 0$.
(b) *Assume that* $\gamma = \alpha(\tau - 1)/d < 2$. *Then,* $\theta(\lambda) > 0$ *for every* $\lambda > 0$, *that is,* $\lambda_c = 0$.

In ordinary percolation, instantaneous percolation in the form $\lambda_c = 0$ can only occur when the degree of the graph is infinite. The *randomness* in the vertex weights facilitates instantaneous percolation in scale-free percolation. We see a similar phenomenon for rank-1 inhomogeneous random graphs, such as the Norros–Reittu model. The instantaneous percolation is related to *robustness* of the random network under consideration. Graph distances in scale-free percolation have been investigated in [84, 149] by identifying the number of edges between x and y as a function of $|x - y|$ for x, y in the infinite component. Again we see that graph distances are rather small if $\gamma \in (1, 2)$, whereas graph distances are much larger for $\gamma > 2$. There is some follow-up work on scale-free percolation. Hirsch [150] proposes a continuum model for scale-free percolation. Deprez, Hazra, and Wüthrich argue that scale-free percolation can be used to model real-life networks in [85]. Bringmann, Keusch, and Lengler [63, 64] study this model on a *torus* and in continuum space and coin the name *geometric inhomogeneous random graphs*. The first author, Hulshof, and Jorritsma [149] establish recurrence and transience criteria. Deprez et al. [85] show that when $\alpha \in (d, 2d)$, then the percolation function is continuous. For long-range percolation, this was proved by Berger [45] (see Sect. 15.4). However, in full generality, continuity of the percolation function as $\lambda = \lambda_c$ when $\lambda_c > 0$ is unknown. Also the existence of critical exponents is in general unknown:

Open Problem 16.8   (Critical behavior of scale-free percolation). *Investigate when the scale-free percolation function $\lambda \mapsto \theta(\lambda)$ is continuous. Identify the critical exponents for scale-free percolation. What is the upper-critical dimension of scale-free percolation and how does it depend on the parameters α and τ?*

We see that $\gamma \in (1, 2)$, where the variance of the degrees is infinite, is special in the sense that instantaneous percolation occurs as for rank-1 random graphs. This raises the questions to which extent the analogy extends. For example, in rank-1 random graphs, the scaling limits within the scaling window are different for random graphs having infinite *third* moments of the degrees than for those for which the third moment is finite (recall Sect. 13.5). This indicates that the critical behavior of scale-free percolation might be different for $\gamma \in (2, 3)$, where the degrees have infinite third moment, compared to $\gamma > 3$ where the degrees have finite third moment, particularly in high dimensions. Indeed, we can think of the Norros–Reittu as a kind of *mean-field model* for this setting, certainly when we restrict scale-free percolation to the torus. It would be of great interest to investigate these models in more detail. Can the lace expansion be applied to investigate the mean-field behavior of scale-free percolation?

16.4   FK Percolation

A key feature in the models considered earlier in this text is the independence of the edge occupation probabilities. FK percolation, also known as the *Random Cluster Model*, generalizes percolation to include a certain class of dependencies that give a highly fruitful connection to other models of statistical mechanics. The letters FK refer to the Dutch scientists Kees Fortuin and Piet Kasteleyn, who invented the model in the early 1970s [110].

The FK model has two parameters, $p \in [0, 1]$ (as in ordinary percolation) and a new parameter $q \in (0, \infty)$. We define the model on a finite graph $\mathcal{G} = (\mathcal{V}, \mathcal{E})$: for any $\omega \in \Omega = \{0, 1\}^{\mathcal{E}}$, we let

$$\mathbb{P}^{\mathrm{FK}}_{p,q}(\omega) = \frac{1}{Z^{\mathrm{FK}}_{p,q}} p^{|\{b \in \mathcal{E} : \omega(b)=1\}|}(1-p)^{|\{b \in \mathcal{E} : \omega(b)=0\}|} q^{k(\omega)}, \qquad (16.4.1)$$

where $k(\omega)$ is the number of connected components in ω, and $Z^{\mathrm{FK}}_{p,q}$ is a normalizing sum to make $\mathbb{P}^{\mathrm{FK}}_{p,q}$ a probability measure, i.e.,

$$Z^{\mathrm{FK}}_{p,q} = \sum_{\omega \in \Omega} p^{|\{b \in \mathcal{E} : \omega(b)=1\}|}(1-p)^{|\{b \in \mathcal{E} : \omega(b)=0\}|} q^{k(\omega)}. \qquad (16.4.2)$$

Mind that for $q = 1$, we observe that $Z^{\mathrm{FK}}_{p,1} = 1$, and $\mathbb{P}^{\mathrm{FK}}_{p,1}$ is the ordinary bond percolation measure on the finite graph $\mathcal{G}$. The parameter q tunes the number of clusters. For $q < 1$, it is favorable to have few clusters (and thus large connected components) compared to ordinary percolation. Indeed, in the extreme limit $q \searrow 0$, the measure $Z^{\mathrm{FK}}_{p,q}$ converges to the uniform spanning tree, uniform connected subgraph, or uniform spanning forest on $\mathcal{G}$ (depending on the behavior of p). For this and many other beautiful results about this model, we refer to the monograph by Grimmett [123].

It is a major challenge to develop a lace-expansion analysis for the FK model for $q \neq 1$, only the case $q = 2$ has been solved by Sakai [234]. If we review our lace-expansion analysis for $q = 1$ in the first sections, then it becomes clear that correlation inequalities, such as the Harris and BK inequality, play a central role in analyzing it, and we therefore focus on them first. One of the first results of Fortuin and Kasteleyn (together with Ginibre) concerns positive association of the FK model as long as $q \geq 1$:

Theorem 16.7 (FKG inequality [111]). *For $q \geq 1$ and any increasing events A, B,*

$$\mathbb{P}^{\mathrm{FK}}_{p,q}(A \cap B) \geq \mathbb{P}^{\mathrm{FK}}_{p,q}(A)\,\mathbb{P}^{\mathrm{FK}}_{p,q}(B).$$

The FK model is thus positively correlated for $q \geq 1$, and this might be used to derive strong influence bounds, cf. [123, Chap. 2]. Theorem 16.7 generalizes (1.3.1). We cannot expect a similar result to hold for $q < 1$.

A bigger challenge is to prove some sort of negative association, which supposedly holds for $q \leq 1$. The weakest possible form of negative association is *edge-negative-association*, that is,

$$\forall a, b \in \mathcal{E} : \quad \mathbb{P}(\omega(a) = 1, \omega(b) = 1) \leq \mathbb{P}(\omega(a) = 1)\,\mathbb{P}(\omega(b) = 1). \qquad (16.4.3)$$

A stronger form of negative association is the disjoint-occurrence property (1.3.4) for *increasing* events (corresponding to "BK" for percolation), and the strongest form is the disjoint-occurrence property (1.3.4) for *general* events (corresponding to the BKR inequality proved by Reimer [226] for percolation).

Open Problem 16.9 (Negative association for FK percolation). *Prove any form of negative association for the FK model for $q < 1$.*

See [123, Sect. 3.9] for a more complete discussion and references to results in the ("extreme") cases of uniform spanning trees and forests.

Connection to Ising and Potts model. Highly notable is the connection between FK percolation and the Potts–Ising model arising for integer values of q. For $q \in \mathbb{N}, q \geq 2$, the q-Potts model on $\mathcal{G} = (\mathcal{V}, \mathcal{E})$ with parameter $\beta > 0$ is given by

$$\mathbb{P}^{\text{Potts}}_{q,\beta}(\sigma) = \frac{1}{Z^{\text{Potts}}_{q,\beta}} \exp \left\{ \sum_{b=\{x,y\} \in \mathcal{E}} \beta \mathbb{1}_{\{\sigma(x)=\sigma(y)\}} \right\}, \quad \sigma \in \Sigma = \{1, \ldots, q\}^{\mathcal{V}}, \quad (16.4.4)$$

where the partition sum $Z^{\text{Potts}}_{q,\beta}$ acts as the normalizing sum. For $q = 2$, this is called the Ising model and classically write $\Sigma = \{-1, +1\}^{\mathcal{V}}$ instead of $\Sigma = \{1, 2\}^{\mathcal{V}}$, which leaves the measure unchanged.

The equivalence between the FK model with parameters (p, q) and the q-Potts model with $\beta = -\log(1 - p)$ (Ising model for $q = 2$) is given by the Edwards–Sokal coupling [97]. Interestingly, the FK model is defined on the *edges* of the graph, whereas the Potts and Ising models are defined on the *vertices*. Indeed, the Edwards–Sokal coupling is at the basis of many breathtaking recent developments for the two-dimensional Ising model, for example, Smirnov's identification of fermionic observables in the Ising model that are conformally invariant in the scaling limit [248].

We pursue a different route here and focus on rigorous results for the high-dimensional Ising model on $\mathbb{Z}^d$. First of all, we consider the Ising model on the infinite lattice, which we achieve as the limiting measure of the Ising model on a sequence of growing boxes. Such *thermodynamic limits* require sufficient care, but are completely standard in the literature. Similarly to percolation, where the triangle condition (4.1.1) implies the existence and mean-field values of numerous critical exponents, we have the *bubble condition* for the Ising model. To this end, denote by β_c the *critical* value for β in (16.4.4) defined by

$$\beta_c = \sup \left\{ \beta > 0 : \sum_x \text{Cov}^{\text{Potts}}_{q,\beta}(\sigma(0), \sigma(x)) < \infty \right\}. \quad (16.4.5)$$

Note the similarity between β_c in (16.4.5) and p_T in (1.1.4). Indeed, the covariance of $\sigma(0)$ and $\sigma(x)$ plays the same role in the Ising model as the two-point function $\tau_p(x)$ in percolation; this correspondence is made explicit through the earlier mentioned Edwards-Sokal coupling. Furthermore, for the Ising model (i.e., $q = 2$) the phase transition is *sharp* in a sense similar to Thm. 3.1, as proven by Aizenman, Barsky, and Fernández [10]. Duminil-Copin and Tassion [93] give a highly simplified proof of this result using arguments like the ones presented in Sect. 3.2. The *bubble condition* for the Ising model is the condition that

$$\sum_{x \in \mathbb{Z}^d} \mathrm{Cov}^{\mathrm{Potts}}_{2,\beta_c} (\sigma(0), \sigma(x))^2 < \infty , \qquad (16.4.6)$$

and this is sufficient for various critical exponents to exist and to take on their mean-field values as shown by Aizenman [7] and Aizenman and Fernandez [11].

In a seminal paper, Fröhlich, Simon, and Spencer [114] use the technical condition of *reflection positivity* to derive an infrared bound for the Ising model for $d > 4$, which in turn implies (16.4.6). Reflection positivity makes clever use of certain symmetries of the Ising model and holds for nearest-neighbor models (which we are considering here) as well as various long-range versions, such as Kac and power-law interactions. On the downside, it immediately breaks down for small perturbations of a model satisfying reflection positivity, and thus is not a very robust or universal property. Sakai [234] uses the lace expansion to give an alternative proof of the infrared bound for the Ising model in high dimension. His proof is more robust in the sense that it is fairly flexible in the interaction, and he achieves highly accurate error estimates. However, in the nearest-neighbor case, it requires the dimension to be large (which can be compensated by considering finite spread-out models, as usual, for which Sakai's results apply to any $d > 4$). Here is Sakai's result:

Theorem 16.8 (Infrared bound for Ising model [234]). *There is $d_0 > 4$ such that for all $d \geq d_0$ there exists a constant $A_2 = A_2(d) > 0$ such that for $\varepsilon > 0$ and all $x \neq 0$,*

$$\mathrm{Cov}^{\mathrm{Potts}}_{2,\beta_c} (\sigma(0), \sigma(x)) = \frac{A_2}{|x|^{d-2}} \left(1 + O(|x|^{-(2(d-4)-\varepsilon \wedge 2)/d})\right) ,$$

where the constant in the O-notation may depend on ε. Consequently, the bubble condition (16.4.6) holds for $d \geq d_0$ and the critical exponents take on their mean-field values.

It appears rather surprising that Sakai successfully derived a lace expansion despite absence of any form of negative association (BK and alike), although this was a crucial ingredient in Chap. 6. Indeed, Sakai uses the random current representation for the Ising model, where one works with a series expansion for the exponentials occurring in (16.4.4), and bounds the lace-expansion coefficients using a so-called *source-switching lemma* due to Griffiths, Hurst, and Sherman [121]. However, this "trick" uses special features of the Ising model (thus $q = 2$ in FK percolation) and is not suited for generalization to other values of q.

Infinite Volume FK Models. We return to FK percolation for general values of q on the infinite lattice $\mathbb{Z}^d$, which we obtain again as a *thermodynamic limit* of FK measures on a growing sequence of finite boxes. Unlike in percolation, where several proofs of a *sharp phase transition* in Thm. 3.1 are known (we have presented two different proofs in Chap. 3), this is an open problem for FK percolation:

Open Problem 16.10 (Sharp phase transition for FK percolation). *For the FK model on $\mathbb{Z}^d$, show the equality*

$$\sup\left\{p : \sum_{x\in\mathbb{Z}^d} \mathbb{P}^{\mathrm{FK}}_{p,q}(0 \leftrightarrow x) < \infty\right\} = \inf\left\{p : \mathbb{P}^{\mathrm{FK}}_{p,q}(0 \leftrightarrow \infty) > 0\right\}$$

for $q > 0$, thereby defining the critical value $p_c(q)$.

We finally end this text with a *big* open problem, that we find highly relevant even though we have no ideas how to tackle it:

Open Problem 16.11 (Lace expansion for FK percolation). *Derive a lace expansion for FK percolation with $q \notin \{1, 2\}$ and use it to derive an infrared bound for the Fourier transform of*

$$x \mapsto \mathrm{Cov}^{\mathrm{Potts}}_{q,\beta_c}\big(\sigma(0), \sigma(x)\big) \,.$$

Bibliography

1. Addario-Berry, L.: Partition functions of discrete coalescents: From Cayley's formula to Frieze's $\zeta(3)$ limit theorem. In: Mena, R.H., Pardo, J.C., Rivero, V., Uribe Bravo, G. (eds.) XI Symposium on Probability and Stochastic Processes (Guanajuato, 2013, Progress in Probability, vol. 68, pp. 1–45. Birkhäuser, Cham (2015). http://doi.org/10.1007/978-3-319-13984-5_1
2. Addario-Berry, L., Broutin, N., Goldschmidt, C.: Critical random graphs: Limiting constructions and distributional properties. Electron. J. Probab. **15**(25), 741–775 (2010). http://doi.org/10.1214/EJP.v15-772
3. Addario-Berry, L., Broutin, N., Goldschmidt, C.: The continuum limit of critical random graphs. Probab. Theory Relat. Fields **152**(3–4), 367–406 (2012). http://doi.org/10.1007/s00440-010-0325-4
4. Addario-Berry, L., Broutin, N., Goldschmidt, C., Miermont, G.: The scaling limit of the minimum spanning tree of the complete graph. Ann. Probab. **45**(5), 3075–3144 (2017)
5. Addario-Berry, L., Broutin, N., Reed, B.A.: The diameter of the minimum spanning tree of a complete graph. In: Fourth Colloquium on Mathematics and Computer Science Algorithms, Trees, Combinatorics and Probabilities (Nancy, 2006, Discrete Math. Theor. Comput. Sci. Proc., vol. AG, pp. 237–248). Assoc. Discrete Math. Theor. Comput. Sci., Nancy (2006)
6. Adler, J., Meir, Y., Aharony, A., Harris, A.B.: Series study of percolation moments in general dimension. Phys. Rev. B (3) **41**(13), 9183–9206 (1990). http://doi.org/10.1103/PhysRevB.41.9183
7. Aizenman, M.: Geometric analysis of φ^4 fields and Ising models. Parts I and II. Comm. Math. Phys. **86**(1), 1–48 (1982)
8. Aizenman, M.: On the number of incipient spanning clusters. Nuclear Phys. B **485**(3), 551–582 (1997). http://doi.org/10.1016/S0550-3213(96)00626-8
9. Aizenman, M., Barsky, D.J.: Sharpness of the phase transition in percolation models. Comm. Math. Phys. **108**(3), 489–526 (1987)
10. Aizenman, M., Barsky, D.J., Fernández, R.: The phase transition in a general class of Ising-type models is sharp. J. Statist. Phys. **47**(3–4), 343–374 (1987). http://doi.org/10.1007/BF01007515
11. Aizenman, M., Fernández, R.: On the critical behaviour of the magnetization in high-dimensional Ising models. J. Statist. Phys. **44**(3–4), 393–454 (1986). http://doi.org/10.1007/BF01011304
12. Aizenman, M., Kesten, H., Newman, C.M.: Uniqueness of the infinite cluster and continuity of connectivity functions for short and long range percolation. Comm. Math. Phys. **111**(4), 505–531 (1987)
13. Aizenman, M., Newman, C.M.: Tree graph inequalities and critical behavior in percolation models. J. Statist. Phys. **36**(1–2), 107–143 (1984). http://doi.org/10.1007/BF01015729
14. Aizenman, M., Newman, C.M.: Discontinuity of the percolation density in one-dimensional $1/|x-y|^2$ percolation models. Comm. Math. Phys. **107**(4), 611–647 (1986)
15. Ajtai, M., Komlós, J., Szemerédi, E.: Largest random component of a k-cube. Combinatorica **2**(1), 1–7 (1982). http://doi.org/10.1007/BF02579276
16. Aldous, D.: Minimization algorithms and random walk on the d-cube. Ann. Probab. **11**(2), 403–413 (1983)
17. Aldous, D.: Brownian excursions, critical random graphs and the multiplicative coalescent. Ann. Probab. **25**(2), 812–854 (1997). http://doi.org/10.1214/aop/1024404421

© Springer International Publishing Switzerland 2017
M. Heydenreich and R. van der Hofstad, *Progress in High-Dimensional Percolation and Random Graphs*, CRM Short Courses, DOI 10.1007/978-3-319-62473-0

18. Aldous, D., Steele, J.M.: The objective method: probabilistic combinatorial optimization and local weak convergence. In: Kesten, H. (ed.) Probability on Discrete Structures, Encyclopaedia of Mathematical Science, vol. 110, pp. 1–72. Springer, Berlin (2004). http://doi.org/10.1007/978-3-662-09444-0_1

19. Alon, N., Benjamini, I., Stacey, A.: Percolation on finite graphs and isoperimetric inequalities. Ann. Probab. **32**(3A), 1727–1745 (2004). http://doi.org/10.1214/009117904000000414

20. Alon, N., Spencer, J.H.: The Probabilistic Method, 2nd edn. Wiley-Interscience Series in Discrete Mathematics and Optimization. Wiley-Interscience, New York (2000). http://doi.org/10.1002/0471722154

21. Angel, O., Goodman, J., den Hollander, F., Slade, G.: Invasion percolation on regular trees. Ann. Probab. **36**(2), 420–466 (2008). http://doi.org/10.1214/07-AOP346

22. Antunović, T., Veselić, I.: Sharpness of the phase transition and exponential decay of the subcritical cluster size for percolation and quasi-transitive graphs. J. Stat. Phys. **130**(5), 983–1009 (2008). http://doi.org/10.1007/s10955-007-9459-x

23. Athreya, K.B., Ney, P.E.: Branching Processes, Grundlehren der Mathematischen Wissenschaften, vol. 196. Springer, New York (1972). http://doi.org/10.1007/978-3-642-65371-1

24. Babai, L.: The growth rate of vertex-transitive planar graphs. In: Proceedings of the Eight Annual ACM-SIAM Symposium on Discrete Algorithms, New Orleans. LA, 1997, pp. 564–573. ACM, New York (1997)

25. Barlow, M.T.: Random walks on supercritical percolation clusters. Ann. Probab. **32**(4), 3024–3084 (2004). http://doi.org/10.1214/009117904000000748

26. Barlow, M.T., Hambly, B.M.: Parabolic Harnack inequality and local limit theorem for percolation clusters. Electron. J. Probab. **14**(1), 1–26 (2009)

27. Barlow, M.T., Járai, A.A., Kumagai, T., Slade, G.: Random walk on the incipient infinite cluster for oriented percolation in high dimensions. Comm. Math. Phys. **278**(2), 385–431 (2008). http://doi.org/10.1007/s00220-007-0410-4

28. Barsky, D.J., Aizenman, M.: Percolation critical exponents under the triangle condition. Ann. Probab. **19**(4), 1520–1536 (1991)

29. Barsky, D.J., Grimmett, G.R., Newman, C.M.: Dynamic renormalization and continuity of the percolation transition in orthants. In: Alexander, K.S., Watkins, J.C. (eds.) Spatial Stochastic Processes, Progress in Probability, vol. 19, pp. 37–55. Birkhäuser, Boston (1991)

30. Barsky, D.J., Grimmett, G.R., Newman, C.M.: Percolation in half-spaces: equality of critical densities and continuity of the percolation probability. Probab. Theory Relat. Fields **90**(1), 111–148 (1991). http://doi.org/10.1007/BF01321136

31. Bauerschmidt, R., Brydges, D.C., Slade, G.: Critical two-point function of the 4-dimensional weakly self-avoiding walk. Comm. Math. Phys. **338**(1), 169–193 (2015). http://doi.org/10.1007/s00220-015-2353-5

32. Bauerschmidt, R., Brydges, D.C., Slade, G.: Logarithmic correction for the susceptibility of the 4-dimensional weakly self-avoiding walk: a renormalisation group analysis. Comm. Math. Phys. **337**(2), 817–877 (2015). http://doi.org/10.1007/s00220-015-2352-6

33. Beffara, V., Nolin, P.: On monochromatic arm exponents for 2D critical percolation. Ann. Probab. **39**(4), 1286–1304 (2011). http://doi.org/10.1214/10-AOP581

34. Ben Arous, G., Cabezas, M., Fribergh, A.: Scaling limit for the ant in a simple labyrinth. arXiv:1609.03980

35. Ben Arous, G., Cabezas, M., Fribergh, A.: Scaling limit for the ant in high-dimensional labyrinths. arXiv:1609.03977

36. Ben-Hamou, A., Salez, J.: Cutoff for non-backtracking random walks on sparse random graphs. Ann. Probab. **45**(3), 1752–1770 (2017)

37. Benjamini, I., Kozma, G., Wormald, N.: The mixing time of the giant component of a random graph. Random Struct. Algorithms **45**(3), 383–407 (2014). http://doi.org/10.1002/rsa.20539

38. Benjamini, I., Lyons, R., Peres, Y., Schramm, O.: Critical percolation on any nonamenable group has no infinite clusters. Ann. Probab. **27**(3), 1347–1356 (1999). http://doi.org/10.1214/aop/1022677450

39. Benjamini, I., Lyons, R., Peres, Y., Schramm, O.: Group-invariant percolation on graphs. Geom. Funct. Anal. **9**(1), 29–66 (1999). http://doi.org/10.1007/s000390050080

40. Benjamini, I., Schramm, O.: Percolation beyond $\mathbb{Z}^d$, many questions and a few answers. Electron. Comm. Probab. **1**(8), 71–82 (1996). http://doi.org/10.1214/ECP.v1-978

41. Benjamini, I., Schramm, O.: Percolation in the hyperbolic plane. J. Am. Math. Soc. **14**(2), 487–507 (2001). http://doi.org/10.1090/S0894-0347-00-00362-3

42. Berestycki, N., Lubetzky, E., Peres, Y., Sly, A.: Random walks on the random graph. arXiv:1504.01999

43. van den Berg, J., Keane, M.S.: On the continuity of the percolation probability function. In: Beals, R., Beck, A., Bellow, A., Hajian, A. (eds.) Conference in Modern Analysis and Probability (New Haven, CO, 1982), Contemp. Math., vol. 26, pp. 61–65. Amer. Math. Soc., Providence, RI (1984). http://doi.org/10.1090/conm/026/737388

44. van den Berg, J., Kesten, H.: Inequalities with applications to percolation and reliability. J. Appl. Prob. **22**, 556–569 (1985)

45. Berger, N.: Transience, recurrence and critical behavior for long-range percolation. Comm. Math. Phys. **226**(3), 531–558 (2002). http://doi.org/10.1007/s002200200617

46. Berger, N., Biskup, M.: Quenched invariance principle for simple random walk on percolation clusters. Probab. Theory Relat. Fields **137**(1–2), 83–120 (2007). http://doi.org/10.1007/s00440-006-0498-z

47. Bezuidenhout, C., Grimmett, G.R.: The critical contact process dies out. Ann. Probab. **18**(4), 1462–1482 (1990)

48. Bhamidi, S., Broutin, N., Sen, S., Wang, X.: Scaling limits of random graph models at criticality: universality and the basin of attraction of the Erdős–Rényi random graph. arXiv:1411.3417

49. Bhamidi, S., Budhiraja, A., Wang, X.: The augmented multiplicative coalescent, bounded size rules and critical dynamics of random graphs. Probab. Theory Relat. Fields **160**(3–4), 733–796 (2014). http://doi.org/10.1007/s00440-013-0540-x

50. Bhamidi, S., van der Hofstad, R., van Leeuwaarden, J.S.H.: Scaling limits for critical inhomogeneous random graphs with finite third moments. Electron. J. Probab. **15**(54), 1682–1703 (2010). http://doi.org/10.1214/EJP.v15-817

51. Bhamidi, S., van der Hofstad, R., van Leeuwaarden, J.S.H.: Novel scaling limits for critical inhomogeneous random graphs. Ann. Probab. **40**(6), 2299–2361 (2012). http://doi.org/10.1214/11-AOP680

52. Bhamidi, S., Sen, S., Wang, X.: Continuum limit of critical inhomogeneous random graphs. Probab. Theory Relat. Fields **169**(1–2), 565–641 (2017)

53. Bingham, N.H., Goldie, C.M., Teugels, J.L.: Regular Variation, Encyclopedia of Mathematics and its Applications, vol. 27. Cambridge University Press, Cambridge (1989)

54. Bollobás, B.: The evolution of random graphs. Trans. Amer. Math. Soc. **286**(1), 257–274 (1984). http://doi.org/10.2307/1999405

55. Bollobás, B.: Random Graphs, Cambridge Studies in Advanced Mathematics, vol. 73, 2nd edn. Cambridge University Press, Cambridge (2001). http://doi.org/10.1017/CBO9780511814068

56. Bollobás, B., Kohayakawa, Y.: Percolation in high dimensions. Eur. J. Combinatorics **15**, 113–125 (1994). http://doi.org/10.1006/eujc.1994.1014

57. Bollobás, B., Kohayakawa, Y., Łuczak, T.: The evolution of random subgraphs of the cube. Random Struct. Algorithms **3**(1), 55–90 (1992). http://doi.org/10.1002/rsa.3240030106

58. Bollobás, B., Riordan, O.: Percolation. Cambridge University Press, New York (2006). http://doi.org/10.1017/CBO9781139167383

59. Borgs, C., Chayes, J.T., van der Hofstad, R., Slade, G., Spencer, J.H.: Random subgraphs of finite graphs. I. The scaling window under the triangle condition. Random Struct. Algorithms **27**(2), 137–184 (2005). http://doi.org/10.1002/rsa.20051

60. Borgs, C., Chayes, J.T., van der Hofstad, R., Slade, G., Spencer, J.H.: Random subgraphs of finite graphs. II. The lace expansion and the triangle condition. Ann. Probab. **33**(5), 1886–1944 (2005). http://doi.org/10.1214/009117905000000260

61. Borgs, C., Chayes, J.T., van der Hofstad, R., Slade, G., Spencer, J.H.: Random subgraphs of finite graphs. III. The phase transition for the n-cube. Combinatorica **26**(4), 395–410 (2006). http://doi.org/10.1007/s00493-006-0022-1

62. Borgs, C., Chayes, J.T., Randall, D.J.: The van den Berg–Kesten–Reimer inequality: a review. In: Bramson, M., Durrett, R. (eds.) Perplexing Problems in Probability, Progr. Probab., vol. 44, pp. 159–173. Birkhäuser, Boston, MA (1999)

63. Bringmann, K., Keusch, R., Lengler, J.: Average distance in a general class of scale-free networks with underlying geometry. arXiv:1602.05712

64. Bringmann, K., Keusch, R., Lengler, J.: Geometric inhomogeneous random graphs. arXiv:1511.00576

65. Broadbent, S.R., Hammersley, J.M.: Percolation processes. I. Crystals and mazes. Proc. Camb. Philos. Soc. **53**, 629–641 (1957)
66. Brydges, D.C., Spencer, T.: Self-avoiding walk in 5 or more dimensions. Comm. Math. Phys. **97**(1–2), 125–148 (1985)
67. Burton, R.M., Keane, M.S.: Density and uniqueness in percolation. Comm. Math. Phys. **121**(3), 501–505 (1989)
68. Cames van Batenburg, W.P.S.: The dimension of the incipient infinite cluster. Electron. Commun. Probab. **20**: 33, 10 (2015). http://doi.org/10.1214/ECP.v20-3570
69. Cerf, R.: Large deviations for three dimensional supercritical percolation. Astérisque **267** (2000)
70. Cerf, R.: Large deviations of the finite cluster shape for two-dimensional percolation in the Hausdorff and l^1 metric. J. Theoret. Probab. **13**(2), 491–517 (2000). http://doi.org/10.1023/A:1007841407417
71. Cerf, R.: The Wulff Crystal in Ising and Percolation Models. Lecture Notes in Math, vol. 1878. Springer, Berlin (2006)
72. Černý, J., Teixeira, A.Q.: From random walk trajectories to random interlacements, Ensaios Mat, vol. 23. Soc. Brasil. Mat, Rio de Janeiro (2012)
73. Chayes, J.T., Chayes, L.: The mean field bound for the order parameter of Bernoulli percolation. In: Kersten, H. (ed.) Percolation Theory and Ergodic Theory of Infinite Particle Systems (Minneapolis, MN, 1984–1985), IMA Vol. Math. Appl., vol. 8, pp. 49–71. Springer, New York (1987)
74. Chayes, J.T., Chayes, L.: On the upper critical dimension of Bernoulli percolation. Comm. Math. Phys. **113**(1), 27–48 (1987)
75. Chayes, J.T., Chayes, L., Newman, C.M.: The stochastic geometry of invasion percolation. Comm. Math. Phys. **101**(3), 383–407 (1985)
76. Chen, L.C., Sakai, A.: Critical behavior and the limit distribution for long-range oriented percolation. I. Probab. Theory Relat. Fields **142**(1–2), 151–188 (2008). http://doi.org/10.1007/s00440-007-0101-2
77. Chen, L.C., Sakai, A.: Critical two-point functions for long-range statistical-mechanical models in high dimensions. Ann. Probab. **43**(2), 639–681 (2015). http://doi.org/10.1214/13-AOP843
78. Clisby, N., Liang, R., Slade, G.: Self-avoiding walk enumeration via the lace expansion. J. Phys. A **40**(36), 10973–11017 (2007). http://doi.org/10.1088/1751-8113/40/36/003
79. Croydon, D.A.: Scaling limits of stochastic processes associated with resistance forms. arXiv:1609.05666
80. Croydon, D.A.: Hausdorff measure of arcs and Brownian motion on Brownian spatial trees. Ann. Probab. **37**(3), 946–978 (2009). http://doi.org/10.1214/08-AOP425
81. Damron, M.: Recent work on chemical distance in critical percolation. arXiv:1602.00775
82. Dawson, D.A.: Measure-valued Markov processes. In: P.L. Hennequin (ed.) École d'Été de Probabilités de Saint-Flour XXI—1991, Lecture Notes in Mathematics, vol. 1541, pp. 1–260. Springer, Berlin (1993). http://doi.org/10.1007/BFb0084190
83. De Masi, A., Ferrari, P.A., Goldstein, S., Wick, W.D.: An invariance principle for reversible Markov processes. Applications to random motions in random environments. J. Statist. Phys. **55**(3–4), 787–855 (1989). http://doi.org/10.1007/BF01041608
84. Deijfen, M., van der Hofstad, R., Hooghiemstra, G.: Scale-free percolation. Ann. Inst. Henri Poincaré Probab. Stat. **49**(3), 817–838 (2013). http://doi.org/10.1214/12-AIHP480
85. Deprez, P., Hazra, R.S., Wüthrich, M.V.: Inhomogeneous long-range percolation for real-life network modeling. Risks **3**(1), 1–23 (2015). http://doi.org/10.3390/risks3010001
86. Derbez, E., Slade, G.: Lattice trees and super-Brownian motion. Canad. Math. Bull. **40**(1), 19–38 (1997). http://doi.org/10.4153/CMB-1997-003-8
87. Derbez, E., Slade, G.: The scaling limit of lattice trees in high dimensions. Comm. Math. Phys. **193**(1), 69–104 (1998). http://doi.org/10.1007/s002200050319
88. Dhara, S., van der Hofstad, R., Leeuwaarden, J.S.H., Sen, S.: Critical window for the configuration model: finite third moment degrees. Electron. J. Probab. **22**(16), 33 (2017)
89. Dhara, S., van der Hofstad, R., Leeuwaarden, J.S.H., Sen, S.: Heavy-tailed configuration models at criticality. arXiv:1612.00650
90. Dodziuk, J.: Difference equations, isoperimetric inequality and transience of certain random walks. Trans. Amer. Math. Soc. **284**(2), 787–794 (1984). http://doi.org/10.2307/1999107

91. Doyle, P.G., Snell, J.L.: Random Walks and Electric Networks, Carus Math. Monogr., vol. 22. Mathematical Association of America, Washington, DC (1984)
92. Drewitz, A., Ráth, B., Sapozhnikov, A.: An Introduction to Random Interlacements. Springer Briefs Math. Springer, Cham (2014). http://doi.org/10.1007/978-3-319-05852-8
93. Duminil-Copin, H., Tassion, V.: A new proof of the sharpness of the phase transition for Bernoulli percolation and the Ising model. Comm. Math. Phys. **343**(2), 725–745 (2016). http://doi.org/10.1007/s00220-015-2480-z
94. Duminil-Copin, H., Tassion, V.: A new proof of the sharpness of the phase transition for Bernoulli percolation on $\mathbb{Z}^d$. Enseign. Math. **62**(1–2), 199–206 (2016). http://doi.org/10.4171/LEM/62-1/2-12
95. Durrett, R.: On the growth of one-dimensional contact processes. Ann. Probab. **8**(5), 890–907 (1980)
96. Dynkin, E.B.: An Introduction to Branching Measure-Valued Processes, CRM Monograph Series, vol. 6. American Mathematical Soc, Providence, RI (1994)
97. Edwards, R.G., Sokal, A.D.: Generalization of the Fortuin–Kasteleyn–Swendsen–Wang representation and Monte Carlo algorithm. Phys. Rev. D (3) **38**(6), 2009–2012 (1988). http://doi.org/10.1103/PhysRevD.38.2009
98. Erdős, P., Rényi, A.: On random graphs. I. Publ. Math. Debrecen **6**, 290–297 (1959)
99. Erdős, P., Rényi, A.: On the evolution of random graphs. Magyar Tud. Akad. Mat. Kutató Int. Közl. **5**, 17–61 (1960)
100. Erdős, P., Spencer, J.H.: Evolution of the n-cube. Comput. Math. Appl. **5**(1), 33–39 (1979). http://doi.org/10.1016/0898-1221(81)90137-1
101. Etheridge, A.M.: An Introduction to Superprocesses, University Lecture Series, vol. 20. American Mathematical Society, Providence, RI (2000)
102. Evans, S.N.: Two representations of a conditioned superprocess. Proc. Roy. Soc. Edinburgh Sect. A **123**(5), 959–971 (1993). http://doi.org/10.1017/S0308210500029619
103. Federico, L., van der Hofstad, R., den Hollander, F., Hulshof, T.: Expansion of percolation critical points for Hamming graphs. arXiv:1701.02099
104. Federico, L., van der Hofstad, R., den Hollander, F., Hulshof, T.: The scaling limit for critical percolation on the Hamming graph. In progress
105. Fernández, R., Fröhlich, J., Sokal, A.D.: Random Walks, Critical Phenomena, and Triviality in Quantum Field Theory. Texts and Monographs in Physics. Springer, Berlin (1992)
106. Fitzner, R.: Non-backtracking lace expansion (2015). http://www.fitzner.nl/noble/
107. Fitzner, R., van der Hofstad, R.: Generalized approach to the non-backtracking lace expansion. arXiv:1506.07969. To appear in Probab. Theory Related Fields
108. Fitzner, R., van der Hofstad, R.: Mean-field behavior for nearest-neighbor percolation in d > 10. Electron. J. Probab. **22**(43), 65 (2017)
109. Fitzner, R., van der Hofstad, R.: Non-backtracking random walk. J. Stat. Phys. **150**(2), 264–284 (2013). http://doi.org/10.1007/s10955-012-0684-6
110. Fortuin, C.M., Kasteleyn, P.W.: On the random-cluster model. I. Introduction and relation to other models. Physica **57**, 536–564 (1972)
111. Fortuin, C.M., Kasteleyn, P.W., Ginibre, J.: Correlation inequalities on some partially ordered sets. Comm. Math. Phys. **22**, 89–103 (1971)
112. Fountoulakis, N., Reed, B.A.: The evolution of the mixing rate of a simple random walk on the giant component of a random graph. Random Struct. Algorithms **33**(1), 68–86 (2008). http://doi.org/10.1002/rsa.20210
113. Frieze, A.M.: On the value of a random minimum spanning tree problem. Discrete Appl. Math. **10**(1), 47–56 (1985). http://doi.org/10.1016/0166-218X(85)90058-7
114. Fröhlich, J., Simon, B., Spencer, T.: Infrared bounds, phase transitions and continuous symmetry breaking. Comm. Math. Phys. **50**(1), 79–95 (1976)
115. Gandolfi, A., Keane, M.S., Newman, C.M.: Uniqueness of the infinite component in a random graph with applications to percolation and spin glasses. Probab. Theory Relat. Fields **92**(4), 511–527 (1992). http://doi.org/10.1007/BF01274266
116. Gaunt, D.S., Ruskin, H.: Bond percolation processes in d dimensions. J. Phys. A **11**(7), 1369–1380 (1978)
117. Gilbert, E.N.: Random graphs. Ann. Math. Statist. **30**, 1141–1144 (1959)

118. Gordon, D.M.: Percolation in high dimensions. J. London Math. Soc. **44**(2), 373–384 (1991). http://doi.org/10.1112/jlms/s2-44.2.373

119. Graham, B.T.: Borel-type bounds for the self-avoiding walk connective constant. J. Phys. A **43**(23), 235001 (2010). http://doi.org/10.1088/1751-8113/43/23/235001

120. Grassberger, P.: Critical percolation in high dimensions. Phys. Rev. E (3) **67**(3), 036101 (2003). http://doi.org/10.1103/PhysRevE.67.036101

121. Griffiths, R.B., Hurst, C.A., Sherman, S.: Concavity of magnetization of an Ising ferromagnet in a positive external field. J. Math. Phys. **11**, 790–795 (1970)

122. Grimmett, G.R.: Percolation, Grundlehren der mathematischen Wissenschaften, vol. 321, 2nd edn. Springer, Berlin (1999). http://doi.org/10.1007/978-3-662-03981-6

123. Grimmett, G.R.: The Random-Cluster Model. Springer, Berlin (2006). http://doi.org/10.1007/978-3-540-32891-9

124. Grimmett, G.R., Hiemer, P.: Directed percolation and random walk. and Out of Equilibrium (Mambucaba. 2000), Progress in Probability, vol. 51, pp. 273–297. Birkhäuser, Boston, MA (2002)

125. Grimmett, G.R., Manolescu, I.: Bond percolation on isoradial graphs: criticality and universality. Probab. Theory Relat. Fields **159**(1–2), 273–327 (2014). http://doi.org/10.1007/s00440-013-0507-y

126. Grimmett, G.R., Newman, C.M.: Percolation in $\infty + 1$ dimensions. In: Grimmett, G.R., Welsh, D.J.A. (eds.) Disorder in Physical Systems, Oxford Sci. Publ., pp. 167–190. Oxford University Press, New York (1990)

127. Hammersley, J.M.: Percolation processes: Lower bounds for the critical probability. Ann. Math. Statist. **28**, 790–795 (1957)

128. Hammersley, J.M.: Comparison of atom and bond percolation processes. J. Math. Phys. **2**, 728–733 (1961)

129. Hara, T.: Mean-field critical behaviour for correlation length for percolation in high dimensions. Probab. Theory Relat. Fields **86**(3), 337–385 (1990). http://doi.org/10.1007/BF01208256

130. Hara, T.: Decay of correlations in nearest-neighbor self-avoiding walk, percolation, lattice trees and animals. Ann. Probab. **36**(2), 530–593 (2008). http://doi.org/10.1214/009117907000000231

131. Hara, T., van der Hofstad, R., Slade, G.: Critical two-point functions and the lace expansion for spread-out high-dimensional percolation and related models. Ann. Probab. **31**(1), 349–408 (2003). http://doi.org/10.1214/aop/1046294314

132. Hara, T., Slade, G.: Mean-field critical behaviour for percolation in high dimensions. Comm. Math. Phys. **128**(2), 333–391 (1990)

133. Hara, T., Slade, G.: On the upper critical dimension of lattice trees and lattice animals. J. Statist. Phys. **59**(5–6), 1469–1510 (1990). http://doi.org/10.1007/BF01334760

134. Hara, T., Slade, G.: The lace expansion for self-avoiding walk in five or more dimensions. Rev. Math. Phys. **4**(2), 235–327 (1992). http://doi.org/10.1142/S0129055X9200008X

135. Hara, T., Slade, G.: Self-avoiding walk in five or more dimensions. I. The critical behaviour. Comm. Math. Phys. **147**(1), 101–136 (1992)

136. Hara, T., Slade, G.: Unpublished note for [136] (1993). http://www.ma.utexas.edu/mp_arc/index-93.html

137. Hara, T., Slade, G.: Mean-field behaviour and the lace expansion. In: Grimmett, G.R. (ed.) Probability and Phase Transition (Cambridge, 1993, NATO Adv. Sci. Inst. Ser. C Math. Phys. Sci., vol. 420, pp. 87–122. Kluwer Academic Publishers, Dordrecht (1994)

138. Hara, T., Slade, G.: The self-avoiding-walk and percolation critical points in high dimensions. Combin. Probab. Comput. **4**(3), 197–215 (1995). http://doi.org/10.1017/S0963548300001607

139. Hara, T., Slade, G.: The scaling limit of the incipient infinite cluster in high-dimensional percolation. I. Critical exponents. J. Statist. Phys. **99**(5–6), 1075–1168 (2000). http://doi.org/10.1023/A:1018628503898

140. Hara, T., Slade, G.: The scaling limit of the incipient infinite cluster in high-dimensional percolation. II. Integrated super-Brownian excursion. J. Math. Phys. **41**(3), 1244–1293 (2000). http://doi.org/10.1063/1.533186

141. Harris, T.E.: A lower bound for the critical probability in a certain percolation process. Proc. Camb. Philos. Soc. **56**, 13–20 (1960)

142. Harris, T.E.: The Theory of Branching Processes, Grundlehren der Mathematischen Wissenschaften, vol. 119. Springer, Berlin (1963)

143. Heydenreich, M., van der Hofstad, R.: Random graph asymptotics on high-dimensional tori. Comm. Math. Phys. **270**(2), 335–358 (2007). http://doi.org/10.1007/s00220-006-0152-8

144. Heydenreich, M., van der Hofstad, R.: Random graph asymptotics on high-dimensional tori. II. Volume, diameter and mixing time. Probab. Theory Relat. Fields **149**(3–4), 397–415 (2011). http://doi.org/10.1007/s00440-009-0258-y

145. Heydenreich, M., van der Hofstad, R., Hulshof, T.: High-dimensional incipient infinite clusters revisited. J. Stat. Phys. **155**(5), 966–1025 (2014). http://doi.org/10.1007/s10955-014-0979-x

146. Heydenreich, M., van der Hofstad, R., Hulshof, T.: Random walk on the high-dimensional IIC. Comm. Math. Phys. **329**(1), 57–115 (2014). http://doi.org/10.1007/s00220-014-1931-2

147. Heydenreich, M., van der Hofstad, R., Miermont, G.: Backbone scaling limit of the high-dimensional IIC (2017). arXiv:1706.02941

148. Heydenreich, M., van der Hofstad, R., Sakai, A.: Mean-field behavior for long- and finite range Ising model, percolation and self-avoiding walk. J. Stat. Phys. **132**(5), 1001–1049 (2008). http://doi.org/10.1007/s10955-008-9580-5

149. Heydenreich, M., Hulshof, T., Jorritsma, J.: Structures in supercritical scale-free percolation. Ann. Appl. Probab. **27**(4), 2569–2604 (2017)

150. Hirsch, C.: From heavy-tailed Boolean models to scale-free Gilbert graphs. Braz. J. Probab. Stat. **31**(1), 111–143 (2017). http://doi.org/10.1214/15-BJPS305

151. van der Hofstad, R.: Spread-out oriented percolation and related models above the upper critical dimension: Induction and superprocesses. In: On the Nature of Isotherms at First Order Phase Transitions for Classical Lattice Models. Spread-Out Oriented Percolation and Related Models above the Upper Critical Dimension: Induction and Superprocesses, Ensaios Mat., vol. 9, pp. 91–181. Sociedade Brasileira de Matematica, Rio de Janeiro (2005)

152. van der Hofstad, R.: Infinite canonical super-Brownian motion and scaling limits. Comm. Math. Phys. **265**(3), 547–583 (2006). http://doi.org/10.1007/s00220-006-0044-y

153. van der Hofstad, R.: Percolation and random graphs. In: Kendall, W.S., Molchanov, I. (eds.) New Perspectives in Stochastic Geometry, pp. 173–247. Oxford University Press, Oxford (2010)

154. van der Hofstad, R.: Critical behavior in inhomogeneous random graphs. Random Struct. Algorithms **42**(4), 480–508 (2013). http://doi.org/10.1002/rsa.20450

155. van der Hofstad, R.: Random Graphs and Complex Networks, vol. 1. Cambridge Series in Statistical and Probabilistic Mathematics. Cambridge University Press, Cambridge (2016)

156. van der Hofstad, R., den Hollander, F., Slade, G.: Construction of the incipient infinite cluster for spread-out oriented percolation above 4 + 1 dimensions. Comm. Math. Phys. **231**(3), 435–461 (2002). http://doi.org/10.1007/s00220-002-0728-x

157. van der Hofstad, R., den Hollander, F., Slade, G.: The survival probability for critical spread-out oriented percolation above 4 + 1 dimensions. I. Induction. Probab. Theory Relat. Fields **138**(3–4), 363–389 (2007). http://doi.org/10.1007/s00440-006-0028-z

158. van der Hofstad, R., den Hollander, F., Slade, G.: The survival probability for critical spread-out oriented percolation above 4 + 1 dimensions. II. Expansion. Ann. Inst. H. Poincaré Probab. Statist. **43**(5), 509–570 (2007). http://doi.org/10.1016/j.anihpb.2006.09.002

159. van der Hofstad, R., Holmes, M.: The survival probability and r-point functions in high dimensions. Ann. Math. (2) **178**(2), 665–685 (2013). http://doi.org/10.4007/annals.2013.178.2.5

160. van der Hofstad, R., Holmes, M., Perkins, E.A.: A criterion for convergence to super-Brownian motion on path space. Ann. Probab. **45**(1), 278–376 (2017). http://doi.org/10.1214/14-AOP953

161. van der Hofstad, R., Járai, A.A.: The incipient infinite cluster for high-dimensional unoriented percolation. J. Statist. Phys. **114**(3–4), 625–663 (2004). http://doi.org/10.1023/B:JOSS.0000012505.39213.6a

162. van der Hofstad, R., Keane, M.S.: An elementary proof of the hitting time theorem. Amer. Math. Monthly **115**(8), 753–756 (2008)

163. van der Hofstad, R., Luczak, M.J., Spencer, J.H.: The second largest component in the supercritical 2D Hamming graph. Random Struct. Algorithms **36**(1), 80–89 (2010). http://doi.org/10.1002/rsa.20288

164. van der Hofstad, R., Nachmias, A.: Unlacing hypercube percolation: a survey. Metrika **77**(1), 23–50 (2014). http://doi.org/10.1007/s00184-013-0473-5

165. van der Hofstad, R., Nachmias, A.: Hypercube percolation. J. Eur. Math. Soc. **19**(3), 725–814 (2017). http://doi.org/10.4171/JEMS/679

166. van der Hofstad, R., Redig, F.: Maximal clusters in non-critical percolation and related models. J. Stat. Phys. **122**(4), 671–703 (2006). http://doi.org/10.1007/s10955-005-8012-z

167. van der Hofstad, R., Sakai, A.: Gaussian scaling for the critical spread-out contact process above the upper critical dimension. Electron. J. Probab. **9**(24), 710–769 (2004)

168. van der Hofstad, R., Sakai, A.: Convergence of the critical finite-range contact process to super-Brownian motion above the upper critical dimension: The higher-point functions. Electron. J. Probab. **15**(27), 801–894 (2010). http://doi.org/10.1214/EJP.v15-783

169. van der Hofstad, R., Sapozhnikov, A.: Cycle structure of percolation on high-dimensional tori. Ann. Inst. Henri Poincaré Probab. Stat. **50**(3), 999–1027 (2014). http://doi.org/10.1214/13-AIHP565

170. van der Hofstad, R., Slade, G.: Convergence of critical oriented percolation to super-Brownian motion above $4+1$ dimensions. Ann. Inst. Henri Poincaré Probab. Statist. **39**(3), 413–485 (2003). http://doi.org/10.1016/S0246-0203(03)00008-6

171. van der Hofstad, R., Slade, G.: The lace expansion on a tree with application to networks of self-avoiding walks. Adv. in Appl. Math. **30**(3), 471–528 (2003). http://doi.org/10.1016/S0196-8858(02)00507-9

172. van der Hofstad, R., Slade, G.: Asymptotic expansions in n^{-1} for percolation critical values on the n-cube and $\mathcal{Z}^n$. Random Struct. Algorithms **27**(3), 331–357 (2005). http://doi.org/10.1002/rsa.20074

173. van der Hofstad, R., Slade, G.: Expansion in n^{-1} for percolation critical values on the n-cube and $\mathcal{Z}^n$: the first three terms. Combin. Probab. Comput. **15**(5), 695–713 (2006). http://doi.org/10.1017/S0963548306007498

174. Holmes, M.: Convergence of lattice trees to super-Brownian motion above the critical dimension. Electron. J. Probab. **13**(23), 671–755 (2008). http://doi.org/10.1214/EJP.v13-499

175. Holmes, M., Perkins, E.A.: Weak convergence of measure-valued processes and r-point functions. Ann. Probab. **35**(5), 1769–1782 (2007). http://doi.org/10.1214/009117906000001088

176. Hughes, B.D.: Random Walks and Random Environments, vol. 2. Random Environments. Oxford Sci. Publ. Oxford University Press, New York (1996)

177. Hulshof, T.: The one-arm exponent for mean-field long-range percolation. Electron. J. Probab. **20**:115, 26 (2015). http://doi.org/10.1214/EJP.v20-3935

178. Hulshof, T., Nachmias, A.: Slightly subcritical hypercube percolation. arXiv:1612.01772

179. Hutchcroft, T.: Critical percolation on any quasi-transitive graph of exponential growth has no infinite clusters. C. R. Math. Acad. Sci. Paris **354**(9), 944–947 (2016). http://doi.org/10.1016/j.crma.2016.07.013

180. Häggström, O.: Problem solving is often a matter of cooking up an appropriate Markov chain. Scand. J. Statist. **34**(4), 768–780 (2007). http://doi.org/10.1111/j.1467-9469.2007.00561.x

181. Häggström, O., Peres, Y.: Monotonicity of uniqueness for percolation on Cayley graphs: all infinite clusters are born simultaneously. Probab. Theory Relat. Fields **113**(2), 273–285 (1999). http://doi.org/10.1007/s004400050208

182. Häggström, O., Peres, Y., Schonmann, R.H.: Percolation on transitive graphs as a coalescent process: relentless merging followed by simultaneous uniqueness. In: Bramson, M., Durrett, (eds.) Perplexing Problems in Probability, Progr. Probab., vol. 44, pp. 69–90. Birkhäuser, Boston, MA (1999)

183. Jagers, P.: Branching Processes with Biological Applications. Wiley Ser. Probab. Math. Statist. Appl. Probab. Statist. Wiley-Interscience, London-New York-Sydney (1975)

184. Janson, S., Knuth, D.E., Łuczak, T., Pittel, B.: The birth of the giant component. Random Struct. Algorithms **4**(3), 231–358 (1993). http://doi.org/10.1002/rsa.3240040303

185. Janson, S., Łuczak, T., Rucinski, A.: Random Graphs. Wiley-Intersci. Ser. Discrete Math. Optim. Wiley-Interscience, New York (2000)

186. Janson, S., Marckert, J.F.: Convergence of discrete snakes. J. Theor. Probab. **18**(3), 615–645 (2005). http://doi.org/10.1007/s10959-005-7252-9

187. Janson, S., Warnke, L.: On the critical probability in percolation. Electronic J. Probab. arXiv:1611.08549

188. Joseph, A.: The component sizes of a critical random graph with given degree sequence. Ann. Appl. Probab. **24**(6), 2560–2594 (2014). http://doi.org/10.1214/13-AAP985

189. Járai, A.A.: Incipient infinite percolation clusters in 2D. Ann. Probab. **31**(1), 444–485 (2003). http://doi.org/10.1214/aop/1046294317
190. Járai, A.A.: Invasion percolation and the incipient infinite cluster in 2D. Comm. Math. Phys. **236**(2), 311–334 (2003). http://doi.org/10.1007/s00220-003-0796-6
191. Kesten, H.: Full Banach mean values on countable groups. Math. Scand. **7**, 146–156 (1959)
192. Kesten, H.: Symmetric random walks on groups. Trans. Am. Math. Soc. **92**, 336–354 (1959). http://doi.org/10.2307/1993160
193. Kesten, H.: The critical probability of bond percolation on the square lattice equals $\frac{1}{2}$. Comm. Math. Phys. **74**(1), 41–59 (1980)
194. Kesten, H.: Percolation Theory for Mathematicians. Progr. Probab. Statist., vol. 2. Birkhäuser, Boston, MA (1982)
195. Kesten, H.: The incipient infinite cluster in two-dimensional percolation. Probab. Theory Relat. Fields **73**(3), 369–394 (1986). http://doi.org/10.1007/BF00776239
196. Kesten, H.: Subdiffusive behavior of random walk on a random cluster. Ann. Inst. H. Poincaré Probab. Statist. **22**(4), 425–487 (1986)
197. Kesten, H.: Asymptotics in high dimensions for percolation. In: Grimmett, G.R., Welsh, D.J.A. (eds.) Disorder in Physical Systems, Oxford Sci. Publ., pp. 219–240. Oxford Univ. Press, New York (1990)
198. Kesten, H.: Some highlights of percolation. In: Proceedings of the International Congress of Mathematicians, vol. I, pp. 345–362. Higher Ed. Press, Beijing (2002)
199. Kipnis, C., Varadhan, S.R.S.: Central limit theorem for additive functionals of reversible markov processes and applications to simple exclusions. Comm. Math. Phys. **104**, 1–19 (1986). http://doi.org/10.1007/BF01210789
200. Kozma, G.: The triangle and the open triangle. Ann. Inst. Henri Poincaré Probab. Stat. **47**(1), 75–79 (2011). http://doi.org/10.1214/09-AIHP352
201. Kozma, G., Nachmias, A.: The Alexander-Orbach conjecture holds in high dimensions. Invent. Math. **178**(3), 635–654 (2009). http://doi.org/10.1007/s00222-009-0208-4
202. Kozma, G., Nachmias, A.: Arm exponents in high dimensional percolation. J. Amer. Math. Soc. **24**(2), 375–409 (2011). http://doi.org/10.1090/S0894-0347-2010-00684-4
203. Kumagai, T., Misumi, J.: Heat kernel estimates for strongly recurrent random walk on random media. J. Theor. Probab. **21**(4), 910–935 (2008). http://doi.org/10.1007/s10959-008-0183-5
204. Lalley, S.P.: Percolation clusters in hyperbolic tessellations. Geom. Funct. Anal. **11**(5), 971–1030 (2001). http://doi.org/10.1007/s00039-001-8223-7
205. Lawler, G.F., Schramm, O., Werner, W.: One-arm exponent for critical 2D percolation. Electron. J. Probab. **7**:2, 13 (2002)
206. Le Gall, J.F.: Spatial Branching Processes, Random Snakes, and Partial Differential Equations. Lectures Math. ETH Zürich. Birkhäuser, Basel (1999). http://doi.org/10.1007/978-3-0348-8683-3
207. Levin, D.A., Peres, Y., Wilmer, E.L.: Markov Chains and Mixing Times. American Mathematical Society, Providence, RI (2009)
208. Łuczak, T.: Component behavior near the critical point of the random graph process. Random Struct. Algorithms **1**(3), 287–310 (1990). http://doi.org/10.1002/rsa.3240010305
209. Łuczak, T., Pittel, B., Wierman, J.C.: The structure of a random graph at the point of the phase transition. Trans. Amer. Math. Soc. **341**(2), 721–748 (1994). http://doi.org/10.2307/2154580
210. Lyons, R., Peres, Y.: Probability on Trees and Networks. Cambridge Series in Statistical and Probabilistic Mathematics. Cambridge University Press, Cambridge (2017)
211. Madras, N., Slade, G.: The Self-Avoiding Walk. Probability and its Applications, Birkhäuser, Boston (1993)
212. Margulis, G.A.: Probabilistic characteristics of graphs with large connectivity (Russian). Problemy Peredači Informacii **10**(2), 101–108 (1974). English transl., Problems Inform. Transmission **10**(2), 174–179 (1974)
213. Mathieu, P., Piatnitski, A.: Quenched invariance principles for random walks on percolation clusters. Proc. R. Soc. Lond. Ser. A Math. Phys. Eng. Sci. **463**(2085), 2287–2307 (2007). http://doi.org/10.1098/rspa.2007.1876
214. Meester, R.W.J.: Uniqueness in percolation theory. Statist. Neerlandica **48**(3), 237–252 (1994). http://doi.org/10.1111/j.1467-9574.1994.tb01446.x

215. Menshikov, M.V.: Coincidence of critical points in percolation problems (Russian). Dokl. Akad. Nauk SSSR **288**(6), 1308–1311 (1986). English transl., Soviet Math. Dokl. **33**(3), 856–859 (1986)
216. Nachmias, A., Peres, Y.: Critical random graphs: Diameter and mixing time. Ann. Probab. **36**(4), 1267–1286 (2008). http://doi.org/10.1214/07-AOP358
217. Nachmias, A., Peres, Y.: Critical percolation on random regular graphs. Random Structures Algorithms **36**(2), 111–148 (2010). http://doi.org/10.1002/rsa.20277
218. Newman, C.M., Schulman, L.S.: Infinite clusters in percolation models. J. Statist. Phys. **26**(3), 613–628 (1981). http://doi.org/10.1007/BF01011437
219. Newman, C.M., Stein, D.L.: Spin-glass model with dimension-dependent ground state multiplicity. Phys. Rev. Lett. **72**(14), 2286–2289 (1994). http://doi.org/10.1103/PhysRevLett.72.2286
220. Nguyen, B.G.: Gap exponents for percolation processes with triangle condition. J. Statist. Phys. **49**(1–2), 235–243 (1987). http://doi.org/10.1007/BF01009960
221. Nguyen, B.G., Yang, W.S.: Triangle condition for oriented percolation in high dimensions. Ann. Probab. **21**, 1809–1844 (1993)
222. Nguyen, B.G., Yang, W.S.: Gaussian limit for critical oriented percolation in high dimensions. J. Statist. Phys. **78**(3–4), 841–876 (1995). http://doi.org/10.1007/BF02183691
223. Nickel, B., Wilkinson, D.: Invasion percolation on the Cayley tree: Exact solution of a modified percolation model. Phys. Rev. Lett. **51**(2), 71–74 (1983). http://doi.org/10.1103/PhysRevLett.51.71
224. Perkins, E.A.: Dawson–Watanabe superprocesses and measure-valued diffusions. In: Bernard, P. (ed.) Lectures on Probability Theory and Statistics Saint-Flour, 1999, Lecture Notes in Mathematics
225. Reed, M.C., Simon, B.: Methods of Modern Mathematical Physics. II. Fourier Analysis, Self-Adjointness. Academic Press, New York–London (1975)
226. Reimer, D.: Proof of the van den Berg-Kesten conjecture. Combin. Probab. Comput. **9**(1), 27–32 (2000). http://doi.org/10.1017/S0963548399004113
227. Riordan, O.: The phase transition in the configuration model. Combin. Probab. Comput. **21**(1–2), 265–299 (2012). http://doi.org/10.1017/S0963548311000666
228. Russo, L.: A note on percolation. Z. Wahrsch. Verw. Gebiete **43**(1), 39–48 (1978). http://doi.org/10.1007/BF00535274
229. Russo, L.: On the critical percolation probabilities. Z. Wahrsch. Verw. Gebiete **56**(2), 229–237 (1981). http://doi.org/10.1007/BF00535742
230. Ráth, B., Valesin, D.: Percolation on the stationary distributions of the voter model. Ann. Probab. **45**(3), 1899–1951 (2017). arXiv:1502.01306
231. Rényi, A.: On connected graphs. I. Magyar Tud. Akad. Mat. Kutató Int. Közl. **4**, 385–388 (1959)
232. Sakai, A.: Mean-field critical behavior for the contact process. J. Statist. Phys. **104**(1–2), 111–143 (2001). http://doi.org/10.1023/A:1010320523031
233. Sakai, A.: Hyperscaling inequalities for the contact process and oriented percolation. J. Statist. Phys. **106**(1–2), 201–211 (2002). http://doi.org/10.1023/A:1013197011935
234. Sakai, A.: Lace expansion for the Ising model. Comm. Math. Phys. **272**(2), 283–344 (2007). http://doi.org/10.1007/s00220-007-0227-1
235. Sapozhnikov, A.: Upper bound on the expected size of the intrinsic ball. Electron. Commun. Probab. **15**(28), 297–298 (2010). http://doi.org/10.1214/ECP.v15-1553
236. Sapozhnikov, A.: Random walks on infinite percolation clusters in models with long-range correlations. Ann. Probab. **45**(3), 1842–1898 (2014)
237. Schonmann, R.H.: Stability of infinite clusters in supercritical percolation. Probab. Theory Relat. Fields **113**(2), 287–300 (1999). http://doi.org/10.1007/s004400050209
238. Schonmann, R.H.: Multiplicity of phase transitions and mean-field criticality on highly non-amenable graphs. Comm. Math. Phys. **219**(2), 271–322 (2001). http://doi.org/10.1007/s002200100417
239. Schonmann, R.H.: Mean-field criticality for percolation on planar non-amenable graphs. Comm. Math. Phys. **225**(3), 453–463 (2002). http://doi.org/10.1007/s002200100587
240. Schramm, O.: Scaling limits of loop-erased random walks and uniform spanning trees. Israel J. Math. **118**, 221–288 (2000). http://doi.org/10.1007/BF02803524
241. Sidoravicius, V., Sznitman, A.S.: Quenched invariance principles for walks on clusters of percolation or among random conductances. Probab. Theory Relat. Fields **129**(2), 219–244 (2004). http://doi.org/10.1007/s00440-004-0336-0

242. Simon, B.: Functional Integration and Quantum Physics, 2nd edn. AMS Chelsea Publishing, Providence, RI (2005)

243. Slade, G.: The diffusion of self-avoiding random walk in high dimensions. Comm. Math. Phys. **110**(4), 661–683 (1987)

244. Slade, G.: Lattice trees, percolation and super-Brownian motion. In: Bramson, M., Durrett, R. (eds.) Perplexing Problems in Probability. Birkhäuser, Boston, MA (1999)

245. Slade, G.: Scaling limits and super-Brownian motion. Notices Am. Math. Soc. **49**(9), 1056–1067 (2002)

246. Slade, G.: The Lace Expansion and Its Applications. Lecture Notes in Math, vol. 1879. Springer, Berlin (2006)

247. Smirnov, S.: Critical percolation in the plane: conformal invariance, Cardy's formula, scaling limits. C. R. Acad. Sci. Paris Sér. I Math. **333**(3), 239–244 (2001). http://doi.org/10.1016/S0764-4442(01)01991-7

248. Smirnov, S.: Conformal invariance in random cluster models. I. Holomorphic fermions in the Ising model. Ann. Math. **172**(2), 1435–1467 (2010). http://doi.org/10.4007/annals.2010.172.1441

249. Smirnov, S., Werner, W.: Critical exponents for two-dimensional percolation. Math. Res. Lett. **8**(5–6), 729–744 (2001). http://doi.org/10.4310/MRL.2001.v8.n6.a4

250. Stauffer, D., Aharony, A.: Introduction to Percolation Theory, 2nd edn. CRC Press, Boca Raton (1994)

251. Sznitman, A.S.: Vacant set of random interlacements and percolation. Ann. Math. **171**(3), 2039–2087 (2010). http://doi.org/10.4007/annals.2010.171.2039

252. Sznitman, A.S.: Topics in Occupation Times and Gaussian Free Fields. Zur. Lect. Adv. Math. Eur. Math. Soc., Zürich (2012). http://doi.org/10.4171/109

253. Tasaki, H.: Hyperscaling inequalities for percolation. Comm. Math. Phys. **113**(1), 49–65 (1987)

254. Timár, Á.: Percolation on nonunimodular transitive graphs. Ann. Probab. **34**(6), 2344–2364 (2006). http://doi.org/10.1214/009117906000000494

255. Turova, T.S.: Diffusion approximation for the components in critical inhomogeneous random graphs of rank 1. Random Struct. Algorithms **43**(4), 486–539 (2013). http://doi.org/10.1002/rsa.20503

256. Uchiyama, K.: Green's functions for random walks on $\mathcal{Z}^n$. Proc. London Math. Soc. **77**(1), 215–240 (1998). http://doi.org/10.1112/S0024611598000458

257. Wu, C.C.: Critical behavior or percolation and Markov fields on branching planes. J. Appl. Probab. **30**(3), 538–547 (1993)

Printed in the United States
By Bookmasters